CONCRETE MATERIALS AND PRACTICE

CONCRETE MATERIALS AND PRACTICE

L. J. Murdock D.Sc. (ENG.), Ph.D., M.I.C.E.
G. F. Blackledge B.Sc. (TECH.), A.M.I.STRUCT.E.

London EDWARD ARNOLD (Publishers) Ltd.

First Published 1948
Second Edition 1955
Third Edition 1960
Fourth Edition 1968

SBN: 7131 3178 0

Printed at The Universities Press, Belfast, N. Ireland

Preface to First Edition

RESEARCH both in Great Britain and abroad has provided an extensive knowledge of the fundamental properties and characteristics of concrete and it is possible in the laboratory to design and prepare concrete of predetermined quality. In the field, however, varying conditions, lack of knowledge, relatively inefficient plant, errors, quantities of material involved, changes in the basic materials, and other difficulties, cause variations in quality of the concrete which can only be minimized by careful control, education, and experience.

The object of this book is to provide a broad outline of the science of concrete making, particular attention being given to the translation of the knowledge gained by testing and research into practice in the field.

The properties of concrete are described early in the book since it is important to recognize that changes in the constituent materials, and the manner in which they are combined and used, may improve some properties at the expense of others, and a compromise must always be made to obtain the special qualities required as economically as possible.

The basic materials and the technique of concrete making under various conditions and for many purposes are discussed in some detail. The method of testing, the design of mixes to meet various requirements, and the methods of control of quality on the site, which are included, have been found by experience to give satisfactory results.

Brief mention is made of the design of reinforced concrete mainly to enable the concrete engineer to recognize errors in bending and fixing the reinforcement. Similarly, examples of formwork are included on account of the importance of careful design and construction to the final result.

The author is indebted to Messrs. G. Wimpey and Co., Ltd., for permission to publish information based on experience on many sites large and small, as well as a number of photographs and charts. Considerable use has been made of the results of researches carried out at the Building Research Station and the Road Research Laboratory, and due acknowledgement is made to the Directors of these establishments for access to this work. Acknowledgment is also made to the Cement and Concrete Association for their generous help in providing photographs; to Mr. R. V. Chate of the Reinforced Concrete Association and Dr. N. Davey of the Building Research Station for their valuable comments and criticisms of the original manuscript; and to Dr. W. MacGregor for reading the proofs.

Preface to Fourth Edition

SUBSTANTIAL revision has again been made to include the latest advances in technology and to describe improvements in plant and equipment. Greater attention has been given to concreting practice in other countries, particularly America, and comparisons are made with British technology and its application on site.

For the fourth edition, Mr. G. F. Blackledge, B.Sc., A.M.I.Struct.E. has become joint author. He has a very considerable knowledge of concrete quality control and the problems which arise in concrete production in many parts of the world.

The authors are indebted to those who have made helpful comments and criticism, including members of the staff of George Wimpey & Co. Limited. Mr. H. Holland in particular has taken a keen interest in the design of concrete mixes, and was responsible for preparing Figs. 7.3, 7.4, 7.5.

Grateful acknowledgment is also made to the following who have given permission for the use of photographs and information:

Cement and Concrete Association. See Preface to First Edition and Figs. 10.3, 10.7, 12.7, 15.1, 15.2, 15.3, 15.4, 15.5, 15.7, 15.14, 15.15, 16.8, 16.10, 16.17, 17.8, 17.9, 19.1, 19.3, 19.6, 19.7, 20.4.
Construction Industry Research and Information Association. Figs. 16.1, 16.2, 16.3.
Institution of Civil Engineers. Figures and information reproduced from Papers to which reference is made in the Bibliography.
Reinforced Concrete Association. Frontispiece and Figs. 17.14, 17.15.
The Controller, H. M. Stationery Office. Figs. 2.5, 2.9, 3.8, 4.11, 4.13, 5.2, 6.2, 10.2, 10.4, 10.5, 12.1.

E. P. Allam Ltd. Figs. 12.6, 21.4.
American Concrete Institute. Fig. 3.7, also parts of Figs. 3.6, 3.14, 9.13, 11.2.
Associated Portland Cement Manufacturers Ltd. Figs. 4.1, 4.6.
Benford Ltd. Figs. 8.7, 9.9, 9.11, 9.12.
Blaw Knox Ltd. Figs. 8.3, 8.5, 9.5, 16.20.
British Steel Piling Ltd. Fig. 20.3.
H. H. Broughton. Fig. 16.19.
Columbus Dixon Ltd. Fig. 23.2.
Concrete Ltd. Figs. 20.5, 21.5.
Concrete Development Co. Ltd. Figs. 18.1, 21.5.
Concrete Pump Co. Ltd. Fig. 9.17.
Fielding & Platt Ltd. Fig. 21.3.
Franki Piling Co. Ltd. Fig. 20.2.
S. V. Gardner. Figs. 17.11, 24.10.
Ham River Grit Co. Ltd. Fig. 3.1.

Holland & Hannen and Cubitts Ltd. Fig. 18.2.
Nordberg Manufacturing Ltd. Fig. 3.3.
Frederick Parker Ltd. Fig. 3.2.
Pegson Ltd. Fig. 21.7.
Road Machines (Drayton) Ltd. Figs. 8.6, 9.16.
Rocoman Plant Ltd. Fig. 9.19.
Stuarts Granolithic Ltd. Figs. 21.1, 23.2.
Trianco Ltd. Fig. 21.8.
Winget Ltd. Fig. 9.8.

Extracts from British Standards and from British Standard Codes of Practice are given by permission of the British Standards Institution, 2 Park Street, London W1, from whom official copies of the Standards and the Codes can be obtained.

Contents

Twickenham Bridge: an example of the pleasing appearance obtainable by bush-hammering gravel concrete, and by the use of inset lines of tiles to mask the lift planes. Concrete which has stood the test of the passing years

CHAPTER I

Introduction

THE USE of some form of cement to bind together stones, gravel and other material has been practised from very early times. The Egyptians certainly used lime mortar in the construction of their pyramids, and analyses have shown that they possessed much practical knowledge of the subject.

The Greeks and Etruscans also used cementitious mortars, but a Roman, Vitruvius, is the first author on the subject whose writings have descended to us. His works show that, although the Romans adopted a different scientific phraseology, they knew as much about the chemistry of cementitious limes as the scientists of the eighteenth century. Vitruvius's work was not, in fact, superseded until M. Vicat made his researches in France in the early part of the nineteenth century.

One of the most notable examples of Roman work which still remains is the Pantheon, the dome of which is 142 ft 6 in. in span and cast in solid concrete. Wooden boards were used as shuttering, and the marks of these can still be clearly seen on the face of the concrete. The calcareous cements used by the Romans were either composed of suitable limestones burnt in kilns or were mixtures of lime and pozzolanic material, which combine to form a fairly hard weather-resistant concrete. Pozzolana was found naturally as volcanic ash or tuff in the district around Puzzuoli in Italy, or was produced artificially by grinding clay tiles and potsherds.

Joseph Aspdin first used the term Portland cement in 1824 to describe a patent cement formed by heating a mixture of finely divided clay and limestone or chalk in a furnace to a temperature sufficiently high to drive off all the carbon dioxide.

It was not until 1845 that Isaac C. Johnson invented the cement which was the prototype of our modern Portland cement, by increasing the temperature at which the limestone and clay were burnt until they formed a clinker. From then onward many refinements and technical changes produced a gradual improvement in the quality of cement, accompanied by a steady expansion in the uses to which concrete was applied.

Improvements in Portland cement production and in the technique of concrete construction have greatly accelerated in recent years, and a great deal of research has been done in this country, in the U.S.A., France and Germany. Concrete, as it is known to-day, is a building and structural material, the properties of which may be governed by careful design and control of the constituent materials. These constituent materials are, a hardened binding medium or matrix formed by a chemical reaction between cement and water, and the aggregate to which the hardened cement adheres to a greater or less degree. The aggregate may be gravel, slag, crushed rock, crushed brick or clinker, sand or other similar material.

The aggregate, the cement and the water are all mixed together, and in this state are plastic and workable, properties which permit moulding into any desired shape. Within a few hours of the preparation of the mixture, the cement and water undergo a chemical combination, generally referred to as hydration, which results in solidification and gradual hardening. The hardening process may continue indefinitely under favorable moisture and temperature conditions, with a general improvement in the quality of the concrete.

Fig. 1.1. A pleasing example of concrete construction. The external structure is built up with I-shaped storey height precast concrete structural units. These were precast in white concrete using capstone aggregate and white Portland cement.

Selection of constituent materials

If concrete is to be suitable for a particular purpose, it is necessary to select the constituent materials and to combine and use them in such a manner as to develop the special qualities required as economically as possible. For instance, the qualities required in an engineering structure are, generally, weather resistance and the strength characteristics on which the designer has based his calculations. In roads, concrete must be strong and wear resisting to sustain the pounding action of heavy and fast-moving traffic. In water towers and reservoirs it must be impermeable. In foundations it may be required to resist the disintegrating action of sulphates in the soil. Similarly, sea structures must be resistant to the chemical attack of sea salts, as well as to abrasive action. Concrete floors, in addition to abrasion by foot traffic and all types of vehicles, are often subjected to chemical attack by various agencies. Other qualities which may be required in certain circumstances include light weight, thermal insulation, and impedance to X-rays or the products of nuclear fission.

Some special types of construction such as piling, dock work, mass concrete dams, tunnel work, and so on, call for a special technique which will be described later. Similarly, the laying of concrete floors such as granolithic, the moulding of cast stone, precast concrete staircase units, concrete railway sleepers, fence posts, paving flags, kerbs, garden ornaments, and other items require individual modifications in procedure. Lightweight concrete and open-textured concrete form another class of concrete product to be treated separately.

The selection of materials and the choice of method of construction is not easy, since many variables will affect the quality of the concrete produced, and both quality and economy must be considered. As an instance of the necessity for compromise, concrete of the same strength may be produced by a lean mix which is difficult to place, or by a somewhat richer and more plastic mix which is easy to place. In such a case it is probable that the increased cost of cement in the richer mix will be offset by the saving in cost of placing the concrete. In addition, the finish of the richer mix will in all probability be better. Other properties of the concrete are, of course, affected by an increased cement content, and must be considered according to their relative importance.

Limitations of concrete

Concrete has limitations and certain disadvantages which should be realized by both the designer and the builder, since recognition of these characteristics will go far towards the elimination of costly difficulties in construction, and of cracking and other structural weaknesses that detract from the appearance and serviceability, and life of the structure.

The principal limitations and disadvantages are:

(1) *Low tensile strength.* Members which are subjected to tensile stress must, therefore, be reinforced with steel bars or mesh.

(2) *Drying shrinkage and moisture movements.* Concrete shrinks as it dries out and even when hardened, expands and contracts with wetting and drying. These movements necessitate the provision of contraction joints at intervals if unsightly cracks are to be avoided.

(3) *Creep.* Concrete gradually deforms under load, the deformation due to creep not being wholly recoverable when the load is removed. Creep is particularly important in relation to prestressed concrete. Creep and shrinkage are difficult to separate in measurements of deformation during testing.

(4) *Thermal movements.* Concrete expands and contracts with changes in temperature at roughly the same rate as steel. Expansion joints must be provided in many types of structure if failures are to be prevented.

(5) *Permeability.* Even the best concrete is not entirely impervious to moisture, and contains soluble compounds which may be leached out to a varying extent by water. Joints, unless special attention is given to their construction, are apt to form channels for the ingress of water.

Impermeability is particularly important in reinforced concrete where reliance is placed on the concrete cover to prevent rusting of the steel.

Variability in quality of concrete

A list of variables likely to occur during the manufacture of concrete is given below, and while this list is not exhaustive it does give some indication of possible sources of variation in strength and other properties.

(A) *Materials*
- (1) Cement: Quality and rate of hardening.
- (2) Fine aggregate: (a) Grading, affecting workability
  - (b) Moisture content, affecting water/cement ratio
  - (c) Silt, affecting strength
  - (d) Cleanliness, affecting strength and durability
- (3) Coarse aggregate: (a) Grading, some effect on strength
  - (b) Moisture content, affecting water/cement ratio
  - (c) Cleanliness, affecting strength and durability
- (4) Water: Quantity affecting most properties, quality affecting setting, strength, durability, etc.

(B) *Methods of batching and mixing*
- (1) By volume: (a) Bulking of sand affects proportions
  - (b) Accuracy of measurement
- (2) By weight: (a) Moisture content of aggregate
  - (b) Accuracy of measurement
- (3) Loss of materials entering mixer
- (4) Efficiency of mixer

(C) *Methods of construction*
- (1) Compaction: Air left entrained reduces strength
- (2) Curing: Necessary to increase strength and improve other properties
- (3) Weather conditions during placing and curing of concrete

Workmanship

In concrete construction it is not enough for the concrete engineer to supervise and criticize the work, without justifying his comments or attempting to educate the foremen, gangers, and workmen. Quality of workmanship is one of the most important factors in the production of good concrete structures, and the key to good workmanship is knowledge and an interest in the work being done. Most workmen are keen to do a job well, subject to the conflicting interest of speed since this affects their earnings, and if the reasons for doing a job in a certain way are explained to them, they will often make an effort to follow out the suggestions made.

The concrete engineer must, therefore, learn to deal tactfully and to encourage an interest in improvements of method by the men actually making and placing the concrete. He must also see that the quality of the work complies with the requirements of the particular purpose for which it was intended, while studying economy and the smooth progress of the work. The concrete engineer must bear in mind that he is responsible for maintaining a reasonable and uniform standard of concrete quality, while in the interests of economy it is obviously unnecessary to produce concrete having a crushing strength far in excess of that required.

CHAPTER 2

Properties of Concrete

THE characteristics of concrete should be considered in relation to the quality required for any given construction purpose. The closest practicable approach to perfection in every property of the concrete would result in poor economy under many conditions, and the most desirable structure is that in which the concrete has been designed with the correct emphasis on each of the various properties of the concrete, and not solely with a view to obtaining, say, the maximum possible strength.

Although the attainment of the maximum strength should not be the sole criterion in design, the measurement of the crushing strength of concrete cubes provides a means of maintaining a uniform standard of quality, and, in fact, is the usual way of doing so. Since the other properties of any particular mix of concrete are related to the crushing strength in some manner, it is possible that as a single control test it is the most convenient and informative.

The testing of the hardened concrete in prefabricated units presents no difficulty, since complete units can be selected and broken if necessary in the process of testing. Samples can be taken from some parts of a finished structure by core borings, but at considerable cost and with a possible weakening of the structure. It is customary, therefore, to estimate the properties of the concrete in the structure on the basis of the tests made on specimens moulded from the fresh concrete as it is placed. These specimens are rammed and cured in a standard manner given in B.S. 1881: 1968[2.1]* as in these two respects it is impossible to simulate exactly the conditions in the structure. Since the crushing strength is also affected by the size and shape of a specimen or part of a structure, it follows that the crushing strength of a cube is not necessarily the same as that of the mass of exactly the same concrete.

Crushing strength

Concrete can be made having a strength in compression of up to about 12,000 lb/in², or even more depending mainly on the relative proportions of water and cement, that is, the water/cement ratio, and the degree of compaction. Crushing strengths of between 3,000 and 7,000 lb/in² at 28 days are normally obtained on the site with reasonably good supervision, for mixes roughly equivalent to 1:2:4 of cement:sand:gravel. In some types of precast concrete such as railway sleepers, strengths ranging from 6,000 to 10,000 lb/in² at 28 days are obtained with rich mixes having a low water/cement ratio.

The crushing strength of concrete is influenced by a numbers of factors in addition to the water-cement ratio and the degree of compaction. The more important factors are:

(1) *Type of cement and its quality.* Both the rate of strength gain and the ultimate strength may be affected.

* The numbers refer to the Bibliography at the end of each chapter.

(2) *Type and surface texture of aggregate.* There is considerable evidence to suggest that some aggregates produce concrete of greater compressive and tensile strengths than are obtained with smooth river gravels. See p. 53.

(3) *Efficiency of curing.* A loss in strength of up to about 40 per cent may result from premature drying out. Curing is therefore of considerable importance both in the field and in the making of tests. The method of curing concrete test cubes given in B.S. 1881 should, for this reason, be strictly adhered to.

(4) *Temperature.* In general, the rate of hardening of concrete is increased by an increase in temperature. At freezing temperatures the crushing strength may remain low for some time.

(5) *Age.* Under normal conditions concrete increases in strength with age, the rate of increase depending on the type of cement. For instance, high-alumina cement produces concrete with a crushing strength at 24 hours equal to that of normal Portland cement concrete at 28 days. Hardening continues but at a much slower rate for a number of years.

Variations in the crushing strength of site-made concrete are discussed in Chapter 25 page 378.

The above refers to the static ultimate load. When subjected to repeated loads concrete fails at a load smaller than the ultimate static load, a fatigue effect. A number of investigators have established that after several million cycles of loading the fatigue strength in compression is 50–60 per cent of the ultimate static strength.

Tensile and flexural strength

The tensile strength of concrete varies from one-eighth of the compressive strength at early ages to about one-twentieth later, and is not usually taken into account in the design of reinforced concrete structures. The tensile strength is, however, of considerable importance in resisting cracking due to changes in moisture content or temperature.

The measurement of the strength of concrete in direct tension is difficult and is rarely attempted. Two more practical methods of assessing tensile strength are available. One gives a measure of the tensile strength in bending, usually termed the flexural strength. B.S. 1881: 1968[2.1] gives details concerning the making and curing of flexure test specimens, and of the method of test. The standard size of specimen is 28 in. × 6 in. × 6 in. for aggregate of maximum size 1½ in. If the largest nominal size of the aggregate is ¾ in., specimens 20 in. × 4 in. × 4 in. may be used.

A load is applied through two rollers at the third points of the span until the specimen breaks. The extreme fibre stresses, that is, compressive at the top and tensile at the bottom, can then be computed by the usual beam formulae. The beam will obviously fail in tension since the tensile strength is much lower than the compressive strength.

Formulae for the calculation of the modulus of rupture are given in B.S. 1881:1968.

Test specimens in the form of beams are sometimes used to measure the modulus of rupture or flexural strength quickly on the site. The two halves of the specimen may then be crushed so that besides the flexural strength the compressive strength can be approximately determined on the same sample. The test is described in B.S. 1881:1968.

Values of the modulus of rupture are utilized in some methods of design of unreinforced concrete roads and runways, in which reliance is placed on the flexural strength of the concrete to distribute concentrated loads over a wide area.

More recently introduced is a test made by splitting cylinders by compression across the diameter, to give what is termed the splitting tensile strength. The method of test under ASTM Designation C496-64T[2.2] utilizes 12 in. × 6 in. (dia.) cylinders, moist-cured for 7 days, and then air-dried until tested at 28 days. The testing machine is fitted with an extra bearing bar to distribute the load along the full length of the cylinder. Plywood strips, 1 in. wide and ⅛ in. thick are inserted between the cylinder and the testing machine bearing surfaces top and bottom. From the maximum applied load at failure the tensile splitting strength is calculated as follows:

$$T = \frac{2P}{\pi l d}$$

where T = splitting tensile strength, lb/in²,
P = maximum applied load in pounds,
l = length of cylinder in inches,
d = diameter in inches.

As in the case of the compressive strength, repeated loading reduces the ultimate strength so that the fatigue strength in flexure is 50–60 per cent of the static strength.

Shear strength

In practice, shearing of concrete is always accompanied by compression and tension caused by bending, and even in testing it is impossible to eliminate an element of bending.

Deformation under load

When concrete is loaded, deformation occurs which increases as the load increases, as with steel and other materials. However, whereas steel deforms elastically at loads below the elastic limit, so that the specimen returns to its original size when the load is removed, concrete deforms partly as a result of elastic strain, and partly as a result of plastic strain or creep. This is illustrated in Fig. 2.1 which shows the stress–strain curve for a continuously increasing load. The relationship for an elastic material, such as steel, is given by the straight line OA, while that for a partly plastic material, such as concrete, is given by the line OB. On unloading a concrete specimen stressed to the point B, the elastic strain disappears, but the plastic strain represented by OX remains. The effect of repeated loading is also shown in which the deformation due to plastic strain XY, YZ diminishes at each repeated loading, although the total plastic deformation or creep OX, OY, OZ is continually increasing. The strain has been exaggerated for clarity; within the normal working range the stress–strain curve for concrete is for practical purposes a straight line.

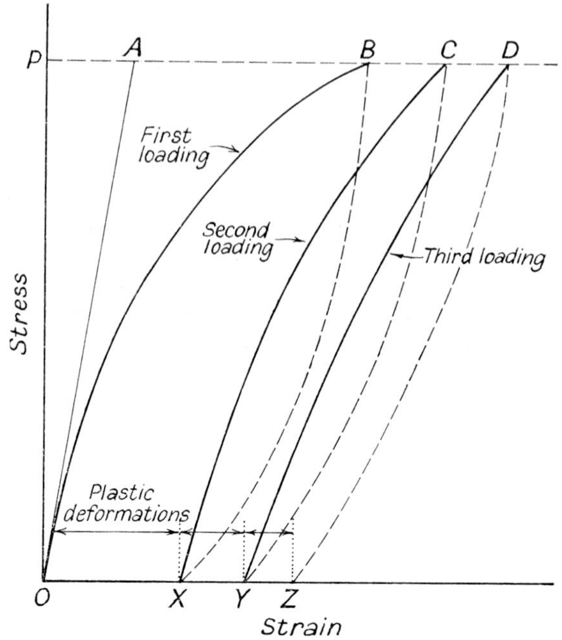

Fig. 2.1. Stress–strain curves for concrete subjected to repeated loading

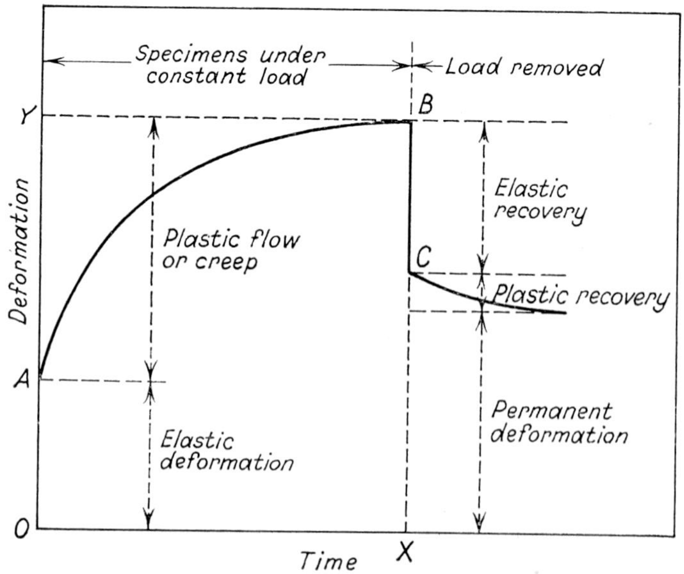

Fig. 2.2. Illustration of deformation of concrete under load

When the load is increased beyond the working range, the stress–strain curve deviates considerably from a straight line, indicating that stress and strain are no longer proportional for practical purposes. This limit of proportionality may be anything from 25 to 75 per cent of the ultimate strength, with 40 per cent as an average value.

When concrete is subjected to a constant sustained load, the deformation produced by the load may be divided into two parts; the elastic deformation which occurs immediately the load is applied, and the plastic flow or creep which begins when the load is applied but continues to increase without further increase in load for the whole time the specimen is loaded. This continuing deformation with time under constant load is illustrated in Fig. 2.2, being represented by line AB. If at time X the load is removed, there is an immediate elastic recovery BC equal to the elastic deformation OA. Plastic recovery which is, however, much smaller than the plastic flow or creep under load AY, then follows slowly.

Modulus of elasticity

The common measure of the elastic property of a material is the modulus of elasticity, which is the ratio of the applied stress to the deformation per unit length resulting from the application of that stress.

Then,
$$E = \frac{f}{e} = \frac{WL}{Ax}$$

where E is the modulus of elasticity

$f = \frac{W}{A}$ is the applied stress, W being the load and A the cross-sectional area

$e = \frac{x}{L}$ is the deformation per unit length or strain, x being the deformation under load W, and L the length of the member.

A, x and L are measured in the same units.

Concrete is not a truly elastic material, as already explained, and the stress–strain curve for continuously increasing loading is of the type shown at OB in Fig. 2.1, the curvature being exaggerated for clarity. If there was no plastic deformation, the curve would become a straight line OA, being the tangent to the curve OB at the origin. The true modulus of elasticity or instantaneous modulus could be calculated from the tangent OA, but for practical purposes it is usual to determine the modulus at an arbitrarily selected stress, applied at a defined rate, which takes into account movement due to creep. The modulus is then

$$\frac{OP}{PB} = \frac{f}{e}$$

It is necessary to load at a defined rate since concrete has more time to creep at a slower rate of loading. Since the creep of concrete is influenced by a number of factors, the modulus of elasticity is also affected.

The modulus of elasticity is not directly related to the other properties of concrete, although higher strengths are usually accompanied by higher values of E. For ordinary concretes the modulus of elasticity lies in the range 2 to 6×10^6 lb/in². As concrete ages, its modulus of elasticity increases, a fact of some importance since a restrained concrete member already in a state of strain due, for instance, to cooling after the initial hardening, becomes subjected to an increasing tensile stress. If completely restrained the stress at any time would be given by the usual formula, $f = Ee$.

Poisson's ratio

When subjected to compression, concrete contracts longitudinally and expands laterally. The ratio of the lateral strain to the longitudinal strain is known as Poisson's ratio, and within the normal working range of loading has a value of about 0·08 to 0·18. Within this range it increases with increase in cement content, an average value for 1:2:4 (520–540 lb cement per cubic yard) concrete being 0·11. The ratio also varies with the factors which affect the other properties of the material.

Creep

Some mention of plastic flow or creep has already been made in connection with the modulus of elasticity. Creep is a non-elastic deformation under load, believed to be due to closure of internal voids, viscous flow of the cement–water paste, crystalline flow in aggregates, and the squeezing of water from the cement gel under load. In practice, creep and drying shrinkage usually occur simultaneously and are often confused; exceptions are dams, tunnels and other underground works, and massive structures where little or no drying out occurs. Creep is useful in that it enables a readjustment of stress to take place when local stresses of high intensity might otherwise result in failure of the structure. Whether creep is beneficial or not depends on circumstances. For instance, when the concrete is reinforced with steel, creep generally leads to a transfer of load from the concrete to the steel. This may relieve high stresses in the concrete but may also result in overstressing of the steel.

In prestressed concrete, creep has an adverse effect resulting in a reduction in the tension in the reinforcement; it is necessary to allow for this by an increase in the initial pretensioning. See also p. 277.

Creep reduces the tendency to crack in restrained concrete members by relieving stresses due to shrinkage. The higher the percentage of reinforcement the greater is the restraint and hence, the tendency to develop cracks, but the restraint has a beneficial effect in that, while cracks are induced at more frequent intervals, they are usually so fine that they pass unnoticed, and are of no importance. See also p. 260.

The creep of unreinforced concrete is approximately proportional to the ratio of stress to strength of concrete within the normal range used in design, but as failure is approached the rate of creep increases rapidly.

The rate of creep is dependent on the following factors apart from the stresses imposed:

(1) *Strength.* An increase in strength usually leads to a reduction in creep, since the stress/strength ratio is reduced. This assumes new conditions are not imposed which also increase the stress; shrinkage of a restrained member for instance.

(2) *Cement.* There is some difference of opinion on the influence of cement type on creep. Generally it is thought that the type of cement is

not of great importance, except that creep is related to the stress/strength ratio for the concrete, and strength development is, of course, related to cement type.

(3) *Proportions of mix.* Creep decreases as the water/cement ratio and the volume of cement paste decrease. The movement in a very wet mix may be as much as twice that in an otherwise similar dry mix.

Neville[2.3] has suggested that the creep of concrete depends mainly on the creep of the cement paste and is reduced by the coarse aggregate. Newman[2.4] suggests that the ultimate creep (C_c) of concrete can be predicted from the equation:

$$C_c = C_p(1 - V_a)^n$$

where C_p is the ultimate creep strain of cement paste, being of the order of 2500–3000 $\times$ 10^{-6} in./in. for a stress level of half the compressive strength,

V_a is the solid volume fraction of total aggregate

n increases from 1·7 to 2·1 with the time under load.

The solid volume of the aggregate includes air voids within the aggregate particles but does not include voids between separate particles. See also p. 50.

(4) *Aggregate.* Creep increases as the aggregate becomes finer; and it is generally greater when porous aggregates are used.

The type of aggregate has a marked influence on creep. R. E. Davis and H. E. Davis[2.5] have given values for the creep of concrete made with different aggregates, as shown in Fig. 2.3.

(5) *Curing.* Creep decreases as cement hydration proceeds, so that concrete kept continuously wet creeps less than that cured in air. Alternate wetting and drying, however, leads to an increase in the magnitude of creep.[2.6]

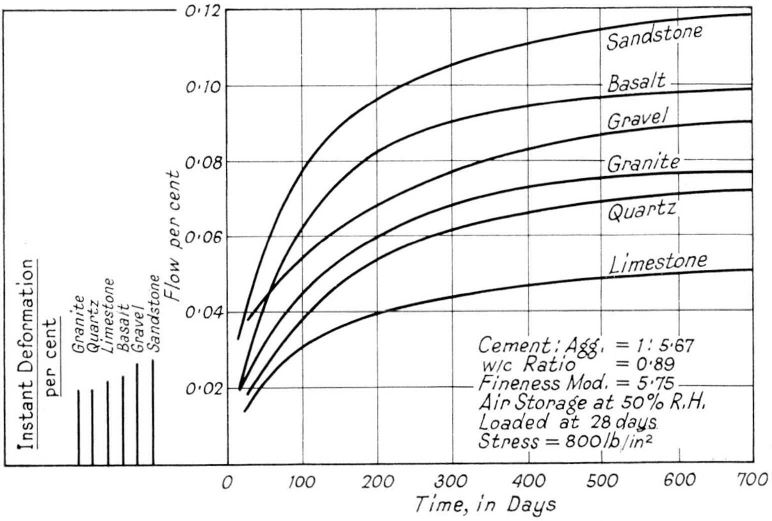

Fig. 2.3. Flow of concrete—effect of mineral character of the aggregate

(6) *Age.* The rate of creep decreases as the concrete ages. The creep at 1 year may be twice that at 28 days, but a further increase in creep of 20 per cent may take 5 years. The ultimate creep after 20 or 30 years may be taken as 30 to 40 per cent above that at 1 year.

Drying shrinkage and moisture movements

Two stages occur in the shrinkage of concrete as it is placed in practice:

(i) An initial shrinkage in the concrete while it is still in a fluid or plastic state, due to the reduction in the volume of the system cement plus water amounting to about one per cent of the absolute volume of the dry cement[(2.7)]; to the loss of water by seepage through the forms; to absorption by the forms, or in the case of roadwork by the formation; and to evaporation.

This shrinkage may be masked by the thermal expansion which accompanies the temperature rise resulting from the chemical reaction in the setting of the cement.

(ii) A further drying shrinkage as the concrete hardens and is allowed to dry.

This may be accompanied by a thermal shrinkage as the concrete cools.

The combined effects are shown in Fig. 2.4, which gives the results of tests made on a column. This diagram shows the effects of both temperature changes in the column and drying shrinkage. While the measurements were commenced at three hours after casting, the point at which the plastic concrete can be considered to have become rigid may not have occurred until some hours later. Expansion in the concrete mass due to the rise in temperature as the cement hydrated may, therefore, have been taken up to some extent by plastic movement. When the concrete became rigid, however, further expansion or contraction of the concrete would be measured as a movement of the rigid structure.

Much depends on the temperature of the concrete at the time it changes from a plastic to a rigid state. If, for instance, this occurs at the peak shown in Fig. 2.4, that is, at about 88°F, then the shrinkage, which will be important from the practical point of view, will be the thermal shrinkage as the concrete cools from 88°F to the surrounding temperature of 64°F, in the test quoted, and—added to this—the drying shrinkage as the concrete dries out.

The shape of the curve on the left-hand side of Fig. 2.4 indicates the occurrence of thermal shrinkage after the temperature peak has been passed.

The extent of the shrinkage which occurs before the concrete sets is dependent on the efficiency of the curing arrangements and on the extent to which water can be absorbed by the shuttering or formation. It can, therefore, be minimized by taking proper precautions. An example of a concrete road cracked by this initial shrinkage shortly after placing is shown in Fig. 2.5.

Dirty and seriously oversanded aggregates which require a high water/cement ratio to obtain satisfactory workability give concrete which is particularly prone to initial shrinkage cracking. Loss of water and consequent shrinkage of the concrete before or during setting is a fairly common cause of cracking in structural concrete, especially at changes of

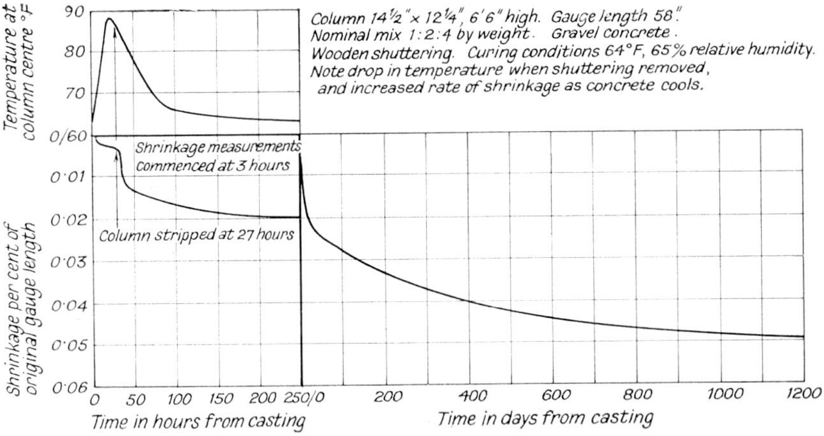

Fig. 2.4. Example of shrinkage of a concrete column from three hours after casting, with indication of temperature rise to be expected

section. Sometimes, incipient cracking occurring at this early stage becomes more obvious later.

The shrinkage which occurs as the concrete hardens and dries out is due to shrinkage of the cement gel and is partly irreversible to the extent of 0·3 to 0·6 of the drying shrinkage[2·6] for ordinary concrete mixes, the lower value being more common. It is only partly irreversible because on subsequent wetting it expands although not sufficiently to recover its original volume. Further drying and wetting produce a practically reversible shrinkage and expansion. These movements are illustrated in Fig. 2.6.

The shrinkage measurements in Fig. 2.4 commenced at three hours so that virtually all the initial drying shrinkage illustrated diagrammatically in Fig. 2.6 was included. Shrinkage test procedures, however, vary considerably. For instance, the Building Research Station in their Digest No. 35,[2·8] specify that after 2 days moist-curing the specimens shall be dried in air for 26 days, then soaked in water for 4 days. The drying

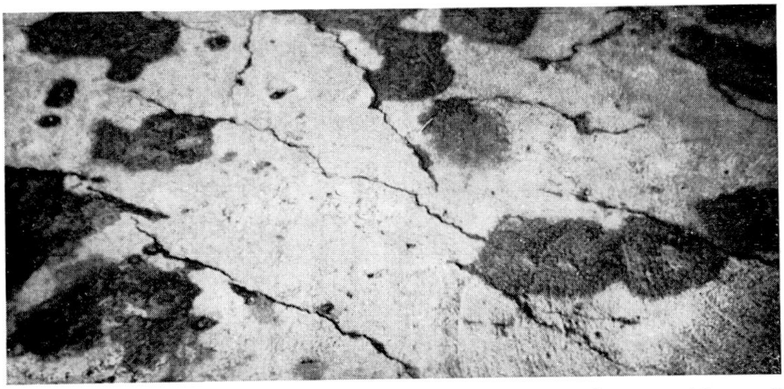

(*Crown copyright reserved*)

Fig. 2.5. Cracks due to initial shrinkage before concrete has hardened

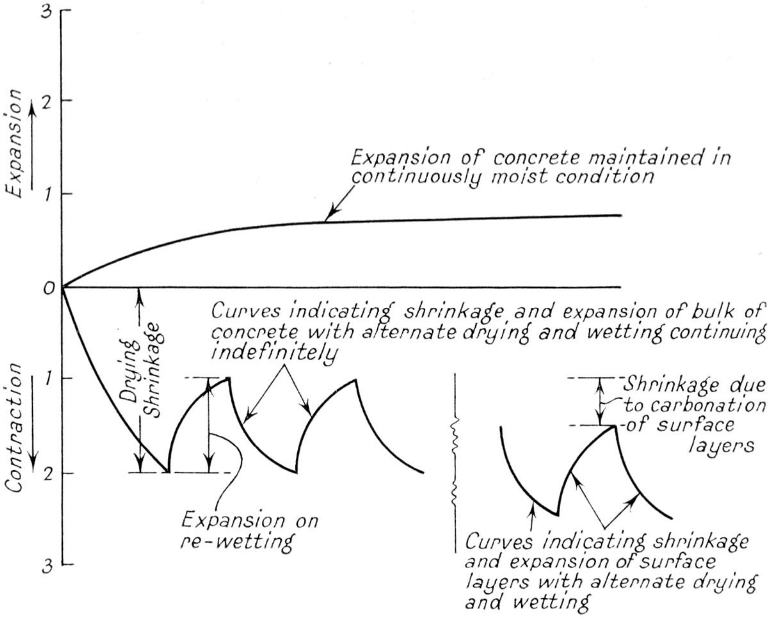

Fig. 2.6. Drying shrinkage and moisture movements with alternate wetting and drying

shrinkage is then measured after drying the specimens at 50°C for a further 3 weeks or so until they attain constant length. By this method the original drying shrinkage appears to be missed and what is measured is the smaller shrinkage on the second cycle, see Fig. 2.6.

The American method, ASTM C157-64T[2.9] is quite different. The specimen is first measured wet about 24 hours after mixing. It is then cured for a further 27 days in lime-saturated water and its length again measured. After that the specimen may either be stored again in lime-saturated water or in air at 23 ± 1·1°C and a relative humidity of 50 ± 4 per cent. The expansion in water or the shrinkage in air is measured at specified intervals up to 64 weeks. It will be noted that this test measures the initial drying shrinkage and, since the temperature is much lower than in the B.R.S. test, an entirely different result is obtained. See also 'Thermal Effects', p. 18.

The variation in methods of measuring shrinkage makes it very difficult to compare the results obtained from different laboratories. While a standard method of measurement would be useful for comparative purposes, the values obtained would not necessarily be a reliable guide to shrinkage in the structure, where conditions have such an important influence.

Continued immersion in water from the time of casting produces a slight expansion of the order of a quarter of the drying shrinkage, but on drying the concrete shrinks as before.

Differences in shrinkage occur between different parts of a structure. For example, foundations may never dry out, and may therefore not shrink at all or even tend to expand, while concrete wall panels may be dried to some extent by the weather outside and more so by heating internally, so that shrinkage certainly occurs. Such differences cause stresses which may lead to cracking. In buildings, for instance, cracking is usually found to be more prevalent between ground level and first floor level than elsewhere in the structure, with the exception of the roof and adjoining members. Drying shrinkage also occurs more quickly at the surface than within a concrete mass, since drying commences from the surface.

An additional shrinkage takes place in the surface layers due to a chemical action between the carbon dioxide in the atmosphere and the lime in the cement. This is known as carbonation, and as may be seen in Fig. 2.6, the total shrinkage at the surface is increased. This shrinkage is of some importance in connection with the crazing of concrete surfaces. In porous concretes the effect of carbonation is more marked than in dense concretes, as might be expected.

The magnitude of the drying shrinkage is more dependent on the water content of the mix than on any other factor. As a result, a rough assessment of drying shrinkage may be obtained by use of the equation

$$S = \frac{3W - 400}{10{,}000}$$

Where S = drying shrinkage, per cent reduction in length,
W = water content expressed in lb/yd^3 of concrete.

This formula takes account of the cement content of the mix, because of its interdependence on the water/cement ratio. Other factors which influence drying shrinkage are the type of cement; the size and shape of specimen; the surrounding atmospheric conditions; age and the amount of reinforcement.

With so many factors, refinements in methods of assessing drying shrinkage are of little practical importance, and there is little point in attempting other than a rough assessment by the simple equation above, or from Fig. 2.7 which for some purposes may be more convenient.

The mineral character of the aggregate affects the drying shrinkage of concrete in which it is used. This has been attributed in the past to greater compressibility or to pronounced moisture movements in the aggregate itself. Carlson[2.10] showed greater shrinkage with the more easily compressible aggregates, while Davis and Troxell[2.11] found that a concrete containing sandstone shrank twice as much as an otherwise similar concrete. More recently it has been suggested[2.12,2.13,2.8] that some dolerites of Scottish origin exhibit high volume changes on wetting and drying and that concrete made from these aggregates has a considerably higher "drying shrinkage" (see p. 12) than that of concrete made from non-shrinking aggregates, e.g. quartzite gravels, limestones and granite. However, aggregates classified by this particular test as having high drying shrinkages have been used in Scotland for many years. There have been a few reports of undue movements and lack of durability, but the evidence is not conclusive that the aggregate was the major cause.

The presence of clay in aggregates or as a coating on aggregates leads to increased shrinkage. Clay coatings, in addition to being undesirable owing to poor bond, can increase shrinkage by up to 70 per cent.[2.14]

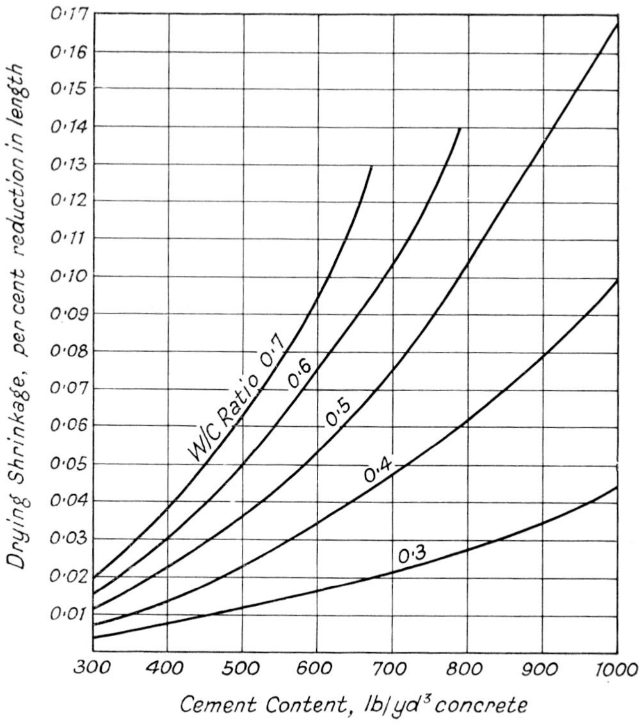

Fig. 2.7. Average values for drying shrinkage of concrete specimens stored in air

The rate at which drying shrinkage occurs decreases as the size of specimen increases. For instance, with a small specimen 2 or 3 in. square the drying shrinkage under constant temperature and humidity conditions is virtually complete in less than a month, whereas a concrete section, say, 2 ft square, continues to shrink for several years under the same conditions.

A further illustration of the influence of size of specimen and age on shrinkage, is given by Atherton[(2·13)] who measured the shrinkage in the members of a turbo-generator foundation, and compared this with the shrinkage of a 3 ft × 4 in. × 4 in. laboratory specimen. The readings were taken in a 5 ft thick wall at the steam end of the turbo-alternator block. The shrinkage in the 5 ft thick wall had reached 0·04 per cent after about a year-and-a-half, by which time the smaller specimen had reached 0·06 per cent. The thick wall was still shrinking after eight years, and had then reached over 0·06 per cent, whereas the smaller specimen was by then shrinking at a much slower rate, and the final shrinkages of the two were about equal. These differences arise because concrete normally only dries from exposed surfaces.

L'Hermite[(2·6)] has observed that desiccation may reach a depth of 3 in. in one month, but 2 ft only after 10 years. This drying of the surface layers gives rise to internal stresses which, if not relieved by creep, may lead to cracking. In practice cracking is not usually noticed until 6 months to 2 years after concreting.

Lea and Davey(2.16) have indicated that the rate of shrinkage is less for concrete exposed to the weather (near London) than for concrete stored under the constant conditions mentioned above. They give shrinkage values for specimens 9 ft 0 in. × 1 ft 8 in. × 2 ft 6 in. (0·4 per cent steel), including the seasonal variation of the member exposed to the weather due to wetting and drying, amounting to nearly 0·01 per cent. Shrinkage may be expected to occur far more rapidly in hot dry atmospheres. An indication of the influence of the relative humidity of the atmosphere is given in Fig. 2.8, based on recommendations by the Comité Européen du

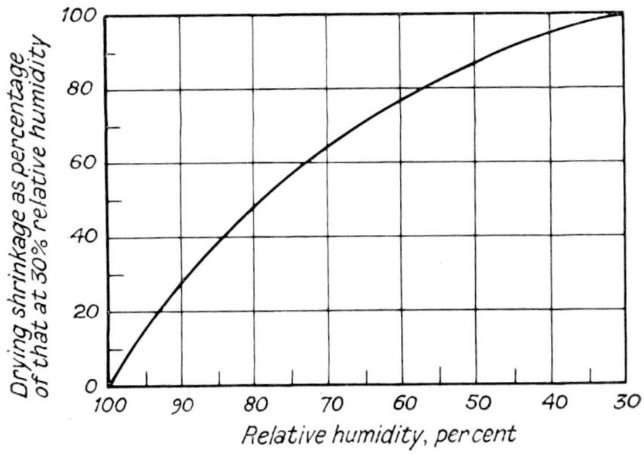

Fig. 2.8. The effect of the relative humidity of ventilated storage conditions on the drying shrinkage of concrete

Béton.(2.17) The diagram is plotted to show the drying shrinkage at any relative humidity as related to that at 30 per cent humidity, since the actual coefficient of shrinkage depends on other factors, as already indicated.

Slow drying of concrete as compared with quick drying tends toward reduced cracking because of the compensating effects of extensibility and plastic flow which are thus given time to take effect, and because of increased strength derived from the longer presence of moisture in the concrete.

In normal reinforced concrete drying shrinkage results in a compressive stress in the bars, which themselves resist and reduce the overall drying shrinkage in the reinforced concrete member. It is generally assumed that the drying shrinkage of a reinforced concrete member is half that of an unreinforced member, although the actual value is obviously dependent on the amount of reinforcement used.

The Comité Européen du Béton suggests(2.17) that the influence of the reinforcement may be taken into account by multiplying the shrinkage coefficients by the term

$$1 - \frac{10 \times A_r}{A_c}$$

where A_r is the cross-sectional area of the longitudinal reinforcement and A_c is the cross-sectional area of the member.

In prestressed concrete, it is necessary to make an allowance for the relief of stress in the steel resulting from creep in the concrete and the steel, and drying shrinkage in the concrete. A good average value for the loss of prestress in the steel is 20,000 lb/in^2, although in poor concrete it may be as high as 50,000 lb/in^2.[2.18] Other authorities assess the loss of prestress at between 12 and 18 per cent of the initial steel stress.[2.19]

Shrinkage and moisture movements are of importance because they result in cracking of the concrete unless the stresses developed can be relieved by creep, or in the case of large areas by suitably placed joints. Thus floors, roads, roofs and long walls are subdivided into panels or sections to control cracking due to drying shrinkage and moisture changes, as well as thermal movements.

Thermal effects

The thermal coefficient of expansion and contraction is the change in a unit length per degree of temperature. Its value varies a little with the richness of the mix and the water content, but for all practical purposes it may be taken as $5{\cdot}5 \times 10^{-6}$ per degree Fahrenheit for unrestrained movement. For estimating the movement likely to occur in large masses of concrete it is usually safe to assume a value of one-half that given above, the other half being absorbed by creep without a serious danger of cracking.

Thermal expansion and contraction are not often uniform throughout a mass of concrete. In the first place, the chemical combination of cement and water is accompanied by the generation of an appreciable quantity of heat, which can only escape by conduction to the surface. It follows that the larger the mass, the higher the internal temperature at an early age compared with that at the surface, with a consequent difference in thermal shrinkage as the mass cools. This leads to the development of tensile stresses conducive to cracking.

When the temperature of concrete is raised by heating, or even by the sun's rays, moisture is lost from the cement gel and a further drying shrinkage results. After the initial thermal expansion, heated concrete therefore tends to shrink again to some extent as this further drying gradually occurs. As a result, the stress conditions in concrete that is subject to temperature changes are complex. Examples are tanks for hot liquids, roads, roofs, and heated floors, particularly when adjoining parts of the structure are not heated so that a thermal gradient is also present. Warping and cracking in these circumstances are not uncommon, and considerable ingenuity in designing expansion joints may be necessary if they are to be avoided. See page 173.

Both shrinkage and creep are affected by temperature. Ross[2.20,2.21] has made extensive studies which have provided much useful data on these relationships.

Durability

The useful life of concrete is ordinarily limited by the disintegrating effects of:

(1) Weathering by the action of rain and frost, and by the expansion and contraction resulting from alternate wetting and drying.

(2) Chemical attack by such agencies as seawater; moorland, marsh and other waters; industrial chemicals and wastes; sewage; animal and vegetable oils, fats, grease; milk; and sugars.

(3) Wear by abrasion from foot and vehicular traffic, by wave action, and by water-borne and wind-borne particles.

Weathering by rain and frost action is chiefly a function of water-tightness or impermeability, since leaching and attack by the carbonic and other acids present in rainwater, and disruption by frost action, depend on the penetration of water into the surface. The mechanism of rain and frost action and the effect of chemical attack by the various agencies listed are not in themselves properties of concrete, and will be described more fully in a later chapter.

The quantity of cement used in concrete does not affect the resistance to weathering to a very large extent unless the cement content is not sufficient to fill the interstices between the aggregate. Glanville[2.22] investigated the effect of cement content on permeability and found it was of no great advantage to use more than about 525 lb of cement per cubic yard provided only sufficient water was used to make the mix workable. To obtain the same degree of impermeability, very wet mixes required a higher cement content, say 625 lb or more of cement per cubic yard.

Permeability

Concrete has a tendency to be porous due to the presence of voids formed during or after placing. In the first place it is usually necessary, in order to obtain workable mixes, to use far more water than is actually necessary for chemical combination with the cement. This water occupies space, and when later it dries out, leaves behind air voids. In addition to these initially water-occupied voids, there is always a small percentage of entrained air voids. Secondly, there is a decrease in absolute volume of the cement and water after chemical reaction and drying out, so that the set cement paste, when dry, occupies less volume than the fresh paste, whatever the water/cement ratio.

If proper care is taken, Portland cement concrete can be made sufficiently impermeable for most purposes without the addition of any special materials. From the above analysis of the cause of voids, it is clear that for the densest and least permeable concrete, the water-cement ratio should be reduced to a minimum consistent with adequate workability for thorough compaction without segregation. A compromise is, therefore, required between a low water-cement ratio and adequate workability, and the best conditions will depend on the type of structure and the method of compaction.

Other factors influencing permeability are:

(1) The soundness and porosity of the aggregate.

(2) Age. The permeability decreases with age, the decrease being greater for wet mixes than for drier ones. Glanville, in Building Research Technical Paper No. 3[2.22] has shown that after a year there is only a small difference between the permeability of mixes of normal consistency and very wet mixes.

(3) The grading of the aggregate should be chosen to give the most

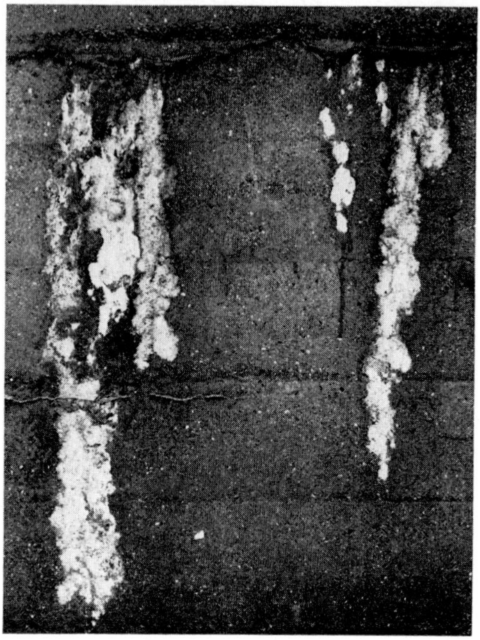

Fig. 2.9. Efflorescence resulting from the percolation of water through a defective construction joint, bringing with it chemical salts

workable concrete with the least water. Harsh gradings, especially as regards the sand, should be avoided.

(4) Curing has an important influence, and it is necessary to keep the concrete moist particularly during the first few days.

Joints are the weakest point in a concrete structure, and an example of weeping through a construction joint is given in Fig. 2.9.[2.23] If impermeability is important, it is wise to provide some form of mechanical seal, as indicated on page 176, since it is difficult to avoid local seepage due to segregation, poor compaction or some other occurrence. Whether such measures are considered necessary or not, it is essential that every effort should be made to obtain a good bond.

Permeability is very difficult to measure accurately, and tests are not often made outside the research laboratory.

Resistance to abrasion

Resistance to abrasion is directly related to crushing strength and in general it is safe to assume that concrete having the highest crushing strength has the greatest resistance to abrasion. Occasionally tests are made on the wear of paving flags, by exposing them to abrasion by steel balls for 48 hours, and finding the loss in weight. The method is described in British Standard 368:1956 for Precast Concrete Flags.

A tough igneous rock aggregate is often preferred to harder and more brittle flint aggregate in paving concrete.

Autogenous healing

Many tests have shown that the finer cracks in fractured concrete will completely heal under moist conditions. Apparently when the fracture occurs, unhydrated cement is exposed, and this in the presence of moisture hydrates and sets.

REFERENCES

2.1 B.S. 1881:1968, *Methods of Testing Concrete*. British Standards Institution.

2.2 ASTM Designation C496–64T, *Tentative Method for Splitting Tensile Strength of Concrete Cylinders*, ASTM Standards, Part 10, 1965. American Society for Testing Materials.

2.3 NEVILLE, A. M., Creep of concrete as a function of its cement paste content, *Mag. Concr. Res.*, **16**, No. 46, March 1964. Cement and Concrete Association.

2.4 NEWMAN, K., Properties of concrete, *Structural Concrete*, **2**, No. 11, Sept./Oct. 1965. Reinforced Concrete Association.

2.5 DAVIS, R. E. and H. E., Flow of concrete under the action of sustained loads, *J. Amer. Concr. Inst.*, March 1931.

2.6 L'HERMITE, R., Volume changes of concrete, *4th International Symposium on the Chemistry of Cement*, Washington D.C., 1960.

2.7 SWAYZE, M. A., Early concrete volume changes and their control, *Proc. Amer. Concr. Inst.*, **38**, April 1942, pp. 425–440.

2.8 *Shrinkage of Natural Aggregates in Concrete*, Building Research Station Digest No. 35, June 1963. H.M.S.O.

2.9 ASTM Designation C157–64T, *Test for Length Change of Cement Mortar and Concrete*, ASTM Standards, Part 10, 1965, p. 107.

2.10 CARLSON, R. W., Drying shrinkage of concrete as affected by many factors, *Proc. Amer. Soc. Civ. Engrs*, **38**, Pt. II, 1938, pp. 419–437.

2.11 DAVIS, R. E., and TROXELL, G. E., Volumetric changes in Portland cement mortars and concretes, *J. Amer. Concr. Inst.*, **25**, 1929, pp. 210–260.

2.12 SNOWDEN, L. C. and EDWARDS, A. G., The moisture movement of natural aggregate and its effect on concrete, *Mag. Concr. Res.*, **14**, No. 41, July 1962. Cement and Concrete Association.

2.13 JENKINS, R. A. SEFTON, PLOWMAN, J. M. and HASELTINE, B. A., Investigation into the causes of the deflection of heated concrete floors, including shrinkage, *Structural Engineer*, **43**, No. 4, April 1965, pp. 105–117.

2.14 POWERS, T. C., Causes and control of volume change, *Portland Cement Assoc. J. Res. and Development Laboratories*, **1**, No. 1, Jan. 1959.

2.15 ATHERTON, A., in Discussion on RAE, F. A., Design and construction of a reinforced concrete foundation block for a 200-MW turbo-generator, *Proc. Inst. Civ. Engrs*, July 1963, pp. 379–380.

2.16 LEA, F. M. and DAVEY, N., The deterioration of concrete in structures, *Paper* No. 5722, Inst. Civ. Engrs, 1949.

2.17 COMITÉ EUROPÉEN DU BÉTON, *Recommendations for an International Code of Practice for Reinforced Concrete*, 1963. Published jointly by the Cement and Concrete Association and the American Concrete Institute.

2.18 FLUCK, P. G. and WASHA, G. W., Creep of plain and reinforced concrete, *J. Amer. Concr. Inst.*, Title No. 54–59, **29**, No. 10, April 1958.

2.19 MAGNEL, GUSTAVE, Creep of steel and concrete in relation to prestressed concrete, *J. Amer. Concr. Inst.*, **19**, No. 6, Feb. 1948.

2.20 ROSS, A. D., The elasticity, creep and shrinkage of concrete, *Proc. Conference on the Mechanical Properties of Non-metallic Brittle Materials*, London, April 1958. Butterworths Scientific Publications.

2.21 ROSS, A. D., ILLSTON, J. M. and ENGLAND, G. L., Short and long-term deformations of concrete as influenced by its physical structure and state, *International Conference on the Structure of Concrete, 1965, Paper* No. H.1. Cement and Concrete Association.

2.22 GLANVILLE, W. H., *The Permeability of Portland Cement Concrete,* Building Research Technical Paper No. 3, revised edition. H.M.S.O.

2.23 DAVEY, N., The surface finishing of concrete structures, *J. Inst. Civ. Engrs*, April 1942.

CHAPTER 3

Aggregates

MOST factors pertaining to the suitability of aggregate deposits are related to the geologic history of the surrounding region. The geological processes by which a deposit was formed or by which it was subsequently modified are responsible for its size, shape and location, the type and condition of the rock, the grading, rounding, and degree of uniformity, and a number of other factors which are pertinent to the question of utilization.

Natural sands and gravels are usually the most economical source of aggregate and are commonly obtained from stream deposits, as in the Thames valley, from glacial deposits, such as occur in Lancashire and Cheshire, and from sand dunes around the coasts. Aggregates are also now being won from the sea bed at an increasing rate. Deposits are being worked in the English Channel, the North Sea, the Mersey, the Bristol Channel, and other estuaries around the coast.

Stream deposits are still the most common and generally the most satisfactory because (1) the individual pieces are usually rounded, (2) streams exercise a sorting action which may improve grading, and (3) abrasion caused by stream transportation and deposition leads to a partial elimination of the weaker materials. Glacial deposits usually contain material having heterogeneous shapes and sizes, ranging widely in quality. Alluvial fans are the deposits, usually in the form of sloping plains, which extend from the feet of mountainous areas into adjacent valleys. The sands and gravels laid down under such conditions are angular, the material is poorly graded, and the deposit is irregular. Marine-dredged aggregates have been used to a limited extent in concrete for many years, but their increasing production and use have given cause for investigation of acceptable limits for shell and salt contents.

In areas where suitable rocks outcrop, crushed quarried stone provides a good coarse aggregate although the cost is higher than a naturally occurring gravel owing to the necessity of blasting the material and then of crushing it to a suitable size. The fines from quarries are usually dusty and the angular shape makes the mix harsher than a similar mix containing rounded sand grains.

Suitability of various types of rock

The suitability of various types of rock is usually as follows:

(1) *Igneous rocks.* These include granites, basalts, dolerites, gabbros and porphyries. Some confusion results from the commercial use of the term "granite" to cover not only most of the igneous rocks, but also some types of limestone. Other local names include: whinstone, toadstone, and greenstone for dolerites and basalts; and elvan for porphyries.

Granite is hard, tough and dense and is an excellent aggregate for concrete. Geologically, granite is coarsely crystalline in structure, free from bedding planes, and is composed of quartz, felspar, mica, and hornblende. In Cornwall, particularly St. Austell, granite sands are found. These have been formed by disintegration of the felspar and hornblende into china clay, leaving the quartz crystals as a sand. In the selection of such material care should be taken to ensure that the sand is washed free from china clay and that excessive amounts of mica are not present, see page 45.

Basalts are igneous rocks similar to granite, but with a much finer grain structure due to more rapid cooling when they were formed. They are excellent aggregates.

Dolerites have a fine crystalline grain structure and contain more felspars and such materials as augite and olivine than do the granites. Some dolerites when used as aggregate for concrete may cause cracking and disruption of the concrete. It has been found that the deleterious rocks expand and contract with changes in humidity. Mineralogical examination has shown that such aggregates contain decomposed ferro-magnesium silicates, such as olivine. The commonest mineral found in these circumstances is montmorillonite. In some parts of the world, the disintegration of some igneous rocks may be far more serious; in Australia for instance, rock which looks as hard as Cornish granite may become a heap of clay in a matter of months simply by exposure to the weather.

(2) *Sandstones.* When hard and dense, most sandstones are suitable as aggregates. The best are those which are composed of quartz grains cemented with hydrated iron oxide or amorphous silica, known geologically as ferruginous and siliceous sandstones. Calcareous sandstones cemented with calcium carbonate are more liable to attack by the carbonic and sulphuric acids present in the atmosphere. Imperfect cementation of the constituent grains makes some sandstones friable and very porous and they are then unsuitable aggregates.

Sandstones range from the harder types composed of very close grains, through the softer and more loosely grained grits, to the sandy shales, in which the presence of clay causes softness, friability and high absorption.

(3) *Limestones* are sedimentary rocks composed chiefly of calcium carbonate. They range from very hard, close-grained crystalline rock to soft chalk. Intermediately, there are the oolitic limestones represented by Portland stone, Bath stone, Clipsham stone, etc., and calcite, such as Derbyshire Spar.

Dolomitic limestones contain a large proportion of magnesium carbonate in addition to calcium carbonate.

The harder and denser types of limestone are very suitable for concrete, but the softer and more porous types should be avoided if possible. The decision will depend on the type of construction; for instance, Bath stone may be used as an aggregate in architectural cast stone, but would be unsatisfactory as a road aggregate because, in addition to poor resistance to wear, it would be liable to become saturated and disintegrated by frost action. Chalk is normally unsuitable.

(4) *Shales* are usually poor aggregates, being soft, weak, laminated and absorptive. In addition, the normal flat shape of the particles makes compaction of any concrete in which they are used, very difficult.

(5) *Metamorphic rocks* are variable in character. On the one hand marbles and quartzites are usually massive, dense, and adequately tough

and strong, providing good aggregates. On the other hand, some schists and slates are often thinly laminated and, therefore, unsuitable.

Artificial aggregates

Blastfurnace slag is probably the most widely used of the artificial aggregates, more so on the Continent and in the U.S.A. than in Britain. It is produced in blast furnaces during the production of pig iron and consists of the silicates and alumino-silicates of lime. Slags vary in hardness, and any which "dust" or disintegrate due to the presence of too much ferrous iron, should be avoided. Many unsound slags may be detected by immersing them in water for a period of two weeks, during which time they should show signs of disintegration. Laboratory methods of testing for soundness of blastfurnace slag are given in B.S. 1047: 1952[3.2]. These are: (i) a chemical analysis for total sulphur, acid-soluble sulphate, silica, alumina, lime, magnesia, iron oxide, and titanium oxide; (ii) tests for stability including iron unsoundness; an analytical test for falling, dusting or lime unsoundness; and a microscopic test; (iii) a water absorption test. A minimum bulk density of 78 lb/ft^3 is also required.

Clean broken-brick of good quality provides a satisfactory aggregate, the strength and density of the concrete depending on the type of brick; engineering and allied bricks when crushed make quite good concrete of medium crushing strength. In using second-hand bricks it is essential that all plaster should be removed, otherwise the calcium sulphate present is liable to prevent or delay setting, and to cause disintegration in a short time. Bricks containing soluble sulphates in excess of $\frac{1}{2}$ per cent should also be avoided. Brick aggregate should be saturated before use, because of its relatively high absorbency. Porous types of brick should not be used as aggregate in reinforced concrete work owing to the danger of penetration of moisture which may lead to serious corrosion of the steel reinforcement.

Glass is not a reliable aggregate, since some varieties possess strong alkali-aggregate reactivity.

Lightweight aggregates

Some aggregates, such as clinker, breeze, pulverized fuel ash (fly ash), pumice, foamed slag, and wood waste are used in the manufacture of lightweight non-structural concrete. Artificial and processed aggregates, notably expanded clays and shales, foamed slag, and sintered fly ash pellets, have been used on a small scale for lightweight structural concrete. Their high absorption, the need for extra cover to the reinforcement, and the relatively low modulus of elasticity of the concrete produced have been the chief factors limiting their usefulness for this purpose.

All these aggregates are discussed in Chapter 22 dealing with lightweight concrete.

Production of sand and gravel aggregates

Sand and gravel are obtained either from pits or by dredging from the bottom of rivers or from the sea bed. Pits are termed "wet" or "dry," according to the water level in the vicinity, although some pits which would otherwise be "wet" are dried out by pumping. The deposits are usually covered with overburden which has to be removed by scrapers, mechanical shovels, draglines or by sluicing. The overburden consists of the soil and stones mixed. Removal is not always straightforward owing to the presence of faults and dips in the gravel seams, and unless care is taken to remove these materials, occasional loads of dirty aggregate occur.

The sand and gravel are then excavated by mechanical shovels in "dry" pits, by draglines in both "wet" and "dry" pits, and by means of crane and grab, or suction pumps, in "wet" pits. Gravel seams often overlie beds of clay, and it is especially necessary that the inclusion of lumps of clay in the gravel should be avoided, since even with vigorous washing the clay lumps are extremely difficult to remove. The danger of the inclusion of clay lumps is greater in "wet" pits than in "dry" pits where draglines or grabs are used, because the driver is unable to see the bottom, and, therefore, to control the depth to which he digs.

The mixed gravel and sand is usually conveyed from the pit to the screening plant by conveyor belt or in light trucks, or by pipe line when the gravel is raised by a suction pump. Conveyor belts are commonly used to raise the materials from ground level to the level of the washing and screening plant.

Screening and washing of the gravel and sand is carried out either by means of cylindrical rotating screens or by the more modern vibrating screens. A recent improvement to vibrating screens has been made by carefully designing them to produce a resonant effect. The sand and dirty water are usually carried away together and are then de-watered by another unit. From the screening unit, the sand and gravel are fed by gravity into suitable hoppers. Oversize material goes to a crusher and is then returned for re-screening. Typical arrangements of vibratory and rotary screens are given in Figs. 3.1 and 3.2. The sand at some pits is

Fig. 3.1. Vibrating screens and overhead storage hoppers at a gravel pit in the London area

Fig. 3.2. A rotary screen

treated further to modify its grading, by means of classifiers, such as those seen in Fig. 3.3.

One method used for obtaining aggregates from the bed of the North Sea 5 miles from the coast in depths up to 55 ft, is by the use of a steel suction tube about 24 in. diameter and 60 ft long. The steel tube is pivoted over the side of the ship until the lower end rests on the sea bed. The aggregate is then pumped up the tube through a large impeller and into the open hold of the ship. Using this method, about 1,500 tons of aggregate can be loaded in about 1 hour.

Fig. 3.3. Overhead classifiers used for improvement of the grading of sands at a gravel pit in the London area

On discharging the aggregate at the plant, the material is processed in the normal manner.

Marine dredged material is usually quite clean apart from the presence of sea salts and sea shells. Washing is likely to remove or reduce the salts. There is, unfortunately, no authoritative guidance as to the permissible limits of salts or shells in fine or coarse aggregates obtained from the sea bed. It is likely, however, that the presence of shells will reduce the compressive strength of concrete. Dredged aggregates, particularly the sands, are apt to vary in grading according to the location from which the material is obtained.

In hot desert countries, aggregate supplies are usually a serious difficulty, particularly the shortage of gravel, and it may be necessary to employ native labour to sweep up the stones by hand from the surface of the desert. There is also always the possibility of quite large areas of desert where the sand, and any gravel present, may be heavily impregnated with salts, particularly gypsum. Such areas are unsuitable, but it is usually necessary to allow the use of aggregates which contain a somewhat higher percentage of these salts than would be considered satisfactory in Great Britain. Although these desert countries consist very largely of sand, it is often quite difficult to find a satisfactory source near the site, since most of the sand is too fine, and frequently contains a considerable proportion of silt. Usual sources are old river valleys or wadis, in which the sand is dug from pits. Since the fine sand and dust on the desert surface is continually drifting in the wind, and more particularly during sand storms, sand pits and stockpiles soon become silted up, causing increased difficulty in obtaining a uniform and satisfactory grading, and cleanliness.

Any question of washing the aggregate may be quite impracticable due to the absence of water, except perhaps for meagre supplies of brackish water, which almost invariably contains appreciable quantities of sulphates and chlorides, and would therefore not be effective in producing a salt-free material.

Production of crushed stone aggregates

The preparation of crushed stone aggregates is not so easy as that of gravel. The overburden must first be removed, the method depending on the formation of the top surface of the rock. The rock is next blasted and conveyed in lumps in trucks, overhead skips, or by belt conveyor to the crushing plant. A primary crusher of the studded roll, gyratory, jaw, or cone type reduces most of the material to less than 3 in. diameter. The crushed rock is taken, usually by belt conveyor, to the primary screen which separates the material above and below the 3 in. size. That passing the primary screen is conveyed to the final screens, which sort it out into the various sizes required. Material retained on the 3 in. primary screen is passed through secondary crushers and again returned to the screens.

Six types of crushers are used for the production of crushed stone aggregate. These are:

(1) Jaw crushers consisting of one or more swinging jaws operating against a fixed jaw, the distance apart and the length of movement

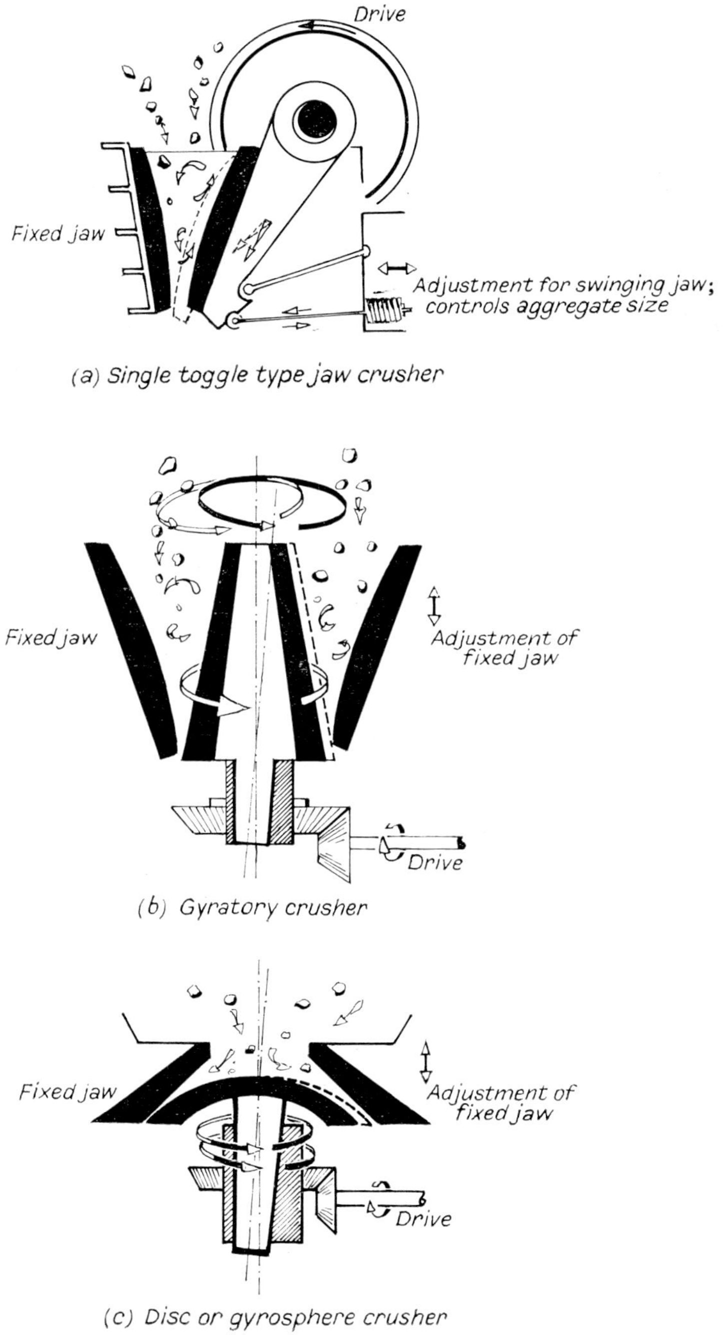

Fig. 3.4. Action of jaw, gyratory and disc crushers

determining the ultimate size of the crushed stone. The single toggle type is used for the softer stones, while a double toggle arrangement gives a more powerful thrust to crush harder stones.

(2) Gyratory crushers in which a crushing head is rocked by an eccentric on the inclined revolving shaft which carries it.

(3) Disc crushers comprising one stationary and one revolving saucer-shaped disc which open and close as the material passes through.

(4) Hammer or impact crushers of various types. Hammer crushers are subject to considerable wear and maintenance costs are liable to be heavy.

(5) Roll crushers in which the crushing is effected by feeding the material between the toothed, serrated or corrugated surface of one roll and the similar or smooth surface of the other. Uneven wear of the rollers causes difficulties in maintaining an even size.

(6) Rod mills are replacing roll crushers for the reduction of fines, being more economical, and giving a more uniform product.

The action of jaw, gyratory and disc crushers is shown diagrammatically in Fig. 3.4. Hammer and gyratory crushers are usually regarded as the most satisfactory since the aggregate produced is more cubical in form.

Further information on factors governing the grading and shape of crushed rock is given in a survey of the literature by Shergold and Greysmith.[3.2]

The screens may be of the revolving type used for sand and gravel, or of one of several types operated by vibration.

The commercial grading of aggregates

Aggregates cannot be screened commercially to the same degree of accuracy as that attained in the laboratory. Several factors are responsible for the variations which normally occur, even with modern screening plants. Perhaps the most important factor is the speed at which the material is passed through the screens, any increase in speed resulting in a greater proportion of undersized material being retained. On the other hand, the presence of oversize material in an aggregate usually results from an increase in the size of the screen meshes due to wear.

Aggregates from properly equipped washing and screening plants may be expected to be of reasonably uniform grading, but it is advisable to inspect each load as it is delivered, and to make check tests on the grading from time to time. An example of the sort of grading variation which may be expected on the site is given in Fig. 3.5. The sieve analyses were made on samples of $1\frac{1}{2}$ in.–$\frac{3}{4}$ in. and $\frac{3}{4}$ in.–$\frac{3}{8}$ in. gravel aggregate delivered during the course of a week on a site where strict supervision was maintained.

In areas where it is not possible to obtain a suitably graded sand it may be necessary to correct the grading by the addition of a suitable blending sand or by removing with classifiers, or other means, portions of sizes present in excessive amounts; for instance, at Parker Dam it was necessary to waste a portion of the sand finer than the American No. 16 sieve and to add a blending sand, in order to obtain closer control of the grading. This practice has also been adopted in Great Britain by some more modern producers.

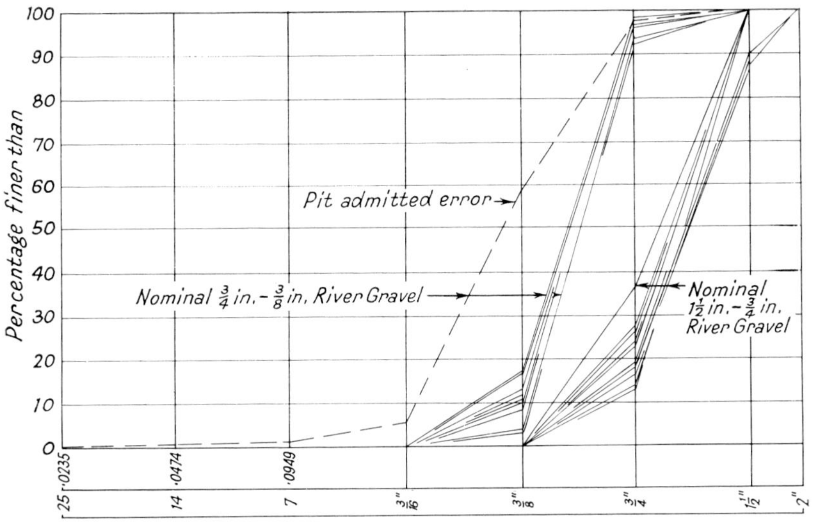

Fig. 3.5. Gradings of a typical day's delivery of $1\frac{1}{2}$ in.–$\frac{3}{4}$ in. and $\frac{3}{4}$ in.–$\frac{3}{8}$ in. river gravel from a pit with a large washing and screening plant

If classifiers are not in use, a considerable variation in the grading of sand as delivered to the site must be expected. A variation from the coarser end to the finer end of any one of the zones given in B.S. 882 : 1965[(3.3)] should not be unexpected, even from one source of supply.

Washed sand is sometimes found to be deficient in material passing a No. 52 B.S. sieve if the washing has been too vigorous. Lack of this fine sand leads to harshness in the concrete mix.

After the material has been washed and screened, the real problem in coarse aggregate grading control begins; namely, prevention of segregation due to handling. Perfectly graded gravel may segregate during a single incorrect stockpiling operation to such an extent that the workability of the concrete in which it is used will be unsatisfactory. Tests[(3.4)] made at Boulder Dam showed that when the fine gravel $\frac{3}{4}$ in. to $\frac{3}{16}$ in. was dropped from a belt into a conical pile and recovered through a gate beneath the centre of the pile, the proportion of $\frac{3}{4}$ in. to $\frac{3}{8}$ in. material varied from 55 per cent to 74 per cent, and the $\frac{3}{8}$ in. to $\frac{3}{16}$ in. material from 42 per cent to 16 per cent. Correct and incorrect methods of stockpiling and handling aggregates are shown in Figs. 3.6 and 3.7.

Sampling of aggregates for testing

Before aggregates are accepted from any particular pit or other source of supply, it is very desirable that a visit should be made in order to inspect the methods of removing overburden, the variations in strata, pockets of unsuitable material, and methods of digging. Furthermore, the final choice of the source of supply of any aggregate should be based not only on the results of tests in the laboratory, but also on the inspection of the method of production and organization at the pit. Only in this way is it possible to eliminate sources of supply likely to give trouble by

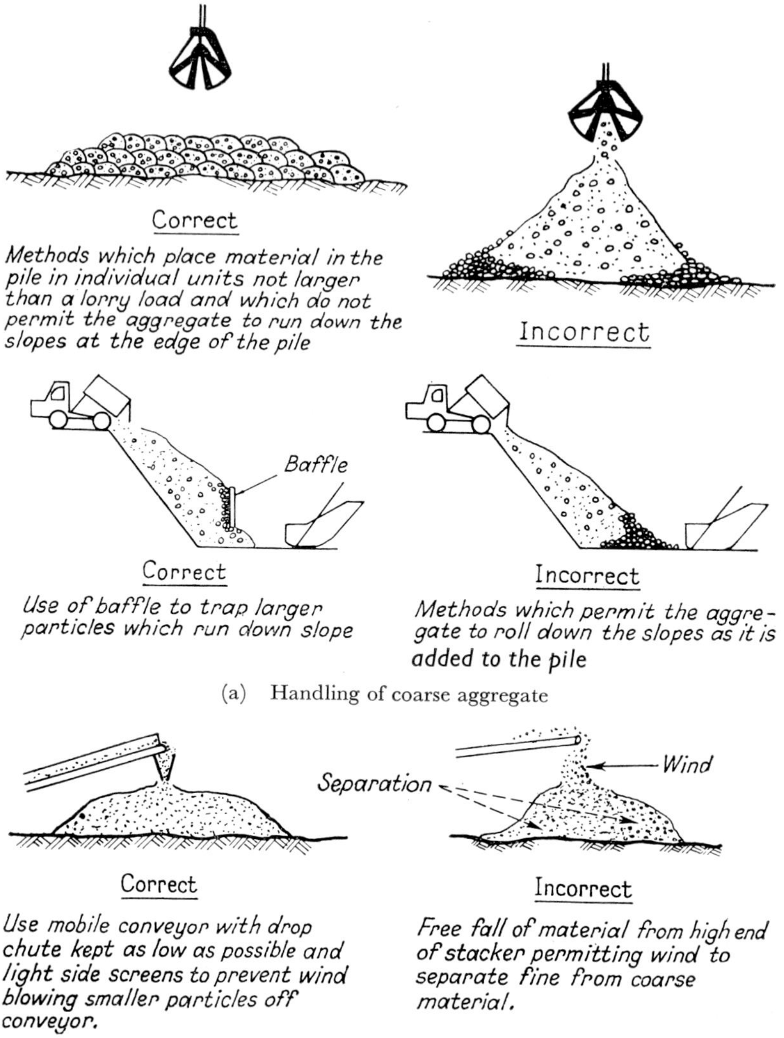

(a) Handling of coarse aggregate

(b) Handling of fine aggregate

Fig. 3.6

large fluctuations in grading, occasional dirty loads, breakdown in continuity of supplies due to failure of worn-out plant at the pit, and other similar causes.

Representative samples of the material as stockpiled or in the bins ready for delivery should be obtained at the same time. The method of taking samples will have an important influence on the accuracy of the results, and reference should be made to B.S. 812:1967[3.5], for the correct procedure.

Important points to note are:

(1) Under favourable conditions at least ten portions (increments) should be drawn from different parts of the bulk. All the increments

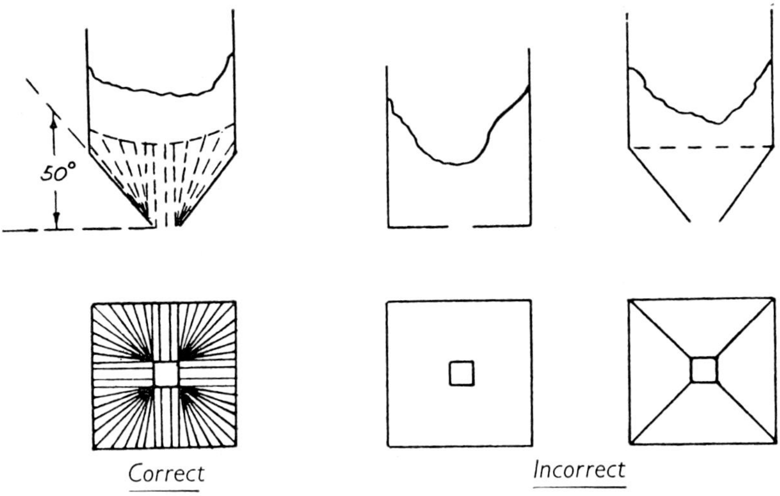

Bin bottom sloping 50° from horizontal in all directions to outlet with corners of bin properly rounded.

Flat bottom bins or those with any arrangement of slopes having corners or areas such that all material in bins will not flow readily through outlet without shovelling.

Slope of aggregate bin bottoms

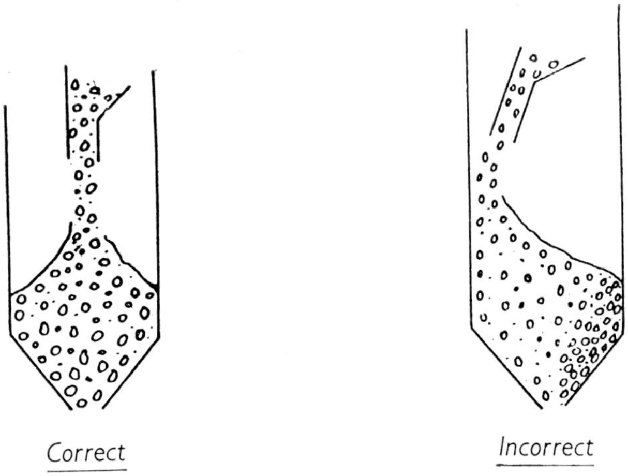

Material drops vertically into bin directly over the discharge opening permitting the discharge of more generally uniform material.

Chuting material into bin at an angle. Material falling other than directly over opening not always uniform as discharged.

Filling of aggregate bins

Fig. 3.7. Correct and incorrect methods of handling aggregate should be combined to form the main sample to be sent to the laboratory. The quantity sent to the laboratory should not be less than that shown in Table 3.1.

(2) In difficult circumstances the quantities shown in Table 3.1 may not be practicable, and a greater number of increments will be required in order that the main sample may be larger and therefore more truly representative of the bulk. It is difficult to sample accurately from stockpiles, where there may be segregation of the coarser material towards

TABLE 3.1 *Minimum Weights for Sampling*

Type of material			Minimum weight of sample despatched for testing	
	Nominal size			
			lb	kg
Aggregate	1 in. and larger	Larger than 25 mm	112	50
	Smaller than 1 in. and larger than $\frac{3}{16}$ in.	Smaller than 25 mm and larger than 5 mm	56	25
	$\frac{3}{16}$ in. or smaller	Smaller than 5 mm	28	13
Filler			10	5

Note 1. Where the material to be sampled consists of a wide range of sizes, larger quantities are likely to be required particularly when the mechanical properties are to be determined.

Note 2. Where it is known which tests are to be carried out on a sample, the sample quantity may be calculated from the data given in the relevant clauses, provided this does not result in quantities smaller than those required by Table 3.1. In addition, larger samples may be required if trial mixes are to be made with the material.

the base and sides. Similarly, "sandwich loaded" vehicles, into which different sizes of aggregate have been loaded one after another, cannot be sampled satisfactorily and a proper sample can only be taken after the load is mixed.

(3) Sampling is best carried out when aggregate is being loaded into or unloaded from vehicles, or when it is being discharged from a conveyor belt.

(4) When sampling rock direct from a natural outcrop or from a pit or quarry face, special care should be taken to ensure that samples are truly representative of the range of variations over the whole of the quarry face, vein or bed, later to become the source of aggregate.

Sample reduction

At the laboratory, the main sample should be reduced to the quantity required for testing, using a sample divider which extracts representative parts of the whole, or by quartering, Fig. 3.8. A suitable form of sample divider is the riffle box shown in Fig. 3.9. For reduction on the sample divider, the material should be surface dry. When reduction is made by quartering, fine and all-in aggregates are most accurately mixed and

Fig. 3.8. Quartering an aggregate

Fig. 3.9. Riffle box for sampling of aggregates

quartered when in a damp condition, if necessary achieved by sprinkling with clean water before mixing. For full details of the method of quartering reference should be made to B.S. 812:1967.

Sieve analysis (grading)

The following definitions have been accepted as Standard in Great Britain:

(a) *Fine Aggregate.* The term fine aggregate shall mean aggregate mainly passing a $\frac{3}{16}$ in. (4·76 mm) B.S. sieve and containing only so much coarse aggregate as is permitted for the various grading zones described in B.S. 882:1965.

(b) *Coarse Aggregate.* The term coarse aggregate shall mean an aggregate mainly retained on a $\frac{3}{16}$ in. (4·76 mm) B.S. sieve and containing only so much finer material as is permitted for the various sizes described in B.S. 882:1965. Coarse aggregate may be described as uncrushed gravel, crushed gravel, crushed stone, or partially crushed gravel when it is a combination of uncrushed and crushed gravels.

(c) *All-in Aggregate.* An aggregate composed of both fine aggregate and coarse aggregate.

Sieve analyses are made by passing the dried aggregate through a series of test sieves beginning with one sufficiently coarse to pass all the material. The diameters of the sieves and the sizes of the mesh openings are given in B.S. 410:1962.[3.6] The sizes of mesh in general use for the sieve analysis of concrete aggregate are given in Table 3.2.

The numbers 7, 14, 25, 52 and 100 referring to the smaller sieves are the numbers of apertures in a 1 in. length of sieve cloth. It will be noted in Table 3.2 that the aperture of each sieve is just half the size of the next larger one.

TABLE 3.2 *British and Equivalent American Sieves*

B.S. sieves			American sieves	
B.S. sieve size	Nominal width of aperture		Equivalent ASTM E11-61 sieve designation	
	in.	mm or microns*	Standard width of aperture mm or microns*	ASTM sieve No.
3 in.	3·0	76·2		
1½ in.	1·5	38·1		
¾ in.	0·75	19·05		
⅜ in.	0·375	9·52		
$\frac{3}{16}$ in.	0·188	4·76	4·76	No. 4
No. 7	0·095	2·40	2·38	No. 8
No. 14	0·047	1·20	1·19	No. 16
No. 25	0·024	*600	*595	No. 30
No. 52	0·012	*300	*297	No. 50
No. 100	0·006	*150	*149	No. 100
No. 200	0·003	* 75	* 74	No. 200

* 1 micron = 0·001 mm.

American sieves have slightly different sizes of aperture from British sieves, a point to be remembered when studying the literature. American sieves should comply with ASTM E.11 and the sieve sizes used for grading analysis of aggregates have been included in Table 3.2.

Sieving by hand is a laborious procedure, and in any laboratory where a considerable amount of sieving is to be done, it is advisable both in order to avoid wastage of labour, and to ensure greater accuracy, to use one of the sieving machines which can deal with a stack of several sieves at a time. A machine of this type is illustrated in Fig. 3.10. Sieves either 8 in. or 12 in. diameter can be used.

An example of the steps in the calculations for the grading of an aggregate is given in Table 3.3. The original sample is assumed to have been 10 lb. in weight after drying. Having completed the sieving, the weight of aggregate retained on each sieve in turn is recorded, giving the figures in column 2. The weight passing each sieve given in column 3 is obtained by the cumulative addition of the weights given in column 2, starting with that material passing 100; thus, the weight of material passing the No. 25 mesh sieve is the sum of the weights retained on sieves Nos. 52 and 100, together with that passing 100. It is then an easy step to convert weight passing into percentage passing.

Grading requirements for aggregates from natural sources, are given in B.S. 882:1965, and have been summarized in Tables 3.4 and 3.5, reproduced by permission of the British Standards Institution.

In the case of fine aggregate, it is required that the grading should lie within one of four zones, subject to certain tolerances. These zones replace the old Class A and Class B classification, which tended to brand as inferior a large proportion of the fine aggregates available. In a note in the Specification it is pointed out that fine aggregate complying with the requirements of any one of the four grading zones is suitable for making concrete, but the quality of the concrete will depend upon a number of factors, including the mix proportions. Generally, the ratio

TABLE 3.3 *Example of Calculation for Grading of Aggregate*

B.S. sieve size	Weight of material retained on each sieve lb	Weight of material passing each sieve lb	Percentage passing each sieve
3 in.	0	10	100
1½ in.	0	10	100
¾ in.	1	9	90
⅜ in.	½	8½	85
$\frac{3}{16}$ in.	1	7½	75
No. 7	½	7	70
No. 14	1	6	60
No. 25	1	5	50
No. 52	1½	3½	35
No. 100	2½	1	10
Passing 100	1		
Total	10		

TABLE 3.4 *B.S. 882:1965 Grading Limits for Coarse Aggregates*

B.S. 410 test sieve	Nominal size							
	Graded aggregate, in.			Single-sized aggregate, in.				
	$1\frac{1}{2}$–$\frac{3}{16}$	$\frac{3}{4}$–$\frac{3}{16}$	$\frac{1}{2}$–$\frac{3}{16}$	$2\frac{1}{2}$	$1\frac{1}{2}$	$\frac{3}{4}$	$\frac{1}{2}$	$\frac{3}{8}$
	per cent by weight passing							
3 in.	100	—	—	100	—	—	—	—
$2\frac{1}{2}$ in.	—	—	—	85–100	100	—	—	—
$1\frac{1}{2}$ in.	95–100	100	—	0–30	85–100	100	—	—
$\frac{3}{4}$ in.	30–70	95–100	100	0–5	0–20	85–100	100	
$\frac{1}{2}$ in.	—	—	90–100	—	—	—	85–100	100
$\frac{3}{8}$ in.	10–35	25–55	40–85	—	0–5	0–20	0–45	85–100
$\frac{3}{16}$ in.	0–5	0–10	0–10	—	—	0–5	0–10	0–20
No. 7	—	—	—	—	—	—	—	0–5

Fig. 3.10. Sieving machine adaptable for 12 in. or 8 in. diameter sieves

TABLE 3.5 *B.S. 882: 1965 Limits of Grading Zones for Fine Aggregates*

B.S. sieve size	Percentage by weight passing B.S. sieves			
	Grading Zone 1	Grading Zone 2	Grading Zone 3	Grading Zone 4
$\frac{3}{16}$ in.	**90–100**	**90–100**	**90–100**	**95–100**
No. 7	**60**–95	75–**100**	85–**100**	95–**100**
No. 14	**30**–70	55–90	75–**100**	90–**100**
No. 25	**15–34**	**35–59**	**60–79**	**80–100**
No. 52	**5**–20	8–30	12–40	15–**50**
No. 100	**0**–10*	**0**–10*	**0**–10*	**0–15***

* For crushed stone sands the permissible limit is increased to 20 per cent. A 5 per cent tolerance is permitted to be applied to the percentages in light type. The tolerance may be split up between two or three sieve sizes if desired.

TABLE 3.6 *ASTM C33-66 Grading Limits for Coarse Aggregates*

ASTM E11-61 sieve size (square openings)	Percentage by weight passing each sieve size (square openings)									
	Nominal size									
	$3\frac{1}{2}$–$1\frac{1}{2}$ in.	$2\frac{1}{2}$–$1\frac{1}{2}$ in.	2 in.–No. 4	$1\frac{1}{2}$ in.–No. 4	1 in.–No. 4	$\frac{3}{4}$ in.–No. 4	$\frac{1}{2}$ in.–No. 4	$\frac{3}{8}$ in.–No. 8	2 in.–1 in.	$1\frac{1}{2}$–$\frac{3}{4}$ in
4 in.	100									
$3\frac{1}{2}$ in.	90–100									
3 in.		100								
$2\frac{1}{2}$ in.	25–60	90–100	100						100	
2 in.		35–70	95–100	100					90–100	100
$1\frac{1}{2}$ in.	0–15	0–15		95–100	100				35–70	90–100
1 in.			35–70		95–100	100			0–15	20–55
$\frac{3}{4}$ in.	0–5	0–5		35–70		90–100	100			0–15
$\frac{1}{2}$ in.			10–30		25–60		90–100	100	0–5	
$\frac{3}{8}$ in.				10–30		20–55	40–70	85–100		0–5
No. 4			0–5	0–5	0–10	0–10	0–15	10–30		
No. 8					0–5	0–5	0–5	0–10		
No. 16								0–5		

TABLE 3.7 *ASTM C33-66 Grading Limits for Fine Aggregates*

ASTM E11-61 sieve size	Percentage by weight passing each sieve size
$\frac{3}{8}$ in.	100
No. 4	95–100
No. 8	80–100
No. 16	50–85
No. 30	25–60
No. 50	10–30
No. 100	2–10

Note 1. The minimum percentages shown above for material passing the No. 50 and No. 100 sieves may be reduced to 5 and 0, respectively, if the aggregate is to be used in air-entrained concrete containing more than 423 lb of cement per cubic yard, or in non air-entrained concrete containing more than 517 lb of cement per cubic yard.

Note 2. It is also required that the fine aggregate shall have not more than 45 per cent retained between any two consecutive sieves of those shown and its fineness modulus shall be not less than 2·3 nor more than 3·1.

of fine to coarse aggregate in a mix should decrease as the fine aggregate becomes finer, i.e. passing from Grading Zone 1 to Grading Zone 4. This is dealt with more fully in Chapter 7, Design of Concrete Mixes, page 106.

The zoning is so arranged that, within the British Isles, the great majority of fine aggregates should fall in one or other of the zones. The user should, therefore, be able to obtain deliveries having gradings lying within the specified zone, so giving reasonable uniformity.

ASTM C33-66 Standard Specification for Concrete Aggregates(3.7) contains similar grading requirements to those in B.S. 882, and these are summarized in Tables 3.6 and 3.7. It will be noted that the Americans specify limits for the fineness modulus of fine aggregates. This is obtained by adding the total percentage shown by the sieve analysis to be retained on each of the following ASTM sieves and dividing the sum by 100: No. 100, No. 50, No. 30, No. 16, No. 8, No. 4, $\frac{3}{8}$ in., $\frac{3}{4}$ in., $1\frac{1}{2}$ in., 3 in. and 6 in.

Combined gradings

If the design of concrete mixes is based on the grading of the aggregates, it is necessary to consider the combined grading of the fine aggregate and one or more coarse aggregates. Given the individual gradings, the theoretical proportions necessary to approximate any desired grading can be determined by trial and error, by arithmetical, or by graphical methods. One graphical method(3.8) applicable to two or more sizes of aggregate is illustrated by the following example, in which the aggregate gradings

TABLE 3.8 *Gradings, percentage passing*

Aggregate	B.S. sieve number or size								
	100	52	25	14	7	$\frac{3}{16}$ in.	$\frac{3}{8}$ in.	$\frac{3}{4}$ in.	$1\frac{1}{2}$ in.
Sand	2	16	51	75	90	99	100	100	100
$\frac{3}{8}$–$\frac{3}{16}$ in.					3	9	94	100	100
$\frac{3}{4}$–$\frac{3}{8}$ in.						2	18	92	100
38% of sand	1	6	19	28	34	38	38	38	38
22% of $\frac{3}{8}$–$\frac{3}{16}$ in.					1	2	21	22	22
40% of $\frac{3}{4}$–$\frac{3}{8}$ in.						1	7	37	40
Combined grading	1	6	19	28	35	41	66	97	100
Required grading	0	5	21	28	35	42	65	100	100

shown in Table 3.8 are to be combined to give a grading approximating to a required grading; this is taken to be grading No. 3 for $\frac{3}{4}$ in. maximum size aggregate as defined in Road Note No. 4.

Rectangular co-ordinates are set up (Fig. 3.11) with a scale of percentage passing (0 to 100%) vertically, and a straight line, *a-h*, is drawn through the origin at any convenient slope.

Vertical lines are drawn through the points between *a* and *h*, where the line *a-h* intersects the horizontal lines, corresponding to the values of the percentage of material required to pass the various sieves—according to the required grading. These lines are the ordinates for the sieve sizes, and are marked accordingly.

By this method the required grading curve has been transformed into a straight line by plotting against an irregular scale.

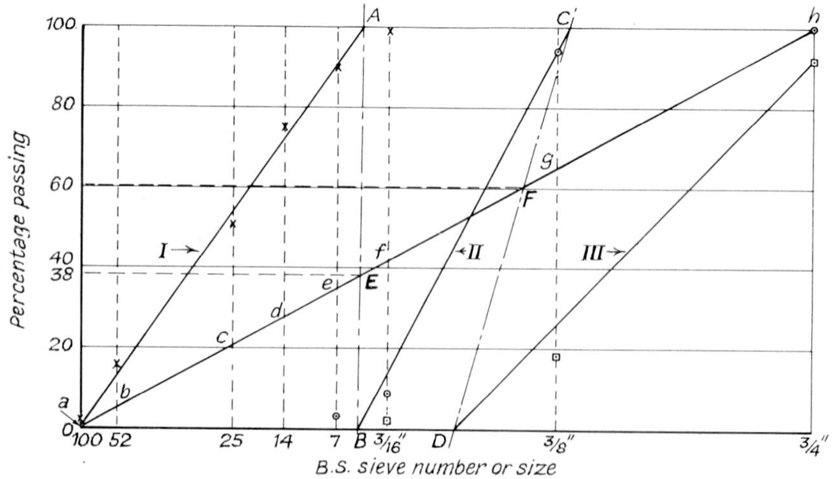

Fig. 3.11. Graphical method of proportioning aggregates approximating to a required grading

The values of actual percentage passing each sieve are then plotted for the separate aggregates and the steepest reasonable straight line (I, II and III) is drawn through each set of points. A straight line is then drawn between the point *A*, where the mean line for the finest material (I) cuts the 100 per cent passing line, and the point *B* where the mean line for the next finest material (II) cuts the 0 per cent passing line. Similarly, a straight line is drawn through the corresponding points *C* and *D* for the coarse aggregates.

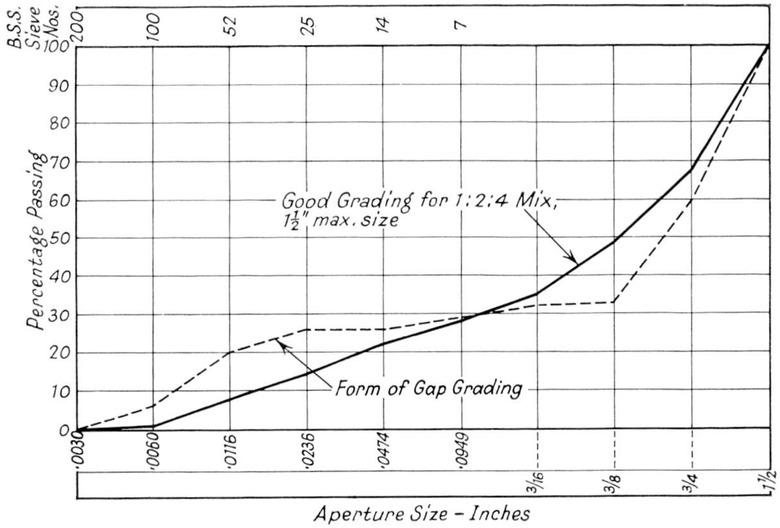

Fig. 3.12. Plotting of grading curves for comparative purposes

The values of percentage passing, corresponding to the intersections *E* and *F* of the straight lines *AB* and *CD* with the initial sloping line *a-h*, are recorded and indicate the proportions in which the aggregates are to be combined.

In the example, the values are 38 per cent and 60 per cent; the proportions are therefore 38 per cent of sand, (60 — 38) = 22 per cent of $\frac{3}{8} - \frac{3}{16}$ in., and (100 — 60) = 40 per cent of $\frac{3}{4} - \frac{3}{8}$ in. by weight. These values are checked in Table 3.8.

Combined gradings, however, can rarely be made to coincide at all sizes with the desired grading.

Comparison of gradings can be made very conveniently by plotting them on a chart as shown in Fig. 3.12. The full curve is a good continuous grading for a 1:2:4 mix using $1\frac{1}{2}$ in. maximum size aggregate and should give a workable concrete at a slump of about 2 in. with a reasonably low water cement ratio. The dotted curve is typical of the combined grading of $1\frac{1}{2}$ in. to $\frac{3}{8}$ in. aggregate and the fine sands common in many parts of Scotland. It is virtually a gap grading.

Grading curves

Mention has been made of a "good grading curve" (Fig. 3.12). Good gradings or "type gradings" have been suggested by various investigators, one of the earliest being that known as Fuller's Best Mix Curve, based on experiments with many gradations of material, from which the one considered the best under normal conditions was chosen. In practice there may be a large number of gradings which may be equally, or more suitable, varying with the cement content of the mix, the nature of the aggregate, the workability required, etc., and Fuller's Best Mix Curve does not necessarily produce the densest and strongest concrete.

Glanville[3.9] made an important contribution concerning grading of aggregates in a series of tests linking grading with workability and water/cement ratio. This work and subsequent advances in knowledge concerning the effect of aggregate characteristics on workability will be discussed more fully in a later chapter dealing with the design of concrete mixes.

Objectionable minerals and salts

Minerals and salts which may occur in natural aggregates and are likely to prove objectionable are:

(1) Coal and coal residues. Coal is objectionable on account of its low strength and its undesirable effect on the appearance of concrete surfaces. Coal residues, as represented by breeze and clinker, are used in certain types of concrete, and give satisfactory results provided the proportion of unburnt coal is small. These materials and fly ash from modern electric power stations are discussed in Chapters 5 and 22.

(2) Certain sulphide minerals such as pyrites readily oxidise through atmospheric attack, causing unsightly rust stains such as that shown in Fig. 3.13, and loss of strength if present in sufficient quantity.

Fig. 3.13. Pyrites stain on a bush-hammered gravel concrete surface
The dampness at the two construction joints is due to seepage of water from the back of the wall

(3) Mica is a contaminating substance sometimes found in aggregates and is undesirable because of its soft, laminated, and absorptive character and its susceptibility to disintegration along cleavage planes. The presence of mica in sand has been found[3.10] to reduce the compressive strength of concrete at constant water/cement ratio, by about 3 per cent for a 1 per cent increase in the mica content.

(4) Soluble sulphates, including gypsum and other similar material.

(5) Sea-shore aggregates contain salts which, while not seriously affecting the strength, are liable to cause unsightly efflorescences on concrete structures.

(6) The presence of sea shells, usually flaky in shape, requires a higher proportion of cement than would otherwise be needed, if adequate workability is to be obtained. The effect on cement content is stated[3.10] to be probably insignificant where up to about 10 per cent of the sand or 2 per cent of the coarse aggregate is shell. Significantly more cement may be necessary where the proportion of shell is higher.

(7) In addition to the presence of sea-shells, sea-dredged aggregate is likely to contain salts, mainly sodium chloride. Although no maximum limits have been agreed by the various authorities in this country, it has been recommended[3.11] that the aggregate should not contain more than 0.4 per cent chloride (Cl) by weight of cement. This amount may still be sufficient to increase the risk of efflorescence and corrosion and is likely to be too high if an admixture containing calcium chloride is used.

Alkali-aggregate reactivity

Serious failures of concrete have been attributed in some parts of Denmark and the United States to expansion arising when the cement and certain minerals in the aggregate react chemically. This reaction occurs with many aggregates and, in fact, there is some aggregate/cement interaction with many aggregates commonly used, although, fortunately, in Great Britain the interaction is insufficient to cause damage to the concrete.

With aggregates in countries abroad the risk is a serious one, in that concrete affected may expand and completely disintegrate in a few years necessitating major repairs. Unfortunately, there are no means of identifying such aggregates from their appearance, and further, aggregates which may be satisfactory with some cements, may be unsatisfactory with others. The reactivity is connected with the alkali content of the cement, and one precaution when making concrete with aggregates not previously employed is to choose a cement with a low alkali content.

Elimination of unsatisfactory aggregates by laboratory tests is difficult, especially as the more reliable tests involving measurement of expansion take a very long time, say, six months to one year. Some indication can, however, be obtained by petrographic examination and other tests which give sufficient indication to eliminate the worst aggregates. Such an examination is, therefore, well worth while, providing it is carried out by an experienced investigator.

In view of the difficulty in identifying other than seriously reactive aggregates by rapid methods of testing, it is necessary to obtain all the evidence possible on the durability of any concrete made in the locality with the particular aggregate to be used. Even this evidence is, however, not completely reliable, since the aggregate may have been used with a low alkali cement, whereas it might cause failure with another cement with a higher alkali content; hence the recommendation given above to use low alkali cements when preparing concrete from unknown aggregates abroad.

The subject is dealt with in more detail in National Building Studies, *Reactions between Aggregates and Cement.*[3.12] A chemical method of test for determining the potential reactivity of aggregates is given in ASTM C.289.

Clay, silt and dust

Clays and silts are commonly present in many gravel and sand deposits. In large quantities they cause a serious reduction in strength, and unsoundness due to their retarding action on the hydration of the cement. They are most dangerous when they form a coating on the particles of aggregate and thus prevent the adherence of the cement. Clays are also deleterious in concrete owing to the swelling and shrinkage which results from alternate wetting and drying; some clays when wet may contain as much as their own weight in water. When clays occur as constituents of other rocks, such as limestones, this absorptive character greatly increases the rock's susceptibility to disruption by weathering.

Stone dust is commonly present in crushed aggregates, but except in the case of certain dolerites, small quantities are not likely to have any deleterious effect on concrete.

Three methods are available for the determination of the clay, silt and dust content, the first of which is suitable for rough estimation in the field.

1. *Field Settling Test (Approximate).* The method as given in B.S. 812:1967 is as follows:

"About 50 ml of a solution of common salt in water (approximately one teaspoonful to one pint of water, or a 1 per cent solution) shall be placed in a 250 ml B.S. measuring cylinder. Sand in the condition as received shall be added gradually until the volume of sand is about 100 ml. The volume shall then be made up to 150 ml by the addition of more salt solution.

The mixture shall be shaken vigorously until adherent clayey particles have been dispersed, and the cylinder then placed on a level bench and gently tapped until the surface of the sand is level.

After three hours standing the height of the silt visible above the sand shall be expressed as a percentage of the height of the sand below.

This is an approximate volumetric method and is intended as no more than a guide to the approximate percentage of silt, loam, clay, etc., since the result will vary according to the fineness and type of clay.

The method is not generally applicable to crushed stone sands or coarse aggregates.

If the amount of clay and fine silt determined by this test is greater than 8 per cent by volume, then the test in B.S. 812 for the determination of the amount of aggregate passing a No. 200 (75 microns) B.S. 410 fine test sieve should be made.

2. *Laboratory Decantation Method.* A method for the determination of material finer than 200 mesh sieve, that is, clay and silt, which has been in use for some time, is described in B.S. 812:1967. Very briefly, the method consists of washing the silt and clay through a 200-mesh sieve, and finding the difference in weight between the original sample and that retained on the sieve. The difference is assumed to be clay and silt. An objection to this method is that any coarse material lost during the test counts as silt and clay. However, the test is simple and, since the sample used in the test is larger than that used in the sedimentation test described below, is less liable to sampling error.

The requirements of B.S. 882:1965 are that unless found satisfactory either as the result of such further tests as may be specified by the purchaser or his representative, or because evidence of general performance is offered which is satisfactory to him, the maximum quantities of clay, silt and fine dust determined in accordance with this test, shall not exceed the following:

(i) Coarse aggregate	1 per cent by weight
(ii) Sand or crushed gravel sand	3 per cent by weight
(iii) Crushed stone sand	15 per cent by weight

ASTM C33 limits the material finer than No. 200 sieve (American) to 3 per cent by weight for concrete subject to abrasion and 5 per cent for all other concrete except for crushed stone sands when the limits may be increased to 5 and 7 per cent respectively.

As the particles in crushed stone dust are not usually as fine as those in natural sand, a somewhat higher quantity is usually tolerable. A figure of 6 per cent is suggested by the Author as a reasonable limit. With some rocks, notably limestone, inclusions of clayey material in the rock as quarried may so increase the proportion of very fine particles as to make the limit of 3 per cent advisable.

3. *Sedimentation Method.* This method of separation of clay and silt is described in B.S. 812:1967. The test is somewhat involved and lengthy, involving pretreatment with sodium oxalate, thorough mixing of the suspension of fine material, and extraction of a small portion by means of a pipette. This portion is dried and weighed.

The test is open to the objection that the pretreatment may radically alter the character of the material being tested, and it is not recommended for general use.

Organic substances

Washing will often remove organic matter which is light in weight, or soluble in water, but will not remove heavy matter such as particles of coal, unless these occur in fine grains. Other organic substances, such as vegetable matter and humus, contain organic acids which inhibit the hydration of the cement, thereby delaying setting and reducing strength.

The colorimetric test for organic impurities is retained in ASTM specification C40–60, but is replaced in B.S. 812:1967 by measurement of the pH value of aggregate–cement mortars under standard conditions. Some cements are not suitable for use in this test, so that the cement to be used should first be tested with standard sand and then with the same sand artificially contaminated with tannic acid.

The sample of sand to be tested is dried and sieved on a $\frac{3}{16}$ in. B.S. sieve, any material retained on the sieve being discarded. Mortar is then prepared in a standard manner with cement checked as above and its pH value determined. If the resulting pH value is 12·25, 12·30 or 12·35, the test is repeated on a further sample heated at 600 ± 25°C for 1 hour.

Fine aggregate may be deemed to be satisfactory for use with Portland cement if the pH value of the mortar is 12·40 or higher or is 12·20 or lower. If in between, the increase in pH as a result of heat treatment should not be more than 0·2.

If lime or calcium chloride are added to the Portland cement B.S. 812 states that these criteria may be unduly restrictive.

Absorption and surface moisture

The following explanations of certain terms used should be clearly understood:

1. *Oven dry.* Completely dry for practical purposes.

2. *Air dry.* Dry at the surface though containing some interior moisture, but less than the amount required to saturate the particles. In this condition an aggregate can absorb more water into itself, and may still appear dry on the surface. Absorbed moisture plays no part in the lubrication of mixed concrete.

3. *Saturated and surface dry.* This is an idealized condition; the aggregate can absorb no more water without a film of water forming on the surface. In this condition the aggregate neither contributes to nor absorbs water from the concrete. If a wet stone is gradually dried or is wiped quickly with an absorbent duster or blotting paper, a stage is reached when the surface moisture disappears, but the stone is still saturated with absorbed water. It is then in a saturated and surface dry condition. Even in the laboratory great care is necessary, particularly with fine aggregate, in achieving the saturated surface dry condition.

4. *Wet.* The aggregate is itself saturated and carries an excess of moisture forming a film on the surface of the particles.

The above conditions are illustrated in Fig. 3.14, which is similar to the representation included in the *A.C.I. Manual of Concrete Inspection.*(3.13)

The absorption of different particle sizes of the same aggregate may vary, as in the example given in Table 3.9, p. 50, so that tests on a single-sized sample should not be taken as representative of the aggregate through the whole range of gradings.

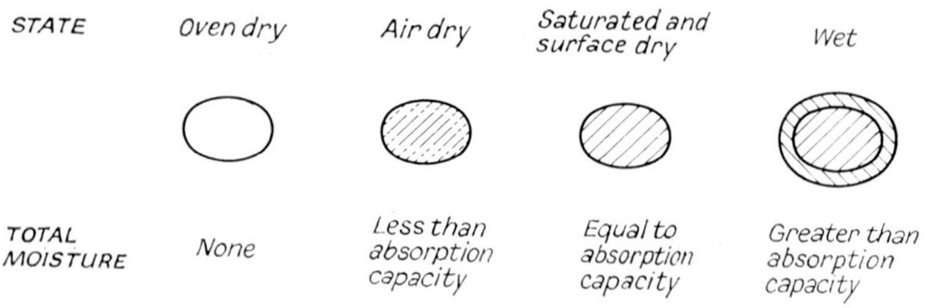

Fig. 3.14. Range of moisture contents as they affect aggregates

The absorption of an aggregate is often a useful guide to the permeability and frost resistance of concrete made with it.

The measurement of the moisture content of an aggregate is described in B.S. 812:1967. Four methods are given, the choice depending on the information required. The most useful of these tests are those giving the absorbed moisture and the surface moisture. These are the most difficult in the field owing to the necessity of drying the aggregate to a saturated surface dry condition. Even in the laboratory great care is necessary, particularly with fine aggregate, in achieving this condition.

In addition to the British Standard tests it is often useful to determine the absorption of an aggregate after only a few minutes soaking, as the rate of absorption provides a guide to the reduction in workability between mixing and placing. Further comment on this point is made in Chapter 6, page 95.

Specific gravity

If a section is cut through any piece of stone and the surface magnified, it will be seen that it is to some extent honeycombed with capillaries and tiny air-holes. For this reason the value obtained for the specific gravity of an aggregate depends on the method of test.

Methods of test and the calculations required for the determination of the specific gravities are given in B.S. 812, ASTM C127.59 and ASTM C128.59.(3.14) The American definitions have been adopted in preference to the British Standard definitions as being more specific.

Whichever test method is adopted the calculations are similar and are based on the following data:

A = weight of the oven dry sample in air

B = weight of saturated surface dry sample in air

C = weight of saturated sample in water

(a) *The Bulk Specific Gravity* referred to in ASTM C127 and 128 may be defined as the ratio of the weight in air of a unit volume of a permeable material (including both permeable and impermeable voids) to the weight in air of an equal volume of water.

$$\text{Bulk specific gravity} = \frac{A}{B - C}$$

In B.S. 812, this value is referred to as "specific gravity."

(b) *The Bulk Specific Gravity* (Saturated Surface Dry Basis) referred to in ASTM C127 and 128 may be defined as the ratio of the weight in air of a permeable material in a saturated surface-dry condition to the weight in air of an equal volume of water.

$$\text{Bulk specific gravity (S.S.D. basis)} = \frac{B}{B - C}$$

No equivalent term or calculation for this value is given in B.S. 812.

(c) *The Apparent Specific Gravity* may be defined as the ratio of the weight in air of a unit volume of a material to the weight in air of an equal volume of water. The volume of the material shall be that of the impermeable portion.

$$\text{Apparent specific gravity} = \frac{A}{A - C}$$

This is also referred to in B.S. 812 as the Apparent Specific Gravity.

In concrete mix calculations for determining the solid volumes of the constituent parts, it is necessary to know the space occupied by the aggregate particles regardless of whether or not pores or voids exist within the particles. The value to be used for this is that referred to above as the bulk specific gravity (S.S.D. basis).

The term "solid volume" is preferred to the older term "absolute volume" as being a more apt description, since the term absolute would

TABLE 3.9 *Apparent Specific Gravity and Absorption for Different Particle Sizes of River Aggregate*

Size of aggregate in.	Apparent specific gravity	Absorption % of dry weight
$1\frac{1}{2}$–$\frac{3}{4}$	2·55	0·3
$\frac{3}{4}$–$\frac{3}{8}$	2·52	0·8
$\frac{3}{8}$–$\frac{3}{16}$	2·45	1·5
$\frac{3}{16}$ down	2·60	1·0

appear to imply that the volume contains only solid matter. This is not true since the volume taken includes voids enclosed within the cement and aggregate particles.

The most difficult part of the determination of the specific gravity is the decision as to when an aggregate is saturated but surface dry. In the field it is usually necessary to drive off the surplus moisture by warming on a Primus stove. Continuous stirring is required until the wet film of water just disappears from the stones, or until sand ceases to stick to the stirrer and becomes free running. In the laboratory a fan blowing over an electric fire (such as a hair drier) is convenient. Considerable care is needed in dealing with sands.

Tests for specific gravity are probably best avoided in the field as far as possible, samples being sent to a central laboratory whenever such tests are required.

At least two determinations should be made, and if these do not agree, further tests should be made until agreement is reached.

Different particle sizes of the same aggregate may have different apparent specific gravities. Average values from a large number of tests on a river aggregate are given in Table 3.9.

Mechanical properties of aggregates

It is required in B.S. 882 that the "ten per cent fines" value when determined by the method given in B.S. 812 shall be not less than 5 tons.

For concrete wearing surfaces, the ten per cent fines value shall be not less than 10 tons. Very briefly, the method consists of placing a quantity of the aggregate, graded between stated sieves, in a cylinder, inserting a plunger, and applying a compression load so as to cause a total penetration of the plunger varying between 0·60 in. (15 mm) and 0·95 in. (24 mm), depending on the type of aggregate. The aggregate is then resieved on the appropriate sieve and the ten per cent fines value calculated from the formula given in B.S. 812.

As an alternative to the ten per cent fines value, the aggregate impact test given in B.S. 812 may be made. The aggregate impact value should not exceed 45 per cent, and for concrete wearing surfaces should not exceed 30 per cent.

For measuring the resistance of aggregates to surface wear by abrasion, an important property for aggregates used in concrete roads, a method of test is given in B.S. 812:1967. American tests for abrasion resistance by use of the Los Angeles machine are given in ASTM C535-65 and C131-66.

Bulk density and voids

The bulk density of a material is the weight of the material held by a container of unit volume when filled or compacted under defined conditions. It is normally expressed in lb/ft^3.

The bulk density of an aggregate is affected by several factors, including the amount of moisture present and the amount of compactive effort used in filling the container.

It is emphasized in B.S. 812 that the methods specified for the determination of the bulk density and voids of an aggregate are laboratory tests intended for comparing properties of different aggregates and are not generally suitable for use in translating proportions by weight into proportions by volume on site. They give the loose weight or the "rodded" or compacted weight per cubic foot, depending on the method of filling the container with dry aggregate.

On the site and for conversion purposes, it is probably best to have the box filled in turn by each of the men filling the mixer hopper, and to take an average value. This average value is likely to lie somewhere between the bulk densities determined by the B.S. 812 methods for compacted and loose weights.

If the aggregate is damp and contains an appreciable amount of material below a $\frac{3}{16}$ in. sieve, it is necessary not only to find the weight per cubic foot dry, but also to make tests for bulking. The weight per cubic foot at any moisture content can be calculated from the dry density and the percentage bulking, not forgetting to allow for the weight of moisture. Normally, the weight per cubic foot of sand or all-in ballast decreases from the dry weight to a minimum value at a certain moisture content, and then increases again until the material becomes saturated.

A method sometimes used in proportioning concrete is based on the void content of the aggregates, and a measurement of voids may be useful in certain cases. B.S. 812 gives a method of measurement for use in the laboratory. As with measurements of the weight per cubic foot, the void content will vary according to the method of filling the gauge-box.

Bulking

When sand is moistened, films of water form on the particles and the surface tension tends to hold them apart, causing an increase in volume or bulking. Fine sand bulks more than coarse sand; aggregate retained on a $\frac{3}{16}$ in. sieve is scarcely affected.

In general, sand bulks rapidly to the extent of 20 or 30 per cent, as the moisture content rises to about 4 to 6 per cent. Further increases in moisture content result in a decrease in bulking, until, when the sand is completely saturated, its volume is practically the same as it was in a dry condition. Curves showing the difference between the bulking of coarse and fine sand plotted against moisture content are given in Fig. 3.15.

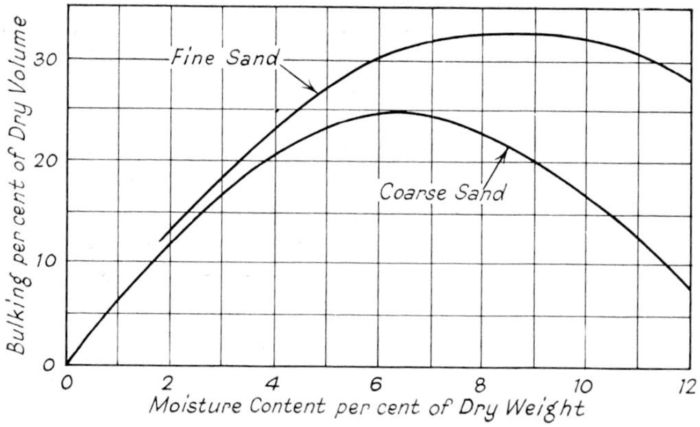

Fig. 3.15. Typical form of bulking curves

In practice it is usual to assume an average value of 20 per cent for the bulking of sand as delivered in a damp condition, unless tests are made to obtain more exact information.

All-in ballast may contain anything up to 60 or 70 per cent of sand, and it is found that bulking occurs roughly in proportion to the amount of sand present.

Bulking can be determined by filling a gauge-box or other container of known volume (A) with damp sand in the manner in which the mixer hopper will be filled. This involves the same difficulties as have been described in the last section. The sand is then either dried and filled back into the gauge-box, or the container is filled with water and the damp sand is poured in to displace the water. Whichever method is adopted, the new depth of aggregate in the container gives the unbulked volume (B).

Then percentage bulking expressed as a percentage of the dry volume

$$= \frac{A - B}{B} \times 100$$

When measuring aggregates by volume it is necessary to make an allowance for bulking in order to obtain the correct proportion of sand particles required to produce good workable concrete. If the cement content is specified as a proportion of the aggregate, omission of an allowance for bulking will also result in the use of too much cement per cubic yard of concrete produced.

Particle Shape and Surface Texture

The particle shape is important since it affects the workability, as is shown in Chapter 7, Design of Concrete Mixes. It may conveniently be assessed by a test for angularity suggested by Shergold,[3.15] which is based on the percentage voids in the aggregate after compaction in a specified manner. The test gives a value termed the angularity number; the method of determination as now given in B.S. 812:1967 is briefly as follows:

A quantity of single-sized aggregate is filled into a metal cylinder of about $\frac{1}{10}$ ft³ (3 litres) capacity having approximate internal dimensions 6 in. (15 cm) diameter and 6 in. high, in three layers, allowing the aggregate to slip in gently from a scoop. Single-sized aggregate is used; in the case of coarse aggregate the size limits specified are $\frac{3}{4}$ in.–$\frac{1}{2}$ in., $\frac{1}{2}$ in.–$\frac{3}{8}$ in., $\frac{3}{8}$ in.–$\frac{1}{4}$ in., and $\frac{1}{4}$ in.–$\frac{3}{16}$ in. Each layer of aggregate is then tamped 100 times by allowing a metal rod $\frac{5}{8}$ in. (16 mm) diameter and 24 in. (60 cm) long, with rounded end, to fall freely from a height of 2 in. (5 cm) above the surface of the aggregate. Further aggregate is then added to fill the cylinder and "rolled off" with the tamping rod. The volume of the cylinder is determined from the weight of water required to fill it (1 cm³ = 1 g).

Then the percentage of voids (V) in the aggregate is:

$$V = \frac{CG_a - W}{CG_a} \times 100 \qquad (1)$$

where C = volume of cylinder in cm³ given by the weight in grammes of water required to fill the cylinder,
G_a = specific gravity of the aggregate calculated on a dry basis,
W = weight of aggregate in cylinder in grammes.

Shergold expressed the voids calculated in this way, less 33, which he regarded as the "optimum," as the angularity number.

Then angularity number

$$f_H = V - 33 \qquad (2)$$

For estimations of workability the author has suggested[3.17] that a different method of expressing the result is more suitable. Thus:

$$\text{Angularity index } f_A = \frac{3f_H}{20} + 1{\cdot}0 \qquad (3)$$

where f_H is the percentage of voids in excess of 33, i.e. the angularity number as expressed by Shergold.

Combining (1), (2) and (3),

$$f_A = 11{\cdot}05 - \frac{15W}{CG_a} \qquad (4)$$

Flaky materials are not satisfactory for concrete work. The surface texture of the aggregate has some bearing on the strength of the concrete produced, in that aggregates with polished surfaces do not produce such strong concrete as those with matt or slightly rougher surfaces. Evidence is accumulating to suggest that the influence of the type of aggregate on the compressive strength, and—more particularly—on the tensile strength or modulus of rupture, is greater than has hitherto been supposed. An analysis of several series of tests made by the Author has indicated that limestone and granite aggregates may give strengths up to 20 per cent higher than river gravel at the same water/cement ratio. Other investigators have reported similar experiences both with limestone and other crushed stones, as compared with river gravels. It is believed that these differences are due mainly to the effects of surface texture and particle shape.

Group classification of rocks

Rocks are classified in B.S. 812:1967, Clause 9. The grouping is a useful system for distinguishing the various rocks such as granite, limestone, quartzite, sandstone, etc.

REFERENCES

3.1 B.S. 1047:1952, *Blastfurnace Slag (Coarse Aggregate)*. British Standards Institution.

3.2 SHERGOLD, F. A. and GREYSMITH, MARY G., Factors governing the grading and shape of crushed rock—a survey of the literature, *Quarry Managers' Journal*, June 1947, pp. 703–712.

3.3 B.S. 882:1965, *Concrete Aggregate from Natural Sources* (includes B.S. 1201, *Aggregates for Granolithic Floor Finishes*). British Standards Institution.

3.4 *Concrete Manual*, 1st ed. 1938, 7th ed. 1963. U.S. Bureau of Reclamation.

3.5 B.S. 812:1967, *Methods for Sampling and Testing of Mineral Aggregates, Sands and Filters*, British Standards Institution.

3.6 B.S. 410:1962, *Test Sieves*. British Standards Institution.

3.7 ASTM C33–66, *Standard Specification for Concrete Aggregates*, ASTM Standards, Part 10, Oct. 1966. American Society for Testing Materials.

3.8 *The Determination of the Proportions of Aggregates approximating to any Required Grading*, Cement and Concrete Association Advisory Note No. 9.

3.9 GLANVILLE, W. H., COLLINS, A. R. and MATHEWS, D. D., *The Grading of Aggregates and Workability of Concrete*, Road Research Technical Paper No. 5, 2nd ed., 1947. H.M.S.O.

3.10 DEWAR, J. D., *Effect of Mica in the Fine Aggregate on the Water Requirements and Strength of Concrete*, Cement and Concrete Association TRA/370, April 1963.

3.11 SHARP, D. R. and SHACKLOCK, B. W., Aggregates in concrete, *Quarry Managers' Journal*, May 1966.

3.12 *Reactions between Aggregates and Cement*. National Building Studies, Parts I, II and III, Research Papers 14, 15 and 17. H.M.S.O.

3.13 *A.C.I. Manual of Concrete Inspection*, 4th ed., revised 1961. American Concrete Institute.

3.14 ASTM C127–59, *Specific Gravity and Absorption of Coarse Aggregate;* C128–59, *Specific Gravity and Absorption of Fine Aggregate*, ASTM Standards, Part 10, Oct. 1966. American Society for Testing Materials.

3.15 SHERGOLD, F. A., The percentage voids in compacted gravel as a measure of its angularity, *Mag. Concr. Res.*, **5**, No. 13, Aug. 1953. Cement and Concrete Association.

3.16 KAPLAN, M. F., The effects of the properties of coarse aggregates on the workability of concrete, *Mag. Concr. Res.*, **10**, No. 29, 1958.

3.17 MURDOCK, L. J., The workability of concrete, *Mag. Concr. Res.*, **12**, No. 36, Nov. 1960. Previously printed in *Research and Application*, No. 11, Jan. 1960 (George Wimpey & Co. Ltd.).

CHAPTER 4

Cement

PORTLAND cement was first produced in 1824 by Joseph Aspdin, who heated a mixture of finely divided clay and limestone or chalk in a furnace at a temperature sufficiently high to drive off the carbonic acid gas. It was not until 1845 that Isaac Johnson burnt the same materials together in a furnace or kiln until they clinkered, thus producing a cement very similar in its essential chemical constituents to modern Portland cement. Since then many refinements, changes in the scale and type of plant, and improvements in methods of control and testing have been made.

The basic materials for the manufacture of Portland cement are usually forms of limestone and clay. Chalk or marl commonly provides the calcareous material in the form of calcium carbonate, while clay or shale provide the argillaceous material, which is mainly alumina and silica. The sites of cement works are normally chosen at places where both these types of material occur adjacent to one another. The method of obtaining the raw material is shown in Fig. 4.1.

Two processes of manufacture are employed, the dry process and the wet process, the latter being the more common in this country. In the dry process the materials are crushed, dried, and then ground in ball mills to a powder which is burnt in its dry condition. In the wet process the materials are first crushed and then ground to form a slurry in the wash mills. After passing through the wash mills and the slurry silos, the slurry passes to the slurry tanks which are shown in Fig. 4.2. Samples of the slurry are taken from these tanks from time to time for testing, and any

Fig. 4.1. Digging limestone for cement manufacture

Fig. 4.2. Slurry tanks

correction in the chemical composition is made by changing the proportions of the calcareous and argillaceous constituents. The slurry is next pumped to the kiln, an example of which is shown in Fig. 4.3. Cement kilns may be as large as 12 ft in diameter and 500 ft in length, and with an output of as much as 700 tons per day. Pulverized coal, oil, or natural gas is used as fuel and is injected into the kiln at the opposite end to that

Fig. 4.3. Rotary kilns

Fig. 4.4. Clinker cooling tubes below rotary kilns

at which the slurry enters. Thus, the slurry enters the kiln at the cooler end, and by the rotation of the kiln in conjunction with its slope, passes down and gradually increases in temperature until it clinkers. The cement clinker then passes through clinker coolers as illustrated in Fig. 4.4. Having cooled sufficiently, the clinker is ground to the required degree of fineness in ball mills shown in Fig. 4.5. While grinding, a "retarder" is

Fig. 4.5. Air-swept cement ball mills. The steel grinding balls of various sizes are seen in the foreground

incorporated, gypsum being the material generally used. An addition of 2 or 3 per cent is usually necessary to slow down sufficiently the chemical action at the time of setting. The friction arising from the grinding operation causes a considerable increase in temperature again, and is the cause of the hot cement sometimes delivered on to a site.

Fig. 4.6. Cement packing machine showing control panel and bag handling

From the mills the cement is conveyed or blown to storage silos, from which the cement is discharged as required, either being fed to a packing plant or direct to bulk-cement lorries. The packing plant, which is shown in Fig. 4.6, automatically fills the paper bags nowadays commonly used. The nominal net weight of cement in each bag is 112 lb, but a variation of a few pounds is likely to occur.

In America bags of cement, referred to as sacks, weigh 94 lb.

Other ancillary plant at the cement works includes dust-extracting equipment, air compressors, and the power generators.

A diagram showing the complete cycle of operations is given in Fig. 4.7.

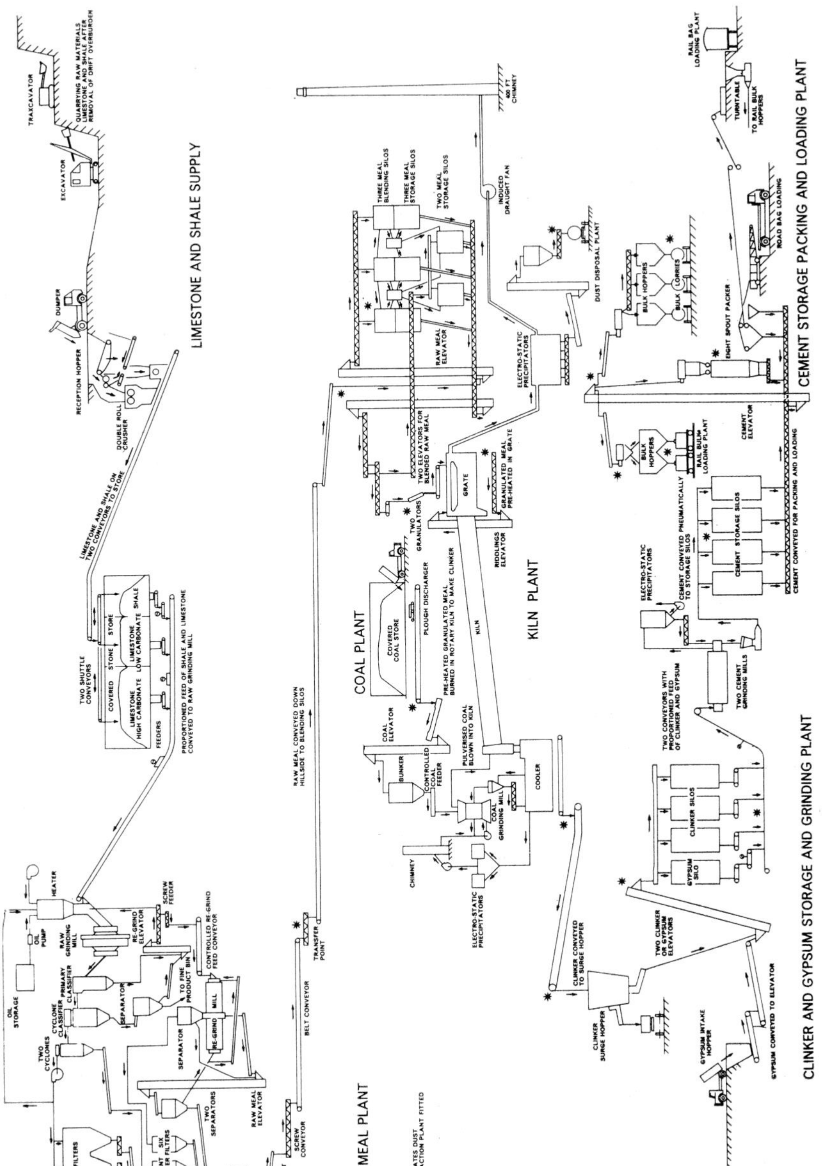

Fig. 4.7. Flow chart

Chemistry of Portland cement

The chemistry of Portland cement is very complicated, and not yet fully understood. It is sufficient to recognize the main constituents and to have some idea of the influence of the four main compounds on the progress of setting and hardening.

The approximate composition of Portland cements is given in Table 4.1.[4.1] Lime constitutes nearly two-thirds of the cement and its proportion has an important effect on the properties. An excess of lime causes unsoundness and disintegration of the cement after setting. A high lime content, but not enough to be excessive, tends to slow down the setting, but produces a high early strength. Too little lime may result in a weak cement, and if underburnt, in one which has a quick set.

TABLE 4.1 *Percentage Composition and Compound Content of Portland Cements*[4.1]

	Rapid-hardening	Normal	Low-heat	"Sulphate resisting"
Analysis				
Lime	64·5	63·1	60·0	64·0
Silica	20·7	20·6	22·5	24·4
Alumina	5·2	6·3	5·2	3·7
Iron oxide	2·9	3·6	4·6	3·0
Compounds				
Tricalcium silicate	50	40	25	40
Dicalcium silicate	21	30	45	40
Tricalcium aluminate	9	11	6	2
Iron compound	9	11	14	9

Silica forms about one-fifth, while alumina is only present to the extent of about one-twentieth of the cement. A high silica content, which is usually accompanied by a low alumina content, produces a slow-setting cement of high strength, and improves resistance to chemical attack. If the reverse is the case, that is, a high alumina and a low silica content, the cement is quick setting and of high strength.

Iron oxide imparts the grey colour to cement, and acts in the same manner as the alumina.

Magnesia is limited by B.S. 12:1958[4.2] to 4 per cent. A greater quantity is liable to cause unsoundness.

The sulphur content is usually calculated as sulphuric anhydride (SO_3) and is limited to 2·5 or 3·0 per cent, depending on whether the tricalcium aluminate content is greater or less than 7 per cent. Excessive quantities cause unsoundness.

The alkalies, soda and potash, are normally carried away in the flue gases when the cement is burnt, and are only present in the finished product in small quantities. If, for any reason, they are present in the cement in excessive quantities, they cause efflorescence, and increased risk of failure due to alkali–aggregate reactivity.

When cement is mixed with water, a chemical action begins between the various compounds and the water. In the initial stages, the small quantity of retarder quickly goes into solution, and is thus enabled to exert its influence on the other chemical reactions which are starting. These reactions result in the formation of the various compounds which cause setting and hardening, the four most important being:

(1) Tricalcium aluminate (three molecules of lime to one of alumina) C_3A.

This compound hydrates very rapidly and produces a considerable amount of heat; it causes the initial stiffening, but contributes least to the ultimate strength, is less resistant to chemical attack, being particularly vulnerable to the disintegrating action of sulphates in ground water, and has the greatest tendency to cause cracking, due to volume changes.

(2) Tricalcium silicate (three molecules of lime to one of silica) C_3S.

This compound jellifies within a few hours, generating considerable heat. The quantity formed in the setting reaction has a marked effect on the strength of concrete at early ages, mainly in the first 14 days.

(3) Dicalcium silicate (two molecules of lime to one of silica) C_2S.

The formation of this compound proceeds slowly with a slow rate of heat evolution. It is mainly responsible for the progressive increase in strength which occurs from 14 to 28 days, and onwards. Cements in which a high proportion of dicalcium silicate is formed, have a relatively high resistance to chemical attack, a relatively low drying shrinkage, and hence, are the most durable of the Portland cements.

(4) Tetracalcium alumino-ferrite (four molecules of lime to one of alumina, and one of iron oxide) C_4AF.

The presence of the alumino-ferrite compound is of little importance, since it has no marked effect on the strength or other properties of the hardened cement.

The chemical reactions outlined above, result in the formation of a mixture of gels and crystals from the solution of cement and water, which, by their adherence and physical attraction to one another and to the aggregate present, gradually set and harden to produce concrete.

It is particularly important to note that the setting and hardening is a chemical reaction, in which water plays an important part, and is not just a matter of drying out, in fact, setting and hardening stop as soon as the concrete becomes dry.

Fineness of grinding

After the chemical composition, the other dominant factor affecting cement properties is the fineness of grinding. Finer grinding increases the speed with which the various constituents react with water, but does not alter their inherent properties. The tendency during recent years has been to grind cements to greater finenesses in order to produce high strengths at early ages. Rapid-hardening Portland cement is an example of this development.

Fineness of grinding is of some importance in relation to the workability of concrete mixes. While claims that an increase in the fineness of cement results in a reduction in the water/cement ratio required to produce a given consistence are open to doubt, it has been established that greater fineness improves the cohesiveness of a concrete mix. Further, the quantity of water rising to the surface of the concrete, known as "bleeding," is reduced; excessive separation is very harmful to the production of sound construction joints. It is particularly apt to occur when vibratory methods of compaction are used, and there is little doubt that serious difficulties would have been encountered in vibrating concrete if modern cements had not been much finer than those in use in the past.

Shrinkage cracking is related to the rate of development of strength of a concrete, and in general, cements which gain strength rapidly are more prone to cracking. Increasing the fineness of any particular cement raises its rate of development of strength, and so indirectly increases the risk of shrinkage crack formation. Some special types of cement gain strength more slowly than normal, although they are finely ground. Since the rate of strength development is again the governing factor they are more resistant to cracking. The shrinkage results from both cooling of the concrete and drying out.

Types of Portland cement

The brief analysis which has been given of the composition of Portland cement and the influence of the various compounds present on its properties, has indicated that a large range in characteristics is possible by varying the proportions of the raw materials, by differences in the fineness of grinding, and by other changes in technique. Since for different purposes it is advantageous to stress one characteristic more than others, various classes of Portland cement have been developed.

Analyses showing the main constituents and compounds of a typical, ordinary, rapid-hardening, low-heat and sulphate-resisting cement are given in Table 4.1. These are average values giving only a broad picture of the make-up of the various classes of Portland cement.

It is interesting to note that the ultimate compressive strengths obtained with all the cements included in Table 4.1, are about the same, and that it is only the rate of hardening which varies to a marked extent.

Rapid-hardening Portland cement has a high tricalcium silicate content which, combined with the fine grinding, contributes towards a high early strength. As with ordinary Portland cement there is a variation in the properties of the rapid-hardening types, and it has been found that the 1 and 3 day compressive strength requirements of B.S. 12:1958 are not always achieved.

It should be noted that the requirements in respect of setting times are the same as those for ordinary Portland cement.

American cement Type III complying with ASTM C150-66 is similar in physical and chemical properties.

The British Standard and ASTM requirements for this and various other types of cement are included in Table 4.2.

Extra-rapid-hardening Portland cement is rapid-hardening cement that has calcium chloride ground with it. The effect of the calcium chloride, as described on p. 81, accelerates the chemical reaction and the evolution of heat so that this type of cement tends to stiffen in 20 to 50 minutes[4.3] after mixture with water, in addition to increasing the early rate of gain of strength.

TABLE 4.2 *Physical Requirements for Cements*

Type of cement	Mortar cube strength lb/in²				Concrete cube strength lb/in²			Fineness cm²/g		Setting time (Vicat)	
	1 day	3 day	7 day	28 day	3 day	7 day	28 day	Air permeability	Turbidimeter	Initial not less than	Final not more than
(a) Ordinary Portland										minutes	hours
B.S. 12:1958	—	2200	3400	—	1200	2000	—	2250		45	10
ASTM Type I	—	1200	2100	3500	—	—	—	2800(1)	1600(2)	45	10
(b) Rapid-hardening Portland											
B.S. 12:1958	—	3000	4000	—	1700	2500	—	3250	—	45	10
ASTM Type III	1700	3000	—	—	—	—	—	—	—	45	10
(c) Low-heat Portland											
B.S. 1370	—	1100	2000	4000	500	1000	2000	3200	—	60	10
ASTM Type IV	—	—	800	2000	—	—	—	2800(1)	1600(2)	45	10
ASTM Type II	—	1000	1800	3500	—	—	—	2800(1)	1600(2)	45	10
(d) Sulphate-resisting Portland											
B.S. 4027:1966	—	2200	3400	—	1200	2000	—	2500	—	45	10
ASTM Type V	—	—	1500	3000	—	—	—	2800(1)	1600(2)	45	10
(e) Portland blastfurnace B.S. 146	—	1600	3000	5000*	800	1600	3200*	2250	—	45	1
(f) High alumina B.S. 915	6000	7000	—	—	—	—	—	2250	—	2 hours and not more than 6 hours	2 hours after the initial set

* Optional
(1) Average value: minimum value any one sample 200 less.
(2) Average value: minimum value any one sample 100 less.

Extra-rapid-hardening cement cannot be stored satisfactorily for more than two or three weeks, since the calcium chloride increases the rate of moisture absorption and consequent deterioration of the cement.

This type of cement is particularly useful for work in cold weather, and when work has to be carried out between tides.

Because of the calcium chloride present it is recommended that this type of cement should not be used in prestressed concrete or reinforced concrete which is to be steam-cured.[4.4]

Low-heat cement has a high proportion of dicalcium silicate mainly at the expense of the tricalcium silicate and is, therefore, slow in hardening and produces less heat than the other cements. This cement was produced mainly as a result of problems which resulted from the temperature rise in mass concrete dams.

Its use is mainly confined to gravity retaining walls, large dams, and similar massive concrete structures where the temperature rise in the mass of concrete, and the subsequent thermal shrinkage on cooling, can be reduced.

It is usually only manufactured in Great Britain for large contracts.

Moderate low-heat cement is manufactured in America in accordance with Type II according to ASTM C150-66.

Sulphate-resisting Portland cement has a better performance in resisting sulphate attack than ordinary Portland cement, due to the reduction in the tricalcium aluminate content. Specification requirements for this type of cement are shown in Table 4.2. It is not resistant to acids. Calcium chloride should not be added as it reduces its sulphate-resisting properties.

As with all concrete, the prime resistance to chemical attack is afforded by the production of a dense well-compacted concrete, and lean mixes that are difficult to compact should be avoided; the concrete should not be leaner than 1 to 6 by weight (525 lb of cement per cubic yard of concrete).

White Portland cement has the same properties as ordinary Portland cement but is manufactured from raw materials containing less than 1 per cent iron. Its cost is some three to four times that of ordinary Portland cement.

Air-entraining cements. These are ordinary Portland cements to which an air-entraining agent has been interground during the manufacturing process. They are available in U.S.A. for Types I, II and III cement, and are required to satisfy the requirements of ASTM C175-66. They are not available in Great Britain where it is found more convenient and satisfactory to vary the amount of air-entraining agent to meet the particular aggregate types, gradings and sizes.

Other types of cement

Portland-blastfurnace cement, complying with the requirements of B.S. 146:1948, is manufactured mainly in Scotland by grinding a mixture of Portland cement clinker and blastfurnace slag in the ratio of about 35:65. On the Continent cements are made with a greater range, 30–85 per cent in the proportion of slag used.

The properties of slag cements vary depending on the amount and characteristics of the slag. They are slower hardening and have reduced 3 and 7 day strength requirements compared with ordinary Portland cement. They are also more affected by low temperatures during the curing period and by inadequate curing[4.5] than is normal cement.

They are particularly useful for mass concrete structures because they have a somewhat reduced heat of hydration.

Portland blastfurnace cement concrete is more resistant to chemical attack, especially in sea water, than ordinary Portland cement.

Pozzolanic cements are produced by grinding together a mixture of 80–60 per cent Portland cement and 20–40 per cent pozzolana, which may be a naturally active material such as volcanic ash or pumice, or an artificial product such as burnt clay or shale. The rate of strength development is slower than that of normal Portland cement, especially at low temperatures. Pozzolanic cements have a higher resistance to chemical disintegration than the base Portland cement which they contain. Their resistance to sulphate attack is similar to that of sulphate-resisting Portland cement. See page 64.

Supersulphated cements are made in Belgium, France and Great Britain by grinding, usually together, but sometimes separately, blastfurnace slag, calcium sulphate and an activator, usually ordinary Portland cement.[4.6]

They have good properties of resistance to attack by sulphates and are stated to withstand attack from acids having pH values as low as 3·5.

They are used in the same way as ordinary Portland cement but it is preferable to increase the time of mixing to about five minutes.

The surface of the concrete may be rather friable and dusty, unless the exposed surfaces are coated with limewash and kept damp during the hardening period after striking the formwork. These precautions need not, of course, be taken when the concrete is subsequently buried or otherwise protected.

They have been found in the past to have rather variable setting and hardening properties, possibly due to temperature effects, so that particular care should be exercised during cold weather.

Supersulphated cement combines chemically with more water than is required for the hydration of Portland cement, and evaporation during curing must be avoided.

As with sulphate-resisting cement, mixes leaner than 1 to 6 by weight (525 lb of cement per cubic yard of concrete) are not recommended.

High-alumina cement is produced by fusing together in a furnace a mixture of limestone and bauxite (aluminium ore).

This cement has an exceptionally high proportion of aluminate (35–44 per cent), resulting in a rapid development in strength; in fact, high alumina cement is as strong, or even stronger, at 24 hours as ordinary Portland cement is at 28 days. Setting and hardening result chiefly from the formation of monocalcium aluminate and are accompanied by a rapid and high evolution of heat, a factor of considerable value when concreting at abnormally low temperatures, but a distinct disadvantage at normal temperatures in all but thin sections from which the heat can escape quickly.

High alumina cement is more resistant than any class of Portland cement to attack by sulphates, soft or peaty water, and dilute acid down to a pH 5 value, but is attacked by caustic alkalis.

A serious disadvantage of high-alumina cement is that the compressive strength of concrete made with it falls if it is exposed to temperatures above about 80°F combined with a high humidity. This applies whether this temperature is attained during setting and hardening, or at any subsequent time after hardening. High alumina cement is therefore unsuitable for concrete requiring a high strength, good wearing resistance, etc., in tropical and sub-tropical regions.

This reduction in strength, known as "conversion," is due to one or more of the calcium aluminate hydrates changing form, and can be recognized sometimes by a change from the characteristic dark grey colour to a chocolate brown colour. This change is generally accompanied by an increase in porosity.

The percentage fall in strength is considerable in high water/cement ratio concrete and may reach half the original strength, but with a water/cement ratio less than 0.4 the concrete will still retain a useful proportion of the original very high strength. Because the degree of loss of strength on conversion can be significant, designers should allow for conversion taking place.

Great care should be taken to keep high-alumina cement apart from Portland cement, as small quantities of either may cause a flash set in case of mixture. High-alumina cement can be bonded to hardened Portland cement concrete that is at least 7 days old. Conversely Portland cement concrete can be bonded to hardened high-alumina cement concrete which is at least 24 hours old.

Hydrophobic cement. This is a Portland cement to which an additive has been introduced, giving the cement particles a protective coating which inhibits hydration of the cement. This enables the cement to be stored in damp or buried conditions for a much longer period than is possible with ordinary Portland cement. The water-repellent film is broken down by abrasion during the concrete mixing process, and normal hydration takes place. The time of mixing should be increased by at least one minute longer than the normal period; hand mixing is unsatisfactory.

Hydrophobic cement is unlikely to provide water-proofing or water-repellent properties in the finished concrete. Hydrophobic cements are produced at present only at two or three places in the world and are likely to be useful and economically worthwhile only in humid and tropical climates.

The properties of hydrophobic cement are stated to comply with the requirements of B.S. 12:1958 for ordinary Portland cement.

The increase of strength with age

Reference has been made to the rate of development of strength of cement and concrete, and it has been shown that this is dependent on the compounds present. The rate of strength increase is greatest during the initial period of hardening and gradually decreases as time goes on. In the early stages of hydration it is only the outer layers of the particles of cement which take part in the chemical reaction, and if a piece of concrete is examined through a microscope, it is seen that particles of unhydrated cement still exist in the hardened paste. These unhydrated particles continue to absorb moisture from the air even after the mixing water has dried out and, by continuing the chemical reaction, gradually increase the strength and density of the concrete, a process which may continue for several years. It is the presence of unhydrated particles of cement which cause the phenomenon of autogenous healing, that is the healing

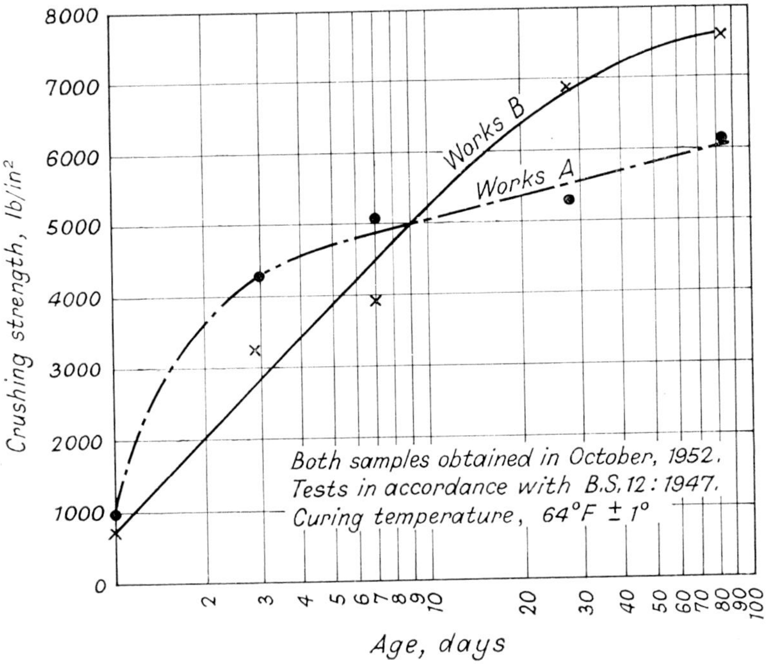

Fig. 4.8. Increase in mortar cube strength with age of two samples of ordinary Portland cement

together of concrete which has cracked. The effect of autogenous healing may be demonstrated by partly crushing a concrete cube, leaving it to weather for a few weeks, and then testing it again. A large proportion of the original strength is recovered.

Cements do not all harden at the same rate; Fig. 4.8[4.7] compares mortar cube strengths at ages up to 90 days, for two samples of ordinary Portland cement. Some authorities suggest that at ages of a year and upwards, the variation in compressive strength, even as between the different classes of Portland cement, is small compared with that at early ages.

The rates of increase in strength of concrete test cubes made with various cements and cured in water at 64°F, are given in Fig. 4.9. The bands are plotted from the results of a large number of cubes made from concrete with mixes between 1:4½ and 1:7 of cement: total aggregate, and show

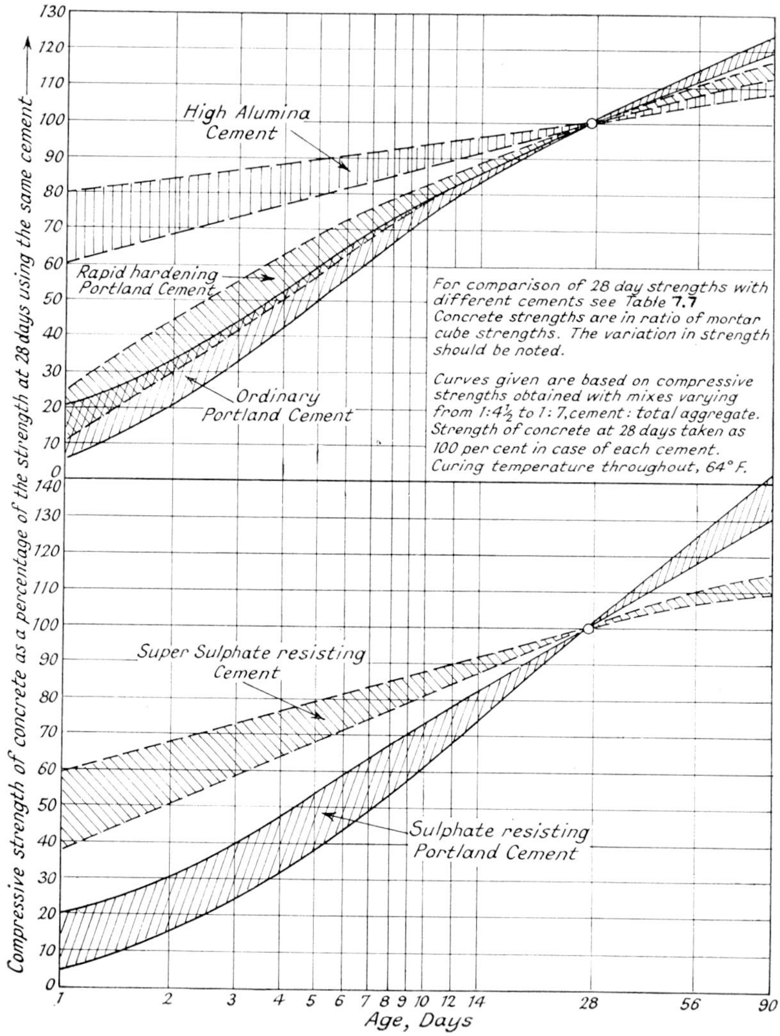

Fig. 4.9. Rate of strength development of different cements

the variation in rate of hardening which may be expected within each type of cement. Values obtained from Fig. 4.9 will be found to be largely in agreement with those given in the proposed International Code of Practice.[2.17]

Effect of temperature

The speed at which the chemical reaction proceeds during the setting and hardening of cement varies with the curing temperature. In small masses of concrete where the heat evolved by the cement is dissipated quickly, the rate of hardening varies with the surrounding temperature in the manner shown in Fig. 4.10. This diagram gives the compressive strength at ages up to 28 days as a percentage of that attained at 28 days

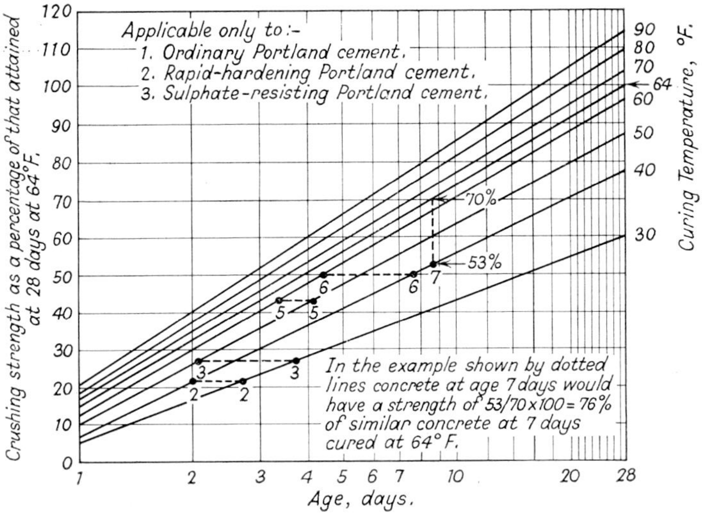

Fig. 4.10. The approximate relationship between curing temperature, crushing strength and age

at 64°F, for a range of temperatures. Thus, concrete cured at 30°F only reaches 37/65, or 57 per cent of the compressive strength of similar concrete cured at 64°F, also for 7 days. The example given in Fig. 4.10 is fully described on p. 376. The relationship should be treated only as a guide since different cements harden at different rates, and an accurate assessment is not possible.

If the temperature is much above 100°F, the paste is likely to set so rapidly that the concrete may stiffen before it is compacted, especially if, in addition, there is an opportunity for water to evaporate from the surface. In large sections, or in sections which are well insulated, the heat evolution of the cement may more than offset the heat losses, and the resultant temperature rise will increase the rate of hardening.

Maturity

The strength of concrete has been shown to depend on both the temperature of curing and age. Several investigators have attempted to express the strength of concrete as a function of the maturity, the summation of the integrals, the product of time and temperature. Perhaps the most easily used and practical "law" is that of Plowman[4.8] in which the percentage of the strength obtained at a maturity of 35600°F hr equals

$$A + B \log_{10} \left(\frac{\text{maturity in °F hr}}{10^3} \right)$$

It should be noted here that 35600°F hr is the maturity at 28 days for curing at 64°F.

The datum temperature for maturity calculation is 11°F. The values of the constants A and B depend on the strength level of the concrete and those suggested by Plowman are shown in Table 4.3.

TABLE 4.3 *Plowman's Constants For Use in the Maturity Equation*

Strength after 28 days at 64°F (maturity of 35600°F hr)lb/in²	Constant A	Constant B
less than 2500	−7	68
2500–5000	6	61
5000–7500	18	54
7500–10000	30	46·5

The values of the constants for the four strength ranges specified are sufficiently accurate for all normal work when the initial temperature of concrete is between 60°F and 80°F, and given the strength at a maturity of 35600°F hr the strength of that grade of concrete at any other maturity may be calculated.

Heat of hydration

The chemical reaction when cement sets and hardens is accompanied by the evolution of heat and, as has already been stated, this is dependent on the relative quantities of the various compounds present in the cement.

For British ordinary Portland cement the heat of hydration averages 70 calories per gramme at 7 days, rising to about 80 calories per gramme at 28 days. A variation of ±10 calories per gramme on these figures may be expected for individual samples. The temperature rise in concrete cured under adiabatic conditions, i.e. when no heat loss can occur, is largest during the first few days of hardening, as is shown in Fig. 4.11,[4.9] gradually becoming less as the concrete ages.

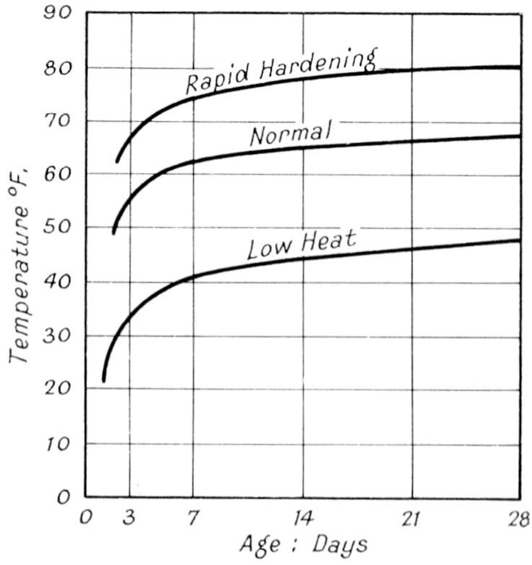

Fig. 4.11 Temperature rise in a typical 1:9 (weight) concrete under adiabatic conditions

The evolution of heat causes an increase in the temperature of the concrete, dependent, in actual practice, not only on the heat evolution from the cement, but also on the rate of placing of the concrete, its thermal conductivity, the type of shuttering, and external atmospheric conditions. The temperature rise will obviously be greatest in a large concrete mass; in dams a maximum rise of as much as 80°F has been recorded, and in most modern dams in which normal Portland cement has been used, it has been at least 40°F. See also p. 305. Dam concrete is usually fairly lean, and since the quantity of heat produced is proportional to the quantity of cement present, richer concretes in less massive structures are often subject to temperature rises of the same order. As an example, Fig. 4.12 shows the average temperature recorded in a part of a large concrete foundation where the concrete had a cement content of 550 lb/yd^3.

Properties of concrete as related to cement content

It is generally known that the strength of concrete increases with the cement content.

It is often argued that an increase in the proportion of cement increases the shrinkage. While this is correct for mixes of the same water/cement ratio, it is not necessarily important in practice, since it is usual to work to a given consistency on the site, as measured roughly by the slump test. As richer concrete requires less water to give a definite consistency, the tendency to shrink due to the larger cement content may be offset to some extent by the lower water content. However, an excess of water combined with a high cement content may cause undue shrinkage with an increased tendency toward cracking.

If the water/cement ratio is maintained constant, an increase in the cement content improves the workability of the mix without affecting the strength.

Transport and storage

Cement for small jobs is usually packed in 1 cwt bags and transported to the site by lorries. Where large quantities are used and portasilos are installed on the site, transport in bulk is more economical.

The storage of cement is entirely a matter of keeping it dry. Cement can be stored in airtight bins indefinitely without deteriorating in any way. However, on the site, except where large plants are installed, with elevators and overhead storage bins, this is not practicable, and it is necessary to stack the bags in a shed or under whatever cover is available. Every effort should be made to prevent moisture from coming into contact with the bags at any point, and it is advisable to provide a raised floor covered with water-proof material such as tarpaulin.

Even when stored under good conditions bagged cement is likely to lose 20 per cent of its strength after 2 months' storage, and 40 per cent after 6 months storage.

Air-set cement results from storage in a damp atmosphere. This is due to the moisture present in the air being absorbed by the cement and causing a partial set. The absorption of moisture from a damp atmosphere is reduced by preventing the movement of air into the store as far

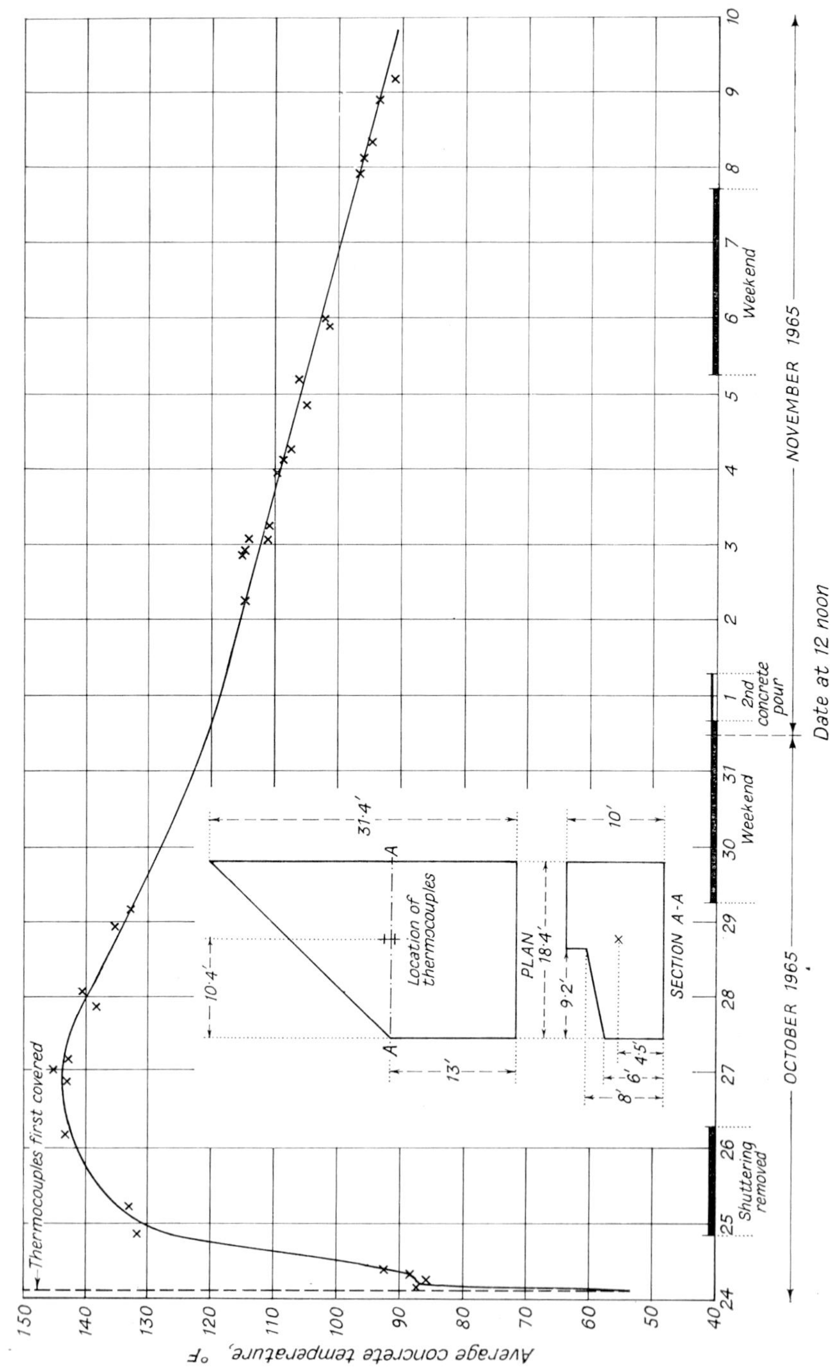

Fig. 4.12 Temperature rise within a large concrete foundation

as practicable. As a rough guide, lumpy cement which cannot be easily crumbled in the fingers is unsatisfactory for general use.

Cement should be stacked in such a way that the cement first delivered can be used first.

Hot cement

The arrival of cement on sites in a warm or hot condition is often the subject of controversy. The high temperature is due to the friction which occurs during grinding, the heat evolved not having been dissipated during storage.

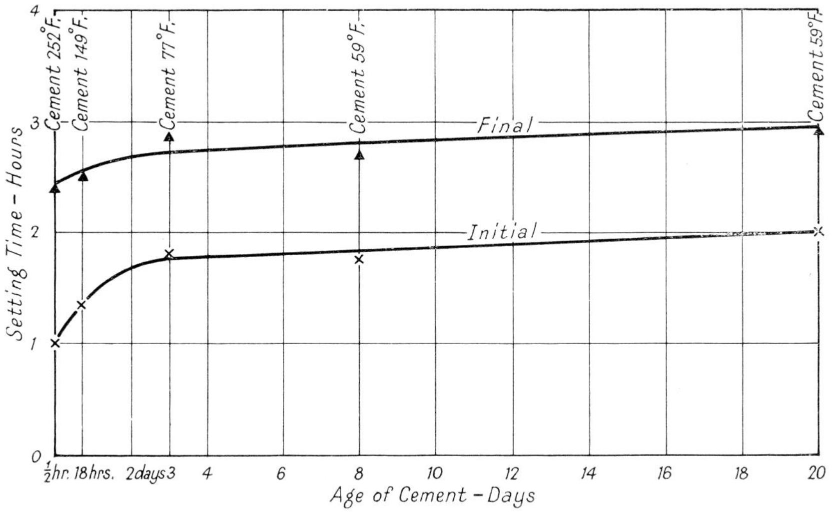

Fig. 4.13. Setting time of hot cement

Tests at the Building Research Station[4.10] have indicated that hot cement has a somewhat shorter setting time than cool cement, as is shown in Fig. 4.13. Since this test is made on neat cement, the effect of the hot cement is probably exaggerated compared with the effect in mixed concrete. Other tests[4.11] have shown that the variation of the temperature of cement over the range 18°–71°C has a negligible effect on the workability and strength of concrete.

Hot cement may be an advantage in cold weather.

Cement tests

Details of the methods of testing are given in the relevant British and ASTM Standards and they should be implicitly followed if reliable results are to be obtained. In general, it is advisable to leave cement testing to an experienced operator in a properly equipped laboratory.

Cements, like other manufactured goods, are subject to some variation in quality, and it is advisable when comparing cement from two sources to obtain at least a dozen samples from different consignments and on different days. The comparison should then be based not only on the standard of quality, but also on the variability or otherwise shown by the tests. An important point which should be remembered in connection with cement tests is that strength and setting times are also affected by the relative humidity of the atmosphere. For both these tests it is essential to maintain the specified temperatures and curing conditions. The tests usually made on cement are:

(1) *Fineness.* All the British Standards for the different cements prescribe a method of test for the determination of the specific surface of cement expressed as the total surface area in cm^2/g based on the Lea-Nurse permeability apparatus. The method is described in detail in each of the British Standards.

ASTM C204-55 gives a similar but modified method of test using the Blaine air permeability apparatus, whereas ASTM C115-58 gives a method of test using the Wagner turbidimeter apparatus.

The Blaine and Lea and Nurse methods give results of specific surface in close agreement with one another, but very much higher values than those obtained from the Wagner method. As a general indication the air permeability methods give results about 1·8 times higher than those from the turbidimeter method.

The relevant specification requirements for the various British and American cements are shown in Table 4.2.

(2) *Chemical Composition.* The analysis of cement is complicated and should be left to an experienced chemist. Limits for the proportions of certain of the constituents are given in the relevant standards.

(3) *Strength.* The strength of cement is determined from compressive tests on mortar cubes or, alternatively, in British Standard tests by 4 in. concrete cubes. Provision is also made for a tensile test when specially desired and provides a guide to the early strength when the speed of demoulding is important.

(*a*) *Compressive strength of mortar cubes.* In Britain the compressive strength is determined on 2·78 in. cubes of a mixture of cement, standard Leighton Buzzard sand and water in certain proportions. These cubes are vibrated in the machine shown in Fig. 4.14, which has a frequency of 12,000 vibrations per minute, with an amplitude of 0·0022 in. In order to obtain the correct amplitude of vibration, the vibrating table is mounted on the springs shown, and the total weight of the machine on its supporting springs is specified.

The cube mould with its base-plate is firmly held on the table of the vibrating machine by screwing down the hopper and by a thumb-bolt on the side. The hopper is filled with mortar and vibrated for 2 minutes at full speed. The mould is then removed and the whole placed in an atmosphere of at least 90 per cent relative humidity and at a temperature of $66° \pm 2°F$ for 24 hours, after which the cubes are removed and immediately submerged in clean water until the time of testing.

This test is not affected by personal error as much as the tensile test, since the compaction is controlled by the machine. Thorough mixing of the materials is important, and a skilled operator should be employed.

The machine requires careful manufacture and it is advisable to purchase from a source known to produce high-class work. Attention to the belt, spring tension, any compression of the carrying springs, and the general cleanliness of the machine, is essential if smooth running is

Fig. 4.14. Mixing slab and mortar cube vibrator. A time switch on the right stops the machine after 2 minutes

to be maintained. The frequency and amplitude should be checked annually. Three cubes are tested in compression at each of the specified ages, generally 3 and 7 days; the minimum strengths which should be obtained are shown in Table 4.2.

For American cements the compressive strength is determined on 2 in. cubes of a mixture of cement, standard Ottawa sand and water but, whereas in the British method a fixed amount of water is added, the amount of mixing water is adjusted to produce a flow of between 100 and 115 when tested on the flow table. Furthermore, the specimens are hand-compacted in a specified manner instead of using the British vibrating machine.

These slight differences in the sand quality and grading, water/cement ratio and method of compaction will result in variations in the results from the two methods and strict comparison between the specification requirements is not possible.

Cements usually have strengths well in excess of the Specification, but a large variation is often observed.

Table 4.4 summarizes a large number of tests at ages from 7 to 90 days on cements delivered to sites in the United Kingdom.

A further illustration of variations both at 3 and 28 days is given in Fig. 4.15. The histograms are based on tests made by the Building Research Station[(4.7)] on 140 batches of cement from a limited number of sources. About half the batches of ordinary and rapid-hardening cements included in the tests overlap in strength. Similar variations have been shown, by Wing and Ruettgers,[(4.12)] to occur with American cements.

The variations in strength are the result of changes in the characteristics of the basic materials, coal quality, plant inefficiencies, and other factors. A small percentage of the variation may be ascribed to errors in the making and testing of mortar cubes. As indicated previously when discussing the increase of strength with age, cements harden at different rates

TABLE 4.4 *Summary of B.S. Vibrated Mortar Cube Tests on a Number of Cements*

Country of manufacture	Type of cement	Compressive strength lb/in²					Standard deviation lb/in²				
		1-day	3-day	7-day	28-day	90-day	1-day	3-day	7-day	28-day	90-day
England	Normal Portland	(14) 1030	(15) 3640	(15) 4840	(14) 6860	(15) 7670	(14) 264	(15) 376	(15) 347	(14) 652	(15) 698
Belgium	Normal Portland	(3) 800	(11) 2630	(11) 4420	(3) 6560	(3) 6960	*	(11) 422	(11) 579	*	*
Germany	Normal Portland	(3) 1370	(8) 3250	(8) 4420	(3) 6720	(3) 7140	*	*	*	*	*
Sweden	Normal Portland	(2) 1970	(3) 3270	(3) 4780	(2) 6400	(2) 7500	*	*	*	*	*
England	Sulphate-resisting	(2) 1310	(6) 3150	(6) 4450	(4) 6510	— —	*	*	*	*	* —
Belgium	Super-sulphate-resisting	(4) 1110	(10) 4170	(10) 5740	(6) 7840	—	*	(10) 1095	(10) 946	*	

(1) The figures in brackets indicate the number of batches tested.
(2) The compressive strengths quoted are the average of the results of tests on the number of batches indicated, three cubes being tested at each age from each batch.
(3) Standard deviation not assessed for less than 10 batches of cement.*

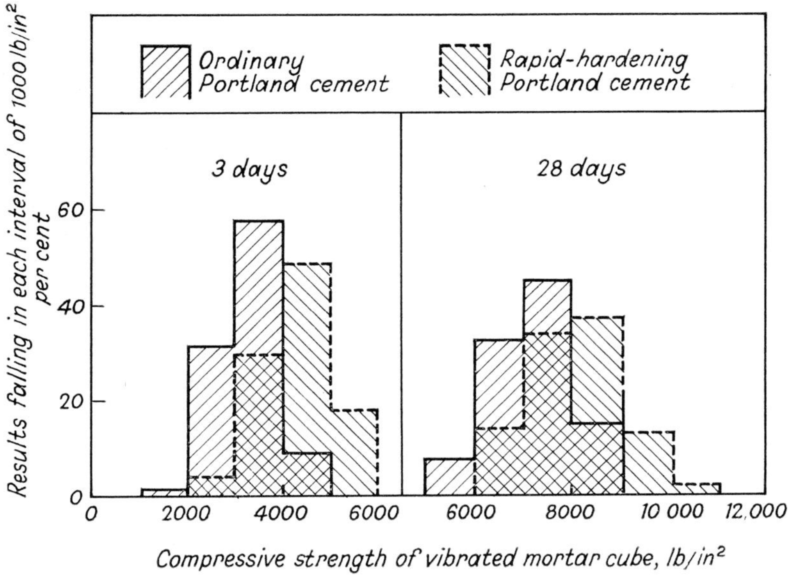

Fig. 4.15. Histograms for compressive strengths of samples of ordinary and rapid-hardening Portland cements

and this accounts for some variation in strength at early ages; the variation in the strength at later ages may be smaller, although no indication of a smaller standard deviation for British cements at later ages was apparent in the tests quoted in Table 4.4.

Mortar cube strengths correlate reasonably well with those obtained with the same cement used in a 1:2:4 concrete mix, and a water/cement ratio of 0·60. Tests made on 220 samples of cement at the Building Research Station(4.13) indicate that the concrete strength using the above mix may be expected to be about 0·7 times the mortar cube strength at the same age. The range in the ratio was about 0·6 to 0·8 in this test series. The mortar cube test, therefore, provides an indication of the compressive strength likely to be obtained in practice with cement similar to that used in the test.

The importance in practice of the variations in the mortar cube strength of the cement is thus demonstrated.

(*b*) *Compressive strength of concrete cubes.* An alternative method included in the British Standard requires the preparation of 4 in. concrete cubes with certain specified types of aggregate, $\frac{3}{4}$ in. maximum size, and sand. The combined grading is adjusted until a mix, in which 325 ± 1 g cement, and 195 ± 1 g water is used to give just sufficient concrete for each 4 in. cube, has a slump of $\frac{1}{2}$–2 in. The cubes are made and stored in water after demoulding at a temperature of 66 ± 2°F, until tested in compression at 3 and 7 days. Three cubes are tested at each age. The required strengths are given in the relevant British Standards and are included in Table 4.2.

(4) *Setting Time.* The test for setting time involves both a determination of the quantity of water necessary to produce a paste of normal consistency, and the test for setting time itself. Figure 4.16 shows the Vicat needle used for the test. The normal consistency is that at which the plunger penetrates to within 5 mm to 7 mm from the bottom of the Vicat mould, $4 \pm \frac{1}{4}$ minutes from the moment of adding water to mix the cement paste. The initial setting time is the period which elapses between the time when the water is added to the cement, and the time at which the thin needle does not penetrate beyond a point 5 mm from the bottom

Fig. 4.16. Measurement of the initial setting time of cement by means of the Vicat needle

of the mould. The cement is considered as finally set when, on applying the needle with the angular attachment, it makes an impression on the block while the attachment fails to do so. It should be noted that the initial and final setting times are a gauge of the period which elapses before a certain consistency of paste is reached; setting and hardening actually commence before the initial setting time and proceed long after the final setting time.

(5) *Soundness.* The Le Chatelier method of test for soundness is a simple little test to determine whether undue expansion is likely to occur. Providing a bath controlled at $66° \pm 2°$F is available, this test can easily be done on the site. British cements rarely give a measurable expansion, however, and the test is for that reason not often made.

REFERENCES

4.1 LEA, F. M., Modern developments in cements in relation to concrete practice, *J. Inst. Civ. Engrs*, Feb. 1943.

4.2 B.S. 12:1958, *Portland Cement (Ordinary and Rapid-hardening)*. British Standards Institution.

4.3 BRITTAIN, K. H., Special cements and their use, *Consulting Engineer*, Oct. 1963.

4.4 Roberts, M. H., Effect of calcium chloride on the durability of pretensioned wire in prestressed concrete, *Mag. Concr. Res.*, **14**, No. 42, Nov. 1962.

4.5 Fulton, F. S., *Concrete Technology*, 1964. The Portland Cement Institute, Johannesburg.

4.6 Baxter, D. J. and Boardman, J. M., The manufacture, properties and use of super-sulphated cement in civil engineering works, *Civ. Eng. Publ. Wks Rev.*, April 1962.

4.7 Snow, F. S., Lea, F. M., Murdock, L. J. and Burke, E., Would the strength grading of ordinary Portland cement be a contribution to structural economy? *Introductory Notes, Proc. Inst. Civ. Engrs*, Part III, Dec. 1953.

4.8 Plowman, J. M., Maturity and the strength of concrete, *Mag. Concr. Res.*, **8**, No. 22, March 1956.

4.9 Davey, N., *Temperature Rise in Hydrating Concrete*, Building Research Technical Paper, No. 15. H.M.S.O.

4.10 Davey, N., *Hot Cement*, Building Research Bulletin No. 7, 1930. H.M.S.O.

4.11 Keene, P. W., *An Investigation of the Effect of Cement Temperatures and Ambient Temperatures on the Workability and Strength of Concrete*, Cement and Concrete Association DN/17, April 1962.

4.12 Wing, S. P. and Ruettgers, A., Variations in strength of Portland cement, *Proc. Amer. Soc. Test. Mater.*, **43**, 1943, p. 917.

4.13 Newman, A. J., Materials for concrete, *Proceedings of a Symposium on Concrete Quality*, Nov. 1964. Cement and Concrete Association.

CHAPTER 5

Admixtures

Introduction

ADMIXTURES, ranging from additions of chemicals to the use of waste materials for which an outlet is being sought, have been advocated for use in concrete almost from the time when cement was first invented.

The use of an admixture should only be considered where it is desired to modify for a particular reason the properties of either fresh or hardened concrete, or both, which cannot be modified or obtained by changes in the composition or proportions of the normal concrete mix. For instance, a harsh mix may be made more plastic and cohesive by the addition of a plasticizer, a pore filler, an air-entraining agent, a change in the proportion of sand to coarse aggregate, a change in the sand grading, or by the use of additional cement.

Because an admixture is generally introduced into a concrete mix in a relatively minute quantity, the degree of control must be higher than is usual for ordinary concrete work to ensure that overdosages are unlikely to occur. There is always a temptation for the operative to add that little bit extra, especially if conditions appear to him to be more adverse than usual. The trouble and expense involved in achieving the required control can often out-weigh any advantages obtainable by using admixtures. Excess amounts of an admixture can be highly detrimental to the strength or other properties of the concrete.

In Great Britain there are about 300 proprietary admixtures available on the market, and the selection of a suitable one for a particular purpose can be confusing. In order to be able to assess the suitability of an admixture the active ingredients must be known, because, while some properties of the concrete may be enhanced by one ingredient, detrimental effects to other properties may be caused by other ingredients. As an example, a plasticizer was being used on site in concrete for airfield runways, the control being exercised over the concreting was good and an automatic dispenser was in use for measuring the correct dosage of admixture. Three days after placing one bay of concrete it had not set and slumped on removing the side forms. It was established that a valve in the dispenser had stuck, resulting in a ten-fold increase in the amount of admixture for each batch of concrete. This overdosage did not result in excessive workability but the small amount of retarder present became the predominating factor.

The Reinforced Concrete Association, now incorporated in The Concrete Society, has recently made a useful start in overcoming the confusion concerning admixtures by asking manufacturers to disclose the main active ingredients of their products. Many manufacturers have since made available standard data relating to the main active ingredient, whether or not calcium chloride is a constituent, and whether or not the admixture entrains air when normally used in concrete.

The importance of testing admixtures, particularly with the cement and aggregates to be used on a site, cannot be over-emphasized. Such tests should be designed to show that no adverse side effects are likely, such as increased drying shrinkage, reduced bond strength, reduced ultimate compressive strength, and reduced modulus of elasticity.

A concrete of fundamentally poor quality will not be converted into a good concrete by any type of admixture.

Accelerators

Almost all proprietary accelerators are based on calcium chloride and are used generally for accelerating the strength development of concrete in cold weather. Calcium chloride is available in two forms, firstly, the more general flake calcium chloride which contains about 75 per cent of $CaCl_2$ and, secondly, the granular calcium chloride containing about 95 per cent of $CaCl_2$. The generally accepted dosage for flake calcium chloride is 2 per cent by weight of the cement. At this concentration, the strength of concrete is increased, particularly at early ages. The setting time is somewhat reduced and Shacklock[5.1] has recommended that concrete containing calcium chloride should be transported, placed and compacted within twenty minutes of mixing, otherwise the mix may stiffen to such an extent that site operatives may be tempted to adopt the bad practice of adding more water, causing a reduction in the compressive strength.

Drying shrinkage generally, but not always, has been found to be increased when calcium chloride is used. Newman[5.2] has shown this to be about 40 per cent when using a 1:2:4 concrete, with a water/cement ratio of 0·60 and ordinary Portland cement. The resistance to sulphate attack given by sulphate-resisting cement may be lost if calcium chloride is used in the concrete.[5.1] Larger quantities of calcium chloride do not cause any appreciable further increase in strength over that obtained with 2 per cent, while the setting time may be unduly reduced and the drying shrinkage is increased. Examination of the stated contents of a number of proprietary accelerators has shown that, at the manufacturers recommended dosage, the amount of calcium chloride is sometimes in excess of the normally accepted 2 per cent by weight of the cement. This applies particularly to accelerators where the manufacturers recommended dosage varies with the temperature of the surrounding air.

Calcium chloride at the normal dosage does not significantly reduce the freezing point of water.

The effect of calcium chloride varies somewhat from one cement to another, even as between different batches of the same type, so that the increase in strength due to calcium chloride cannot be determined precisely. The curve in Fig. 5.1, however, relates the ratio of strength of concrete in proportions approximating to 1:6 cement:total aggregate cured at 64°F, with and without calcium chloride at 2 per cent, to the age in days. Thus, at 7 days, the strength of concrete in which 2 per cent of calcium chloride has been used, may be expected to be $1\frac{1}{4}$ times the compressive strength of similar concrete without the addition, both being cured at 64°F. The ratio in strengths is roughly the same at other curing temperatures, down to freezing point, and the curve in Fig. 5.1,

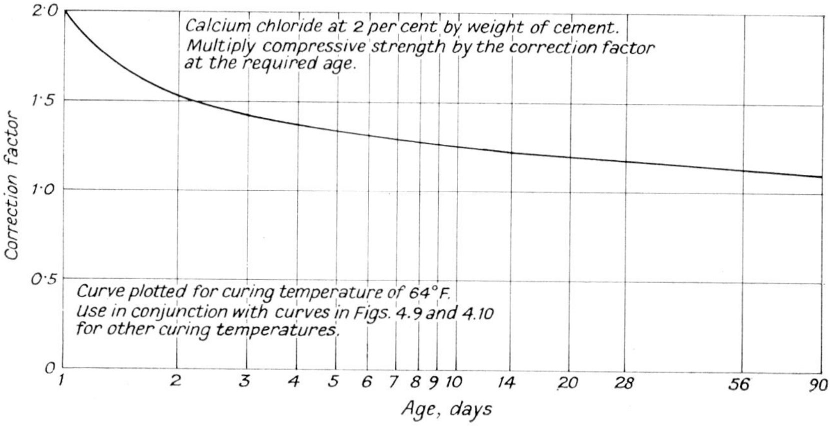

Fig. 5.1. The effect of calcium chloride on the compressive strength of Portland cement concrete

may therefore be used in conjunction with Figs. 4.9 and 4.10, in obtaining a rough estimate of the strength of concrete with calcium chloride addition at various ages and curing temperatures.

The use of calcium chloride is now generally accepted as increasing the risk of corrosion of steel reinforcement and it should be noted that C.P. 115[5.3] prohibits the use of calcium chloride in prestressed concrete. For good quality reinforced concrete, however, the normal dosage of calcium chloride is unlikely to have a significant effect on the corrosion of the reinforcement. Overdosages will increase the risk; in one instance, corrosion of mild steel bar reinforcement, resulting in spalling of the concrete within 12 months, was found to be due to a twelvefold dosage of calcium chloride.

Aluminium chloride has a similar, but more marked, effect in accelerating the setting of cement than calcium chloride, and is sometimes used when concrete is required to set in a few minutes in order to stop water seepage under pressure.

Retarders

Retarders, as this name implies, delay the beginning of the setting and hardening of concrete, and under normal conditions in the United Kingdom their use is seldom necessary. They are most useful in hot countries where concrete has to be transported for some distance, and for concreting in large quantities in very hot weather, when it is important to avoid the formation of "cold joints."

Overdosage is likely to affect seriously the setting of concrete. If deep pours are envisaged the delayed setting is likely to increase formwork pressures considerably.

Pore fillers and pozzolanas

A number of materials in powder form are available which are claimed to increase workability, and often to improve various other properties. Some of these powders act as pore fillers and others react with the cement, as in the case of pozzolanas. These materials include quick lime, slaked lime, diatomaceous earth, bentonite, kaolin, rock flour, ground blastfurnace slag, and fly ash. These are discussed briefly under the various headings below.

(*a*) *Lime.* Slaked lime, has in the past, been used extensively as an admixture on the theory that it improved workability and water-tightness. It is, however, of doubtful value and there seems no advantage in adding lime rather than using a richer mix.

(*b*) *Pore fillers.* Pore fillers, such as diatomaceous earth, bentonite, kaolin and rock flour, are of very doubtful value and, while they provide increased workability and cohesiveness, an increased water/cement ratio is required which reduces the strength.

The drying shrinkage is increased by these materials, which are, in effect, little different from the clay, silt and stone dust so carefully excluded from concrete aggregates.

(*c*) *Pozzolanas.* Pozzolanas are those materials which react with any lime set free during the setting of cement, and so improve the durability of concrete, including its resistance to attack by sulphates, peaty waters, etc., Callet,[(5.4)] however, instances a failure of pozzolanic cement at Le Havre. See also page 351. Pozzolanas are used either in addition to, or to replace up to 40 per cent of the cement. They reduce the rate of hardening of concrete, and this is one of the principal objections to their use. The evidence available suggests that the ultimate strength for replacements of up to at least 20 per cent of the cement by pozzolanas is not very different from that when cement alone is used.

(*d*) *Ground blastfurnace slag.* Portland blastfurnace cement has been in production for many years and is described briefly on page 64. A newer development known as the Trieff process is to transport ordinary Portland cement and blastfurnace slag to the site separately, and then to grind them together wet immediately before mixing with the aggregate to form concrete. It is claimed that up to 60 per cent of blastfurnace slag can be used in this process. The results may be expected to be similar to those obtained with Portland blastfurnace cement. The plant needed to grind the two materials together limits the use of this process to larger works.

(*e*) *Fly ash.* Fly ash is produced during the burning of powdered coal, the chief source of supply being power stations. It contains some unburnt coal, the proportion of which should not exceed 8 per cent if the fly ash is to be used as an admixture in concrete. The proportion of this unburnt material is apt to vary from time to time, being highest when the burners are working inefficiently immediately after being lit, or when the boilers are not working under full load, and frequent testing is important if the occasional use of material with a high carbon content is to be avoided. Fly ash varies in properties and it is necessary to determine independently, the quality of material from any particular source.

In general, fly ash has a somewhat similar effect to pozzolanas, in that the initial setting and hardening may be delayed, but the ultimate strength may be as high as that of concrete made with cement alone. This applies only for proportions of fly ash up to about 20 per cent used as a replacement for cement.

There have also been suggestions that fly ash might be used to replace a proportion of the sand in the mix. Tests have indicated that replacement of up to 20 per cent of sand by fly ash results in an increase in workability in harsh mixes, but not in mixes which already contain an adequate proportion of fines. Some increase in strength occurs under the same conditions, providing the workability is maintained constant. The drying shrinkage is not affected by its use in the proportions stated above and the temperature rise in mass concrete can be reduced. Fly ash has been used in the concrete of several dams.

Water-repellent admixtures

A large number of water repellent admixtures have been placed on the market, their function being to prevent the absorption of rain into the concrete. They are of no use against water under pressure. The water-repellent material, i.e. that which causes the water to run off, as "off a duck's back," is usually a metallic soap or vegetable or mineral oil. Gradual leaching by rainwater reduces their effectiveness in time.

The value of those admixtures which are claimed to waterproof concrete is doubtful, even if they do possess some waterproofing effect, since they are completely ineffective against water penetration through cracks, joints, honeycombed areas, etc., where most leakage is likely to occur. Good concrete is practically impermeable in itself without any admixture.

Workability aids

A number of chemicals have been marketed which, added in very small percentages to the mix, give some increase in workability. They are known as wetting agents, dispersing agents, water-reducing agents, or plasticizers. Their effect is usually marginal and there may be little measurable difference in workability between a concrete without the admixture and one containing the admixture and having the same compressive strength.[5.1] Little appears to have been published on the long-term effects of any of these materials.

The use of workability aids should be limited to concrete made with "difficult" aggregates and so liable to harshness or segregation of the mix, or to special and unusually rich mixes where very high strengths are needed in conjunction with good workability.

Air entrainment

Air-entraining agents are used widely in the United States where it is now generally accepted that the durability of air-entrained concrete, particularly as regards frost resistance and resistance to de-icing salts, is considerably improved. A reduction in strength is, however, accepted, and the American Concrete Institute recommended practice for selecting proportions for concrete, A.C.I. 613–54[5.5], allows for a 20 per cent reduction in strength when air entrainment is adopted. For example, the probable compressive strength at 28 days at a water/cement ratio of 0·54 is given as 4,000 lb/in^2 for non-air-entrained concrete, as compared with 3,200 lb/in^2 for air-entrained concrete.

This reduction in strength is equivalent to an increase in total air content of 3 to 4 per cent over the amount of air normally entrapped in non-air-entrained concrete. A similar decrease in strength would be obtained if the water/cement ratio were increased by 0·07–0·08. On the other hand, the workability is increased by an amount roughly equivalent to the same change in water/cement ratio. If the reduction in strength is acceptable, greater workability is therefore obtained without the use of an unduly high water/cement ratio, which would give rise to risk of frost failure, see also p. 349.

A reduction in the tendency of concrete to bleed is also usually obtained. As bleeding results in the formation of fine cracks below the larger particles of aggregate, which provide paths for the penetration of water, a reduction in bleeding is beneficial from the point of view of permeability and durability.

Air-entraining agents have not been used to the same extent in Great Britain, mainly because the incidence of frost damage has been so much lower than has been the case in the United States. The customary use of lower water contents in Great Britain may be an important factor. They are, however, being used more extensively in concrete for road and runway constructions, particularly to provide increased resistance to the combined effects of frost and the salt used for de-icing.

Air entrainment should not be used unless site control is good, since the amount of air entrained may vary considerably with changes in sand grading, errors in proportioning, workability of the mix, and temperatures. Checks on the air void content should be made at frequent intervals using apparatus such as that shown in Fig. 5.2,[5.6] since each one per cent

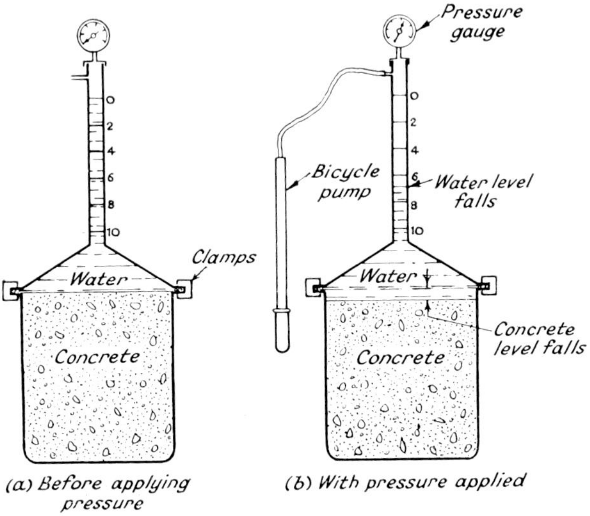

Fig. 5.2. Diagram of pressure-type apparatus for determining the air content of air-entrained concrete

When the air pressure is increased, the air in the concrete is reduced in volume; this slightly lowers the surface of the concrete and causes the level of the water in the tube to fall increase in air voids is likely to result in a loss in strength of between 5 and 6 per cent, see Fig. 12.1.

Careful consideration should be given to the relative economies of using an air-entraining agent, instead of the addition of extra cement; changes in the resulting strength, and the desirability of avoiding complication especially on small sites, should be borne in mind.

REFERENCES

5.1 SHACKLOCK, B. W., The use of admixtures in concrete, *Concr. Constr. Eng.*, Sept. 1962.

5.2 NEWMAN, A. J., Effects of the addition of calcium chloride to Portland cements and concretes, *Concr. Constr. Eng.*, May 1943.

5.3 C.P. 115, *The Structural Use of Prestressed Concrete.* British Standards Institution.

5.4 CALLET, M. P., Concreting in sea-water. Notes from a paper published in *The Dock and Habour Authority, Concr. Constr. Eng.*, Nov. 1949, p. 364.

5.5 A.C.I. 613–54, *Recommended Practice for Selecting Proportions for Concrete, J. Amer. Concr. Inst.*, **26,** No. 1, Sept. 1965, pp. 49–64, or American Concrete Institute Standards, 1965.

5.6 WRIGHT, P. J. F., Entrained air in concrete, *Proc. Inst. Civ. Engrs*, Part 1, **2,** No. 3, May 1953.

CHAPTER 6

Water/Cement Ratio, Workability and Segregation

WATER has two functions in a concrete mix, firstly, to enable the chemical reactions which cause setting and hardening to proceed, and secondly, to lubricate the mixture of gravel, sand and cement in order to facilitate placing.

As in other chemical reactions, the cement and water combine in definite proportions, and for normal Portland cement 1 part by weight of cement requires about 0·25 parts by weight of water for hydration. Concrete containing such a small proportion of water would, however, be very dry and exceedingly difficult to compact. This is discussed more fully in the next Chapter. Extra water is, therefore, added to lubricate the mix, that is, to make it workable, and since all this water evaporates when the concrete dries, leaving voids, it is important that the water used to provide lubrication should be kept to a minimum.

If concrete is not fully compacted, numerous bubbles of air may be trapped, resulting in further voids. There are, then, two main sources of voids in concrete, entrapped air bubbles, and the water required for lubrication, which later evaporates. Air bubbles are more easily eliminated from wet mixes than from dry mixes; the latter require a considerable amount of work to liberate the air. It follows that for any given method of compaction representing a certain amount of work done there is an optimum water content at which the volume of entrapped air plus the volume of water voids is a minimum. Concrete with the minimum volume of total voids is the densest and strongest and therefore, normally the most desirable.

The conclusion to be drawn from the above is that the densest and strongest concrete is obtained by using the minimum quantity of water consistent with the degree of workability required to give maximum density. The degree of workability must be considered in relation to the method of compaction and the type of construction, in order to avoid the necessity for an excessive amount of work to obtain maximum density. This statement can be illustrated by comparing the stiff concrete placed in roadwork by heavy vibratory equipment with the more fluid concrete necessary for thin reinforced concrete wall sections rammed by hand. If stiff concrete were used in the second case, honeycombing would almost certainly occur, and the density and strength would be far less than in the roadwork.

The definition of the term water/cement ratio needs clarification, since there is as yet no definite agreement among the various authorities on concrete. The difficulty arises from the presence in a batch of concrete of water from three possible sources:

(1) Water absorbed into the aggregate, (w_a)
(2) Surface water on the aggregate, (w_s)
(3) Water added during mixing, (w_m)

Water from sources (2) and (3) together provides what might be termed the free water in the mix and the author has throughout this book adopted the basis that:

$$\text{Water/cement ratio} = \frac{w_s + w_m}{W_c} = \frac{w}{W_c} \qquad (1)$$

where W_c denotes the weight of cement.

In this equation it is assumed that the aggregate is wet; if it should be dry the water to be added at the mixer,

$$w_{md} = w_a + w \qquad (2)$$

The relationship between the water/cement ratio, the richness of the mix, the grading of the aggregate, and the workability and strength of concrete was first studied by Professor Abrams in America. The conclusions drawn from this work led to the formulation of Abrams' water/cement ratio law. This reads as follows:

> "With given concrete materials and conditions of test, the quantity of mixing water used determines the strength of the concrete, so long as the mix is of workable plasticity."

This law implies that for fully compacted concrete, sound aggregates and a given cement the strength depends only on the ratio of water to cement. Research in recent years has shown that there is some variation from this law, and this is dealt with in Chapter 7, Design of Concrete Mixes. The strength: water/cement ratio relationship given in Chapter 7 is quoted for an age of 28 days only. Reference should be made to Chapter 4, Cement, for information on the effects of age and temperature.

Increases in the water/cement ratio have an adverse effect on such properties of concrete as permeability, resistance to frost action and weathering, resistance to abrasion, tensile strength, creep, modulus of rupture, and shrinkage. See Chapter 2.

It may be concluded that for nearly all purposes concrete having the least water/cement ratio necessary to give the degree of workability required for full compaction without an undue amount of work, will give the best results.

Workability

The term workability is difficult to define precisely, and Newman[6.1] has proposed that it should be defined in at least three separate properties:

(i) Compactibility, or the ease with which the concrete can be compacted and the air voids removed.

(ii) Mobility, or the ease with which concrete can flow into moulds, around steel and be remoulded.

(iii) Stability, or the ability of concrete to remain a stable coherent homogeneous mass during handling and vibration without the constituents segregating.

No one test gives complete guidance on the degree to which these properties are developed in a mix. The compacting factor, the V-B test and the slump test provide the most useful general guide at present. Of these the slump test is probably the best known in the field, both in this country and in the United States, while the V-B test and the compacting factor test have been used to an increasing extent in the laboratory. The slump test is a good measure of mobility and stability in concretes of medium or high workability. The compacting factor test was devised at the Building Research Station and has certain advantages in that it measures the degree of compaction for a certain amount of work done. It is much more sensitive to changes in the water/cement ratio of rich mixes than of lean mixes where particle interference occurs[6.1]. Cusens[6.2] concluded that the small amount of work applied in the compacting factor test did not provide an accurate measure of the work required to compact dry, harsh mixes with compacting factors less than 0.80. The V-B test is more satisfactory for rich or dry mixes which are to be vibrated to obtain high strengths.

Other methods attempt to measure the consistency of the concrete while still in the mixer. They depend on measurements of torque, or overturning forces. Although they indicate changes in the consistency of concrete mixes, their use must be supplemented by frequent slump tests.

Slump test

The slump test was devised in America nearly forty years ago and has been universally adopted as a check of the consistency of concrete in the field. While it has been shown that with all except rich workable concretes it is liable to random variations, the slump test does provide a simple practical method for maintaining reasonable uniformity in the consistency of concrete produced in the field.

The method of making slump tests is described in B.S. 1881:1968. The test is made with a metal cone 12 in. high, having a bottom diameter of 8 in. and a top diameter of 4 in., on a representative sample of the concrete, the procedure being as follows.

The slump cone should first be inspected to make sure the internal surface is clean, dry, and free from set cement.

The cone is then placed on a smooth, flat, non-absorbent surface, preferably a steel plate, and the operator holds the mould firmly in place, while it is being filled, by standing on the foot pieces. The mould is filled to about one-fourth of its height with the concrete which is then tamped, using 25 strokes of a $\frac{5}{8}$ in. rod, 2 ft long, rounded at the lower end, Fig. 6.1 (*a*). The filling is completed in successive layers similar to the first and the top struck off so that the mould is exactly filled. The feet should still remain on the foot pieces until surplus concrete has been cleaned from round the base of the cone. The mould is then removed by raising vertically, immediately after filling, Fig. 6.1 (*b*). The moulded concrete is finally allowed to subside and the height of the specimen measured after coming to rest, Fig. 6.1 (*c*).

The American slump test as given in ASTM C143-66 is similar except that the mould should be dampened and that it should be filled in three layers, each approximately one third the volume of the cone.

The British Standard requires that the slump of concrete sampled in the field shall be measured within a period of 2 minutes after sampling. In the case of samples prepared in the laboratory the slump is measured 6 minutes after adding the water for mixing. This requirement arises from the change in slump which occurs with time.

Fig. 6.1 (*a*). Tamping concrete in slump cone
Note that the cone is held down by keeping the feet on the small lugs at the bottom

Fig. 6.1 (*b*). Lifting the cone vertically

Fig. 6.1 (*c*). Method of measuring slump
The horizontal arm is 12 in. above the base plate. Alternatively, the slump cone is placed alongside and used as a basis for measuring the slump.

The consistency is recorded in terms of inches of subsidence of the specimen during the test, which is known as the slump.

The random variations which occur between slumps are in some measure due to the fact that three distinct types of slump may occur. These are illustrated in Fig. 6.2.[3.9] The first consists of a general and uniform subsidence without breaking up, and may be termed a "true" slump. The second is a "shear" slump which occurs when the top

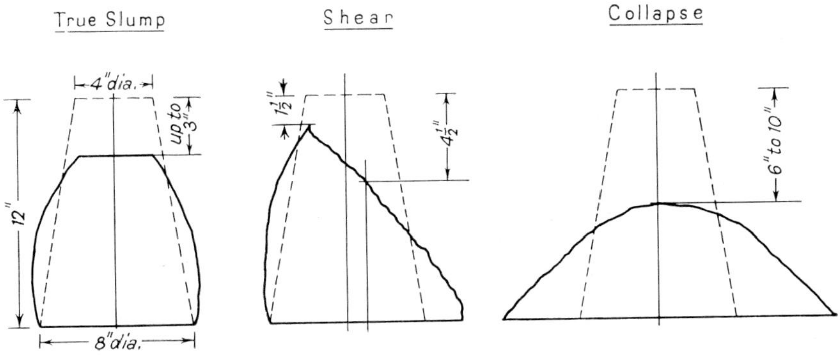

Fig. 6.2. Types of slump

half of the cone shears off and slips down an inclined plane. The third type results from a complete collapse of the cone. The second type of slump occurs frequently with normal concrete mixes, as, for instance, 1:2:4 mixes, and it is difficult to decide just how to measure the slump. In the illustration given, some operators might measure the minimum subsidence of $1\frac{1}{2}$ in. while others might measure the mean value of $4\frac{1}{2}$ in. When this type of slump occurs, the test should be repeated, in order to obtain a shape more nearly the "true" slump. If the repeat test still gives a shear slump, the fact should be recorded on the test report.

The workability of concrete is related to the amount of applied work required to compact it to its maximum density, and is, therefore, not

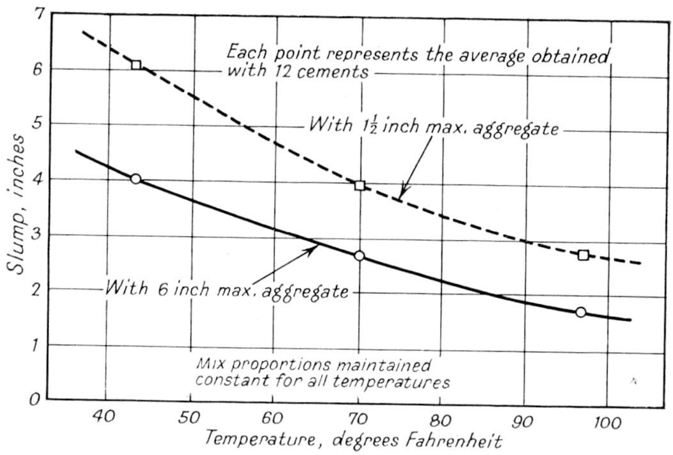

Fig. 6.3. Effect of temperature on slump of concrete

related to the slump. It is, in fact, possible to obtain the same slump with mixes requiring very different amounts of work to compact them.

This constitutes the chief disadvantage to the test.

Tests in the United States have shown that as the temperature of the mixed concrete increases, the slump decreases. Fig. 6.3, reproduced from the Concrete Manual,(3.4) shows the general trend.

Compacting factor test

The compacting factor test was devised to measure the degree of compaction obtained by doing a standard amount of work on the concrete, and therefore bears a close relation to workability.

A full description of the development of the test is given in Road Research Technical Paper, No. 5.(3.9) The method of test is described fully in B.S. 1881:1968, to which reference should be made.

The apparatus used for the compacting factor tests is shown in Figs. 6.4 and 6.5. It consists essentially of two inverted cone-shaped hoppers and a cylindrical container. In the original form the height was 12 in. but it is suggested that this be changed to 12·3 in. in order to provide an

Fig. 6.4. The compacting factor test. The concrete is about to be dropped into the bottom cylinder

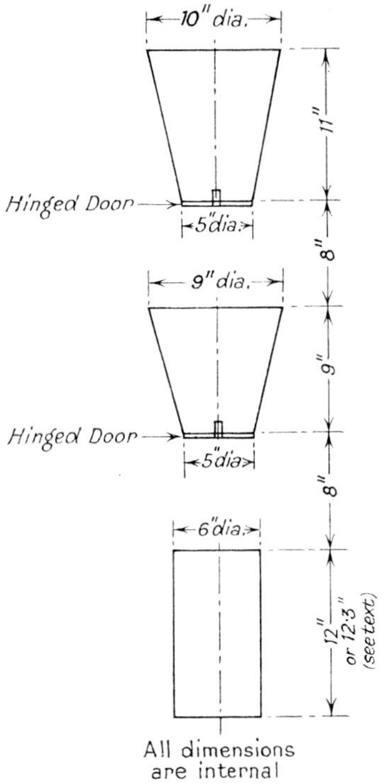

Fig. 6.5. Dimensions of compacting factor apparatus

internal volume of 0·2 ft^3 (0·0057 m^3). In view of the impending change to the metric system in Great Britain and possible further revision, the example of the calculation of the compacting factor given below is still based on a cylinder 12 in. high. In making the test the top hopper is filled with a representative sample of the concrete. When completely filled, a hinged door at the bottom is released and the concrete allowed to fall into the second hopper. The filling of the second hopper is thus effected by a standard method, and personal differences are minimized. The concrete is similarly released from the second hopper and falls into the cylindrical container. Surplus concrete is struck off by simultaneously working two steel floats from the outside to the centre. The cylinder contains a certain volume of concrete which has been subjected to a definite amount of work in falling from the second hopper. The contents of the cylinder are then weighed to the nearest 10 g giving the weight of partially compacted concrete (W_p).

The cylinder is then refilled from the same sample in layers approximately 2 in. deep, the layers being heavily rammed or preferably vibrated so as to obtain full compaction. The top surface is again struck off level with the top of the cylinder and the weight of concrete contained again determined. This is known as the weight of fully compacted concrete (W_f).

The compacting factor is then the ratio (W_p/W_f) of the weight of partially compacted concrete to the weight of fully compacted concrete.

The weight of fully compacted concrete can also be calculated theoretically on the basis of the solid volumes occupied by the various materials, assuming no air voids. The data required are:

(1) Mix proportions by weight.

(2) Water absorbed by the aggregate, since this will also be included in the weight of the partially compacted concrete. Note that in the definition of water/cement ratio (p. 88) the water absorbed by the aggregate is not included. In Table 6.1 the absorbed water provides weight but does not occupy volume, and is therefore not calculated under M/G.

TABLE 6.1 *Method of Calculation of Compacting Factor*

Material	Mix proportions by weight M	Specific gravity G	$\frac{M}{G}$
Coarse aggregate	4·0	2·58	1·55
Fine aggregate	2·1	2·63	0·80
Cement	1·0	3·15	0·32
Water:			
(1) effective	0·5	1·0	0·50
(2) absorbed by aggregate	0·06	—	—
Sum	7·66		3·17

$$\text{If } W_p = 12{,}000 \text{ g}$$

$$V = 5{,}450 \text{ cm}^3$$

$$W_f = \frac{\sum M \times V}{\sum\left(\frac{M}{G}\right)}$$

$$= \frac{7{\cdot}66 \times 5{,}450}{3{\cdot}17}$$

$$= 13{,}200 \text{ g}$$

$$\text{Compacting factor} = \frac{W_p}{W_f}$$

$$= \frac{12{,}000}{13{,}200} = 0{\cdot}91$$

(3) Water/cement ratio.

(4) Specific gravities of all the materials. Cement may be taken as 3·15 and water as 1·0. Bulk specific gravity in case of aggregates. (See p. 49).

(5) The volume of the cylinder. (V cm³).

From this information it is necessary to calculate the weight of concrete (W_f) the cylinder would contain if all the air voids were eliminated. This may be conveniently done by the method given in Table 6.1.

The compacting factor is the ratio of the observed weight of concrete deposited in the cylinder to the calculated weight without air voids.

The result will differ from that obtained by refilling the cylinder if full compaction was not achieved, and provides a useful check.

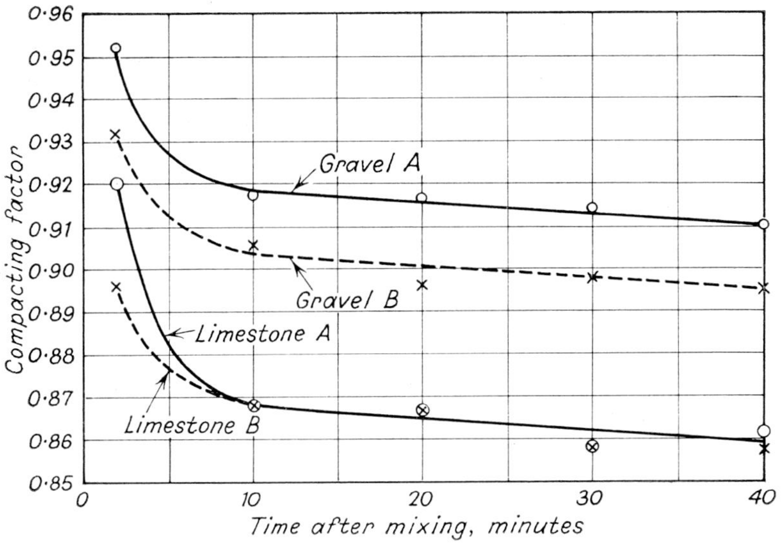

Fig. 6.6. Workability of concrete with gravel and limestone aggregates

Mix in each case was 1:6 (cement:aggregate) and river sand was used for all mixes.

All mixes: effective water/cement ratio 0·6.

Allowance for absorption of aggregate was made by:

A Extra water added at time of mixing (simulating the use of a dry aggregate).

B Sufficient water added to dry aggregate to give saturation ½ hour before mixing.

Mixing and curing temperature: 64°F ± 2° (18°C).

Evaporation was prevented before and after mixing.

It will be noted that, whichever method of determining the compacting factor is adopted, the difference in the two weights is due to air voids, and the closer the values, the less the air voids and the higher the compacting factor. The workability therefore increases as the compacting factor approaches unity.

The workability of concrete decreases considerably during the first few minutes after mixing as is shown in Fig. 6.6. The change is greater when using a dry aggregate (curve *A*) than when using one which has been previously saturated (curve *B*), due to the rapid initial absorption of the dry aggregate. Since the interval between mixing and placing is likely in practice to average 10–15 minutes, it is suggested that for test purposes a time interval of 10 minutes is allowed between the time of discharge from the mixer and the time of testing by the compacting factor method. This time interval also has the advantage that the change in compacting factor with time is relatively small compared with that in the first few minutes. This recommendation differs from that given in B.S. 1881 : 1968, which provides for only six minutes delay between adding the water to dry mixed materials in the laboratory and completing the workability test. Six minutes is, in the author's opinion, too short.

The test is not suitable for use generally on the site, but on larger jobs where a site laboratory is established there appears no reason why the test should not be used as a check on workability from time to time. In the central laboratory, the compacting factor test provides a reliable guide to the workability of a concrete mix. It is used to ensure that a mix designed for use in the field has the required workability for the type of work to be done with the compacting equipment available, while containing the minimum quantity of mixing water.

A general indication of the compacting factors suitable for different types of structure and methods of compaction is given in Table 7.9. The slump which may be expected is also given.

V-B consistometer test

The V-B test is particularly suitable for the measurement of the workability of concrete mixes of very low workability which are compacted by vibration, and of air-entrained concrete mixes. The accuracy decreases with increasing sizes of aggregate, and for aggregates of maximum size over $\frac{3}{4}$ in. the results are of doubtful value.

The method of test is described in B.S. 1881 : 1968, to which reference should be made. The consistometer consists essentially of a cylindrical container clamped on a small vibrating table, details of which are given in the specification. A cone of the same internal diameter as a slump cone is placed concentrically in the container and a funnel fixed on top. The cone is filled with concrete in four layers, each tamped 25 times as in the slump test. The funnel is removed and the top levelled off. The cone is then lifted out and a transparent disc lowered on to the subsided cone. The slump may be measured if required.

Vibration is then started and the remoulding of the concrete in the container is observed through the transparent disc. The time in seconds is recorded until the whole surface of the transparent disc is covered with cement grout, which is taken to be the moment when full compaction is attained. Recording mechanisms or other means may also be used to record the moment of full compaction. The workability by this method is defined as the time in seconds measured above, and reported as V-B degrees.

Cleanliness is of vital importance when using this type of apparatus, and care is necessary to avoid errors due to distortion of the central plunger.

An indication of the relation between slump, compacting factor and V-B time is given in Fig. 6.7.[6.3] This should only be regarded as a generalized relationship since the type of fine aggregate and other factors can have a significant effect.

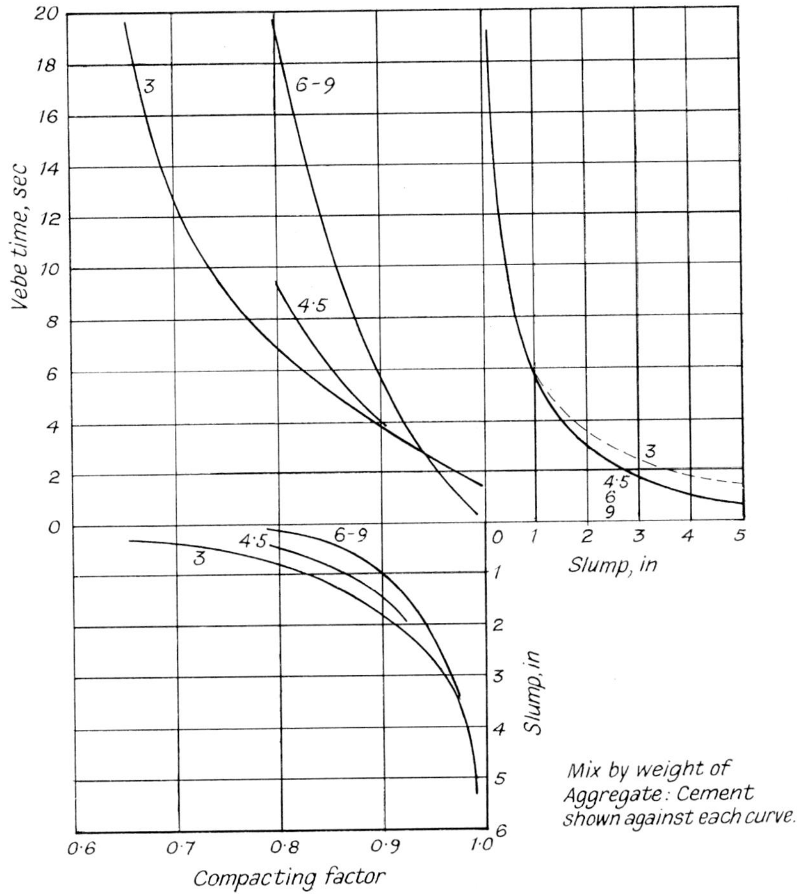

Fig. 6.7. Relationship between slump, compacting factor and V–B time

Factors governing workability

The compacting factor may be shown(3.17) to be related mainly to the mix proportions, to the water content, and to the grading, shape and surface texture of the aggregate. The relationship is discussed under the following headings.

(a) Aggregate Grading

It has been evident for some time that workability was linked in some way with the surface area coefficient of the aggregate, and an approximate method based on research up to 1955, was published in the 2nd edition of this book. It was evident, however, that the surface area coefficient did not apply to material finer than about 100 mesh sieve, nor to the cement. The new evidence now available indicates that other factors intervene; the surface area coefficient, for instance, did not provide fully for the greater compacting factor obtained by increasing the maximum size of the aggregate, e.g., from $\frac{3}{4}$ in.–$1\frac{1}{2}$ in. The reason for this lies in the fact that, in addition to having a smaller surface area requiring wetting, a more dense packing is possible. If the aggregate is assumed, for simplicity, to consist of spheres, it can be shown that the overall volume occupied by a large sphere is only some 70 per cent of the overall volume occupied by eight times the number of spheres of half the diameter, although the nett volume of solids is, in both cases, the same. In the case of the smaller spheres, the interstices must be filled with mortar, which, in the normal course of events will contain voids. There is, therefore, an advantage in using larger aggregate, although the actual increase in density which results cannot be calculated theoretically.

TABLE 6.2 *Surface Index Calculations*

Sieve size within which particles lie	Surface index for particles within sieve sizes indicated	Example of calculation for grading No. 1, 1½ in. max Road note No. 4,	
		Percentage of particles within sieve size	f_s
3 in.–1½ in.	−2½	—	—
1½ in.–¾ in.	−2	50	−100
¾ in.–⅜ in.	−1	14	−14
⅜ in.–3/16 in.	1	12	12
3/16 in.–No. 7	4	6	24
No. 7–14	7	6	42
No. 14–25	9	5	45
No. 25–52	9	4	36
No. 52–100	7	3	21
Passing 100	2	—	—
			66
	Add constant		330
			396
	× 10⁻³		0·396

As already stated, the particles finer than 100 mesh sieve and cement were previously neglected when considering surface area coefficients. It is now evident that, not only is the surface area coefficient of no importance for these finer particles and the cement, but that the value also requires modification for other sizes. This is because the smaller sand sizes in themselves provide some lubrication in the mix without requiring wetting to the same extent. As a result of detailed analysis[3.17] it is suggested that the aggregate particle size is assessed on the basis of a surface index f_s, as given in Table 6.2. It should be noted that the surface index is empirical, based on test evidence, although to some extent related to the surface area coefficient.

The surface index for any aggregate or mixture of aggregates is then calculated by multiplying the individual figures given in Table 6.2 by the percentage of particles of the aggregate within the sieve sizes indicated, summing the results allowing for positive and negative values, and adding the constant indicated. An example is given in Table 6.2 for Grading

No. 1 as shown in Road Note No. 4 for 1½ in. maximum size aggregate; to reduce the figure to a convenient size it is divided by 1,000. To be precise, the surface index should be calculated on a solid volume basis (see p. 100), since it is shown later that workability is dependent on proportions by volume. However, the error in using a weight basis for the sieve analysis is very small unless the specific gravity of different sizes of particle is markedly different.

(b) Particle shape

Various attempts have been made to measure the shape and angularity of aggregates. Perhaps the simplest is that suggested by Shergold[3.15] and termed the "angularity number" of the aggregate. This is described fully on p. 53. For purposes of mix design it has been found more suitable

TABLE 6.3 *Examples of Angularity Indices*

Type of coarse aggregate	Shape	Angularity number	Angularity index (f_A)
Quartzite gravel, natural	Rounded	1	1·15
Flint gravel, natural	Irregular	6	1·9
Limestone, crushed	Angular	7	2·05
Crushed granites and basalts, and crushed flint gravel	Angular	8 to 10	2·2 to 2·5

In the absence of test data natural uncrushed sands may be taken at $f_A = 1{\cdot}8$ to 1·9.

to use the value obtained by calculating the angularity index, as also described. This is in effect the proportion of voids above 33 relative to the "optimum" volume of solids, i.e. 67 per cent.

Examples of angularity indices for a number of aggregates are given in Table 6.3.

Another method of expressing particle shape is the shape index.[6.4] This expresses the particle shape in terms of the specific surface and number of particles. The shape index may vary with the particle size, and it follows that the angularity of the sand sizes is likely to vary in a similar manner. The angularity of the sand sizes is therefore as important as that of the coarse aggregates.

(c) Combining effects of grading and shape

Analysis has shown that the angularity index, as calculated above, can be used to modify the surface index simply by multiplying them together ($f_S f_A$). If aggregates of different angularity indices are used in the same mix, then the average angularity index is calculated in proportion to the percentage of each type of material in the mix.

(d) Influence of mix proportions

It is the proportions by volume of the various constituents of a concrete mix which affects workability, so that the method of assessment of compacting factor described here is based on proportions by volume.

It needs little thought to realise that as a mix becomes richer, the grading and angularity of the aggregates become of less importance, until with high proportions of cement they are little other than plums floating in cement paste. An analysis of a large number of test results(3·17) has shown that when the aggregate/cement ratio (A_v) is reduced to 2, the effects of grading and angularity become negligible (see Fig. 7.2, p. 118). The combined surface and angularity indices ($f_S f_A$) therefore require modification, depending on the aggregate to cement ratio based on solid volumes. The solid volume is obtained by dividing the dry weight of each material by the respective specific gravity. A modified index is therefore obtained as follows:

$$f_S f_A (A_v - 2) \qquad (1)$$

where A_v = volume of aggregate to one volume of cement on the basis of solid volume of particles.

The value of A_v is calculated from the mix parts by weight, taking into account the specific gravity of the materials.

$$A_v = \frac{3{\cdot}15\, A_w}{G_A C_w}$$

where A_w = total weight of dry aggregate
C_w = weight of cement
G_A = average bulk specific gravity of aggregate. (See p. 49 for definition).

If the aggregate varies in specific gravity with particle size, the average specific gravity should be determined. London river gravel, for instance, varies in specific gravity from 2·55 to 2·45 between the $1\frac{1}{2}$ in. and $\frac{3}{16}$ in. sieves, and London area sand usually has a specific gravity of about 2·60. Many hard limestones and aggregates of metamorphic or igneous origin are in the range 2·65 to 2·70.

(*e*) *Water content*

It has been shown previously that the compacting factor is related to the water content of the concrete when expressed as a percentage by volume of the materials in the mix. Recent work(3·17) has confirmed this concept of the water content by volume in the mix as being important, although an important modification is required in assessing the quantity of water available to influence workability. An interesting experiment may be made with neat cement, which demonstrates that the consistency of a cement water paste changes quite suddenly with an increase in water/cement ratio, as measured by weight, of only about 1 per cent. The test can be done quite simply using the consistency plunger of the Vicat needle apparatus. Depths of penetration similar to those shown in Table 6.4 are obtained for ordinary Portland cement. This sudden changing of a cement paste from a stiff silt-like material almost to a fluid, with such a small change in water content, greatly simplifies the problem of determining the water content available for lubrication of the mix. It is generally found that this change in consistency occurs at water percentages of between about 23 and 28, so that excess water beyond these percentages is available for lubrication of the mix.

TABLE 6.4 *Values of Penetration (Vicat Apparatus) Obtained with Neat Cement Paste for Varying Water Contents*

Water percentage	Depth of penetration mm	Remarks
18	Nil	
19	1	Caused subsidence but no actual penetration up to 9 mm
20	3	
21	5	
22	7	
23	9	
24	12	
25	24	
26	40+	Total penetration

Then water content available for lubrication is

$$w/c - w_v \tag{2}$$

where w_v is the water/cement ratio in the Vicat needle test at which a sudden change in consistency occurs, being generally between 0·23 and 0·28.

For most practical purposes the value of w_v may be assumed to be 0·25; at the time a mix is designed the actual cement to be used will often not be known. In the remainder of this chapter this value of 0·25 is used. Where any doubt exists, or where other types of cement are being used, the simple Vicat test may be used to obtain a more precise value for the reduction in water/cement ratio required.

Calculation of compacting factor

If the various influences discussed above and summarised on lines (1) and (2) above, are brought together, it would appear that the compacting factor is a function of:

$$\frac{w_c - 0{\cdot}25}{f_S f_A (A_\nu - 2)}$$

The necessary constants have been calculated from numerous test results,[3.17] and the equation giving the compacting factor then becomes:

$$\text{c.f.} = 0{\cdot}74 \left\{ \frac{10(w/c - 0{\cdot}25)}{f_S f_A (A_\nu - 2)} + 0{\cdot}67 \right\} \tag{3}$$

The basic principles described above have been applied particularly to the assessment of the compacting factor of concrete mixes. They are, however, applicable to workability whether measured by this method or other methods.

Segregation

Segregation is the separation of the various constituents of a concrete mix due to differences in particle size and specific gravity. There is a tendency for the coarser and heavier particles to settle and for the finer and lighter materials, particularly the water, to rise to the surface.

A close relation exists between segregation and many of the imperfections which occur in concrete construction such as honeycombing, weak and porous layers, surface scaling and sand streaks. Segregation is usually a result of using too much mixing water, badly graded aggregate, especially if the quantity of sand is deficient, or unsatisfactory methods of handling.

Segregation has been shown to be a serious risk where the surface index (f_S) falls below about 0·45 for 1½ in. max. sized aggregate and 0·5 for ¾ in. max. sized aggregate. The proportion of fines may then be insufficient adequately to fill the voids between the coarse particles. In no-fines concrete, which is made with ¾ in.–⅜ in. coarse aggregate only, with a surface index of 0·23, a honeycombed effect is obtained.

Control of consistency in the field

The aim in controlling consistency and workability in the field is to produce uniform concrete and to maintain a reasonably constant water/cement ratio. The slump test is the usual method of control apart from the maintenance of a constant water/cement ratio by making tests on the surface moisture content of the aggregates, and allowing a calculated quantity of water at the mixer.

The slump test is useful in two ways:

(1) Provided the grading of the aggregate, the proportioning, and the temperature of the materials are reasonably uniform, the slump indicates variations in the water/cement ratio. Even when tests are made on the moisture content of the aggregate, and an amount of water is added at the mixer calculated to give the correct water/cement ratio, the consistency and slump of the concrete must be the final guide. In practice, the moisture content of the aggregate varies throughout the stockpile, and tests can only give an average value.

On the other hand, the slump may vary by an inch or more when only a 1 per cent variation in the moisture content occurs in the sand used in the mix, and any change of slump should therefore be corrected by an adjustment in the water added at the mixer. It is usual when close control of the water/cement ratio is required, to allow a tolerance on the calculated quantity of water to be added at the mixer, the final adjustment depending on observations of the consistency of the concrete as it is made. Further reference will be made to this point in connection with the control of water content at the mixer.

TABLE 6.5 *Effect of Change of Water Content on Slump*

Change in slump in inches		Percentage change in water content
From	To	
1	2	5
2	3	4½
3	4	3
4	5	2½
5	6	2

NOTE.—Slump tests are somewhat unreliable and the figures given may not apply to individual slump tests.

Fig. 6.8. Mix too harsh

(2) If the quantity of water added at the mixer is constant, and the moisture content of the aggregate is reasonably uniform, the slump test should indicate changes in the grading of the aggregate or errors in the proportioning of the materials.

On the site it often happens that the initial batches of concrete are a little too stiff or too plastic, and the question arises as to how much water to add or subtract to obtain the correct consistency, as measured by the slump test. Values from which the required percentage change in water content for any change in slump may be calculated are given in Table 6.5.

Fig. 6.9. Oversanded mix

Fig. 6.10. Mix of satisfactory consistency

A further method of estimating consistency in order to maintain uniformity from batch to batch, is to watch the shape and appearance of the concrete as deposited in the transport container.

Examples of mixes of varying consistency are given in Figs. 6.8 to 6.11.

The slump may be reduced during transportation of the concrete, and when the length of haul is considerable it is advisable to make tests on the same concrete both at the mixer and at the point of deposition, in order to determine the extent of the loss. When the distance from the mixer to the point of deposition is considerable, a loss of several inches may be recorded. This is exceptional and should not be permitted if over-wet mixes, which are liable to segregate during transport, have to be produced at the mixer to give the correct consistency at the point of deposition.

Fig. 6.11. A concrete mix of satisfactory consistency

REFERENCES

6.1 NEWMAN, K., Properties of concrete, *Structural Concrete*, **2**, No. 11, Sept./Oct. 1965. Reinforced Concrete Association.

6.2 CUSENS, A. R., The measurement of the workability of dry concrete mixes, *Mag. Concr. Res.*, **8**, No. 22, March 1965. Cement and Concrete Association.

6.3 DEWAR, J. D., *Relations between Various Workability Control Tests for Ready-mixed Concrete*, Technical Report TRA/375, Feb. 1964. Cement and Concrete Association.

6.4 *Variation of Specific Surface and Particle Shape with Aggregate Particle Size*, Report TRA/121, July 1952. Cement and Concrete Association.

CHAPTER 7

Design of Concrete Mixes

CONCRETE mixes are usually designed to meet the following requirements:

(1) A minimum compressive strength at 28 days based on the value assumed by the designer of the structure.

(2) The provision of adequate workability to enable the site to obtain full compaction with the compacting equipment available. It is also sometimes necessary to take into account the type of mixer, the method of transport, and the method of handling and placing the concrete, in assessing the workability requirement.

(3) The mix should be cohesive in order to avoid the possibility of honeycombing, water gain (collection of water under aggregate particles resulting from bleeding) and other troubles associated with segregation. In order to satisfy this requirement, and to give the dense concrete necessary to protect reinforcement, an adequate cement content is essential. Suggested minimum cement contents are given on p. 124.

(4) Durability is usually associated with compressive strength; the greater the strength the more durable the concrete. Concrete exposed to frost action should not have a water/cement ratio of more than 0·5 to 0·6 (see p. 349), depending on the severity of exposure.

Other requirements to meet special circumstances, as for instance, low heat of hydration in dams, are dealt with in other Chapters.

Methods of designing concrete mixes

Numerous methods have been suggested for designing concrete mixes. Perhaps the best known in Britain for many years has been that given in Road Note No. 4.[7.1] This is based on grading curves for the combined coarse and fine aggregates, and relationships established by tests[3.9] between mix proportions, water/cement ratio, and compacting factor. There are, however, limitations to this method:

(1) The grading curves are based on the grading of sands commonly available in the London area, and it is not always possible to obtain a good fit for sands from other areas.

(2) Other gradings, e.g. gap gradings, in which the middle sizes of aggregate are absent, can produce good concrete, although extra care may be necessary to avoid segregation.

(3) Particle shape, particularly that of the fine aggregate, is not adequately dealt with.

The American Concrete Institute Committee on the selection of mix proportions[5.5] takes the view that, while established basic relationships and laboratory tests provide guides in approaching optimum combinations, final proportions should be established by actual trial and adjustment on site. The recommended procedure is to make a preliminary estimate of water requirements in the mix from the data given in Table 7.1; these

TABLE 7.1 *Approximate Mixing Water Requirements for Different Slumps and Maximum Sizes of Aggregates**

Slump, in.	Water, lb/yd³ of concrete for indicated maximum sizes of aggregate				
	$\frac{3}{8}$ in.	$\frac{3}{4}$ in.	$1\frac{1}{2}$ in.	3 in.	6 in.
	Non-air-entrained concrete				
Very stiff, no slump	317	283	250	—	—
0 to 1	333	300	267	—	—
1 to 2	350	308	275	242	208
3 to 4	383	342	300	267	233
6 to 7	408	358	317	283	250
Approximate amount of entrapped air in non-air-entrained concrete, per cent	3	2	1	0·3	0·2
	Air-entrained concrete				
Very stiff, no slump	283	250	225	—	—
0 to 1	300	267	233	—	—
1 to 2	308	275	242	208	183
3 to 4	342	300	267	233	200
6 to 7	358	317	283	250	217
Recommended average total air content, per cent	8	6	4·5	3·5	3

* These quantities of mixing water are for use in computing cement factors for trial batches. They are maxima for reasonably well-shaped angular coarse aggregates graded within limits of accepted specifications.

If *more* water is required than shown, the cement factor, estimated from these quantities, *should* be increased to maintain desired water/cement ratio, except as otherwise indicated by laboratory tests for strength.

If *less* water is required than shown, the cement factor, estimated from these quantities, *should not* be decreased except as indicated by laboratory tests for strength.

are based on averages to be expected for fairly well shaped aggregates graded within the limits of the usual accepted specifications.†

In a supplement the A.C.I. have introduced a procedure for designing mixes having slumps in the range of zero to 1 in. for aggregates up to $1\frac{1}{2}$ in. maximum size. Some of the data extracted from this supplement for very stiff and 0–1 in. slump mixes are included in Table 7.1.

† The Tables give extracts; water requirements are given in lb/yd³ to avoid confusion between American and English units of measurement.

TABLE 7.2 *Maximum Permissible Water/Cement Ratios for Different Types of Structures and Degrees of Exposure*

Type of structure	Exposure conditions[1]					
	Severe wide range in temperature or frequent alternations of freezing and thawing (air-entrained concrete only).			Mild temperature rarely below freezing, or rainy, or arid.		
		At the water line or within the range of fluctuating water level or spray			At the water line or within the range of fluctuating water level or spray	
	In air	In fresh air	In sea-water or in contact with sulphates[2]	In air	In fresh air	In sea-water or in contact with sulphates[2]
Thin sections, such as railings, curbs, sills, ledges, ornamental or architectural concrete, reinforced piles, pipes, and all sections with less than 1 in. concrete cover over reinforcement	0·49	0·44	0·40[3]	0·53	0·49	0·40[3]
Moderate sections, such as retaining walls, abutments, piers, girders, beams	0·53	0·49	0·44[3]	†	0·53	0·44[3]
Exterior portions of heavy (mass) sections	0·58	0·49	0·44[3]	†	0·53	0·44[3]
Concrete deposited by tremie under water	—	0·44	0·44	—	0·44	0·44
Concrete slabs laid on the ground	0·53	—	—	†	—	—
Concrete protected from the weather, interiors of buildings, concrete below ground	†	—	—	†	—	—
Concrete which will later be protected by enclosure or backfill but which may be exposed to freezing and thawing for several years before such protection is offered	0·53	—	—	†	—	—

1. Air-entrained concrete should be used under all conditions involving severe exposure and may be used under mild exposure conditions to improve workability of the mixture.

2. Soil or ground water containing sulphate concentrations of more than 0·2 per cent.

3. When sulphate-resisting cement is used, maximum water/cement ratio may be increased by 0·04.

† Water/cement ratio should be selected on the basis of strength and workability requirements.

The water/cement ratio is next selected from Table 7.2. The cement requirement for the concrete can then be calculated from this and the water requirement given by Table 7.1.

The volume of coarse aggregate to be used is chosen from Table 7.3 being based on the fineness modulus of the fine aggregate. The fineness modulus is calculated by the method given on p. 41.

For those cases where the mix is to be designed to give a specified strength the water/cement ratio may be determined by reference to Fig. 7.1, p. 110.

TABLE 7.3 *Volume of Coarse Aggregate per Unit of Volume of Concrete**

Maximum size of aggregate in.	Volume of dry-rodded coarse aggregate per unit volume of concrete for different fineness moduli of sand			
	2·40	2·60	2·80	3·00
$\frac{3}{8}$	0·46	0·44	0·42	0·40
$\frac{3}{4}$	0·65	0·63	0·61	0·59
$1\frac{1}{2}$	0·76	0·74	0·72	0·70
3	0·84	0·82	0·80	0·78
6	0·90	0·88	0·86	0·84

* Volumes are based on aggregates in dry-rodded condition as described in Method of Test for Unit Weight of Aggregate (ASTM Designation C 29). These volumes are selected from empirical relationships to produce concrete with a degree of workability suitable for usual reinforced construction. For less workable concrete such as required for concrete pavement construction they may be increased about 10 per cent.

The Author has devised a method of mix design based on the methods of assessing the effects of grading and angularity of aggregates described in Chapter 6. The method is described in detail, but for many purposes may be reduced to a simple selection of mixes from charts such as those given towards the end of the chapter.

Mix design

The method of mix design described in this chapter is based on the assessment of workability from the characteristics of the aggregate, the mix proportions, and the water content. The limitations of Road Note No. 4,(7.1) which gave the workability for four aggregate gradings only, have been largely overcome. The method of assessing compacting factors has been described in Chapter 6, and may be applied to all types of aggregates of widely varying grading, including gap gradings. The latter can be used satisfactorily provided extra care is taken to avoid segregation, the risk of which is increased by the absence of the middle sizes of aggregate.

The diagrams in this Chapter are based on the compacting factor as a measurement of workability. The important influences are:

(*a*) Water content of the mix available for lubrication: $w/c - w_v$. In Chapter 6 it is indicated that water equivalent to a water/cement ratio of about 0·25 is virtually absorbed by the cement, before fluidity is obtained. The effective water content is then: w/c—0·25.

(*b*) Surface index, which takes account of the grading of the aggregate.

(*c*) Angularity index, which allows for the shape of the aggregate.

(*d*) Mix proportions. Up to a 1:2 mix of cement: aggregate by solid volume, the grading of the aggregate and its shape have little influence on workability. For practical design purposes the effect of the mix proportions is therefore related to: $A_v - 2$, where A_v is the aggregate: cement ratio by solid volume. See definition of solid volume of an aggregate on p. 100.

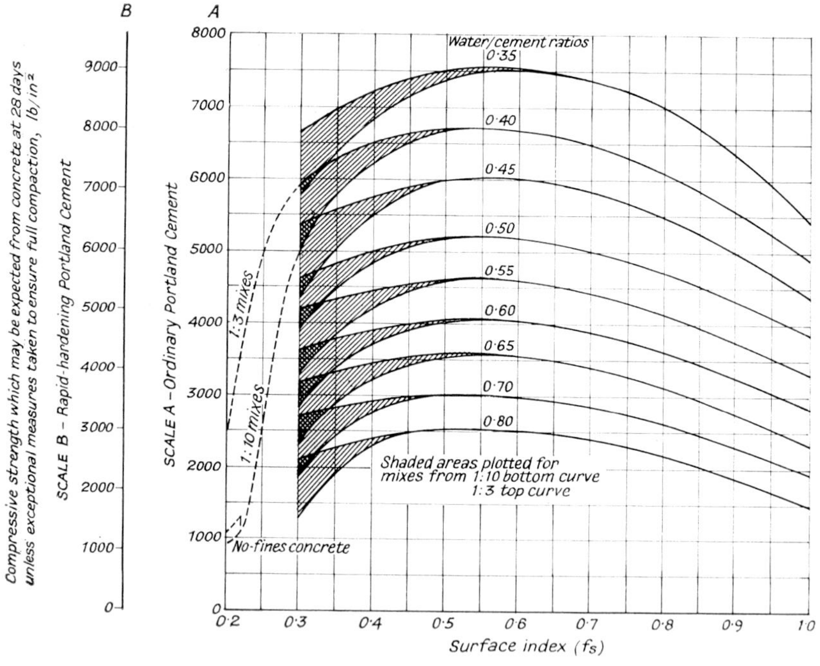

Fig. 7.1. Relationship between compressive strength, water/cement ratio and surface index

Notes.—Scale *A*—Ordinary Portland cement having mortar cube strength of 7000 lb/in² at 28 days.

Aggregate—Irregular river gravel and sand (for other aggregates see Table 7.8 for adjustments).

Compaction—Complete elimination of air voids should give strengths equal to that at peak of curves. If compaction is sufficiently vigorous, there will be less drop in strength above about $f_s = 0{\cdot}6$ than is indicated by the curves. Below about $f_s = 0{\cdot}45$ the strength will not be so much improved by vigorous compaction because there are insufficient fines to fill voids. The curves are based on test cube results with average site compaction, the wetter mixes by hand, the drier mixes by vibration.

The above applies for all mixes in which there are sufficient fines to fill the voids between the larger particles. When the total fines becomes small, as indicated by a small surface index, segregation is liable to occur, and the compacting factor is reduced.

In assessing compressive strengths, whereas the strength is related to the water/cement ratio as defined on p. 88, the surface index again has an influence, causing a gradual decrease above a factor of about 0·6, and a much more rapid decrease, particularly in the case of the leaner mixes below a factor of about 0·5. This rapid decrease arises from the presence of insufficient fine materials to fill the voids between the larger particles, a point well illustrated in Fig. 7.1 by the extension of the dotted curves for mixes with a water/cement ratio of 0·4 to include no-fines concrete, which normally has a surface index of a little over 0·2. No-fines concrete contains about 30 per cent air voids.

Compressive strength required

Concrete mixes are usually designed to provide a minimum compressive strength at 28 days after mixing, based on the value assumed by the designer of the structure. In this Chapter the 28 day strength only is considered. Reference should be made to Chapter 4, Cement, for information on the effects of age and temperature on crushing strengths.

For various reasons listed in Chapter 25, dealing with quality control on the site, considerable variation occurs in the crushing resistance of concrete, and it is therefore necessary to aim at an average strength considerably in excess of the minimum specified. Suggested allowances for these variations are given in Table 7.4, the choice of the value to be used in designing a mix being based on a forecast or knowledge of the methods to be used in construction and the standard of quality control.

Table 7.4[7·2] is based on the author's analyses of cube results from many sites. Views differ[7·3,7·4] on the allowances to be made, but for practical purposes Table 7.4 still gives a convenient and fair assessment. For average strengths below 3,000 lb/in² the variation may be taken as:

$$\frac{\text{average strength expected}}{3{,}000} \times \text{variation shown in Table 7.4}$$

The degree of supervision has a much more important effect than the type of equipment employed.

In Table 7.4 the provision made for the acceptance of results falling below the specified requirement is 2 per cent. Opinions vary on the percentage of results permitted to fall below the specified minimum. Some authorities insist on a 1 per cent limit, but the newer B.S. Codes and Standards have adopted 2½ per cent. If a comparison is made between the figures in Table 7.4 and C.P. 114[7·5] the latter would appear to allow about 4 per cent of failures. The A.C.I. recommends that the permissible percentage of defectives should vary according to the nature of the work. The Comité Européen du Béton[2·17] suggests 5 per cent.

Example of use of Table 7.4

If, as an example, nominal mix 1:1½:3 concrete, having a required minimum works crushing strength of 3,750 lb/in² at 28 days, is to be

TABLE 7.4 *Suggested Allowances for Variations in Cube Strengths for Use in Quality Control*[7.2]

Method of proportioning, type of concrete, and control measures	Variations above or below average strength beyond which investigation is advisable (for average strengths above 3,000 lb/in^2)
All materials by weight. High quality concrete of uniform workability in large quantities. Competent supervision by a qualified engineer.	1,000
All materials by weight. High quality concrete in members of varying sections requiring slight changes in workability. Competent experienced supervision.	1,200
All materials by weight. General concreting where constant supervision is not justified. Lean mixes.	1,500
Aggregates by volume using deep bins for measurement. Cement by full bag or by weight. Good control.	1,200
Aggregates by volume. Cement by full bag or by weight. Measurement of aggregates by marks in loading hopper or by gauge boxes, without constant supervision.	1,500
Aggregates by volume. Cement by bag or part bag. Little control.	1,800
All materials by volume.*	2,100–2,400

NOTE.—Total variation is twice the value quoted. Allowances are intended to avoid more than 2 per cent of cubes failing to reach the required standard.

* Cement should never be measured other than by weight. The figures are given to indicate the extent of the variation which results if weight measurement is not adopted.

produced on a site where all materials are batched by weight, where the concrete is to be used in several different parts of a structure, but where constant supervision cannot be given, this minimum value should be increased by 1,500 lb/in^2 to give an average required crushing strength of 5,250 lb/in^2. The mix should then be designed, in this example, to produce concrete with an intended average strength of 5,250 lb/in^2. This is seen to be slightly more conservative than C.P. 114, if the average strength is regarded as comparable with preliminary test results.

Summarizing the above, the compressive strength on which the mix should be designed is the average strength determined by adding the differences given in Table 7.4 to the minimum strength required.

Code of practice

The requirements of the Code of Practice for Reinforced Concrete C.P. 114:1957 (Amendment 1965),[7.5] for preliminary and works tests on nominal concrete mixes are summarised in Table 7.5.

TABLE 7.5 *Proportions and Strength Requirements for Portland Cement Concrete and Portland Blast-furnace Cement Concrete with Aggregates Complying with* B.S. 882 *or* B.S. 1047
Requirements of Code of Practice, C.P. 114 (1957) *for Nominal Concrete Mixes*

(1)	(2)		(3)		(4)	
Nominal mix	Cubic feet of aggregate per 112 lb of cement		Cube strength within 28 days after mixing		Alternative cube strength within 7 days after mixing	
	Fine	Coarse	Preliminary test	Works test	Preliminary test	Works test
			lb/in²	lb/in²	lb/in²	lb/in²
1:1:2	$1\frac{1}{4}$	$2\frac{1}{2}$	6,000	4,500	4,000	3,000
$1:1\frac{1}{2}:3$	$1\frac{7}{8}$	$3\frac{3}{4}$	5,000	3,750	3,350	2,500
1:2:4	$2\frac{1}{2}$	5	4,000	3,000	2,700	2,000

It will be noted that the preliminary test requirements are one third above the works tests, and the assumed variation if the preliminary test is taken as the "average strength" is not entirely in agreement with Table 7.4. For further details reference should be made to the Code, which also gives strength requirements for high alumina cement concrete.

C.P. 114 also makes provision for "designed concrete mixes," i.e. mixes specially designed to satisfy any specified works cube strength requirements. The range in concrete mixes allowed is given in Table 7.6. The mix limitations should be noted. Other requirements are briefly:

(*a*) Supervision by a competent person of all stages of preparation, placing, site tests, and test cube making.

TABLE 7.6 *Strength Requirements for Designed Concrete Mixes. Requirements of Code of Practice,* C.P. 114 (1957), *Amendment No.* 1, 19th *February,* 1965.

Mix limitations	Specified works cube strength
(1)	(2)
	lb/in²
For Portland cement or Portland blast-furnace cement concrete, not less than 400 lb nor more than 900 lb of cement per cubic yard of finished concrete; and designed for the required concrete strength at 28 days,* within the range given in column (2).	2,250 to 7,500 at 28 days
For high alumina cement concrete, not less than 450 lb nor more than 700 lb of cement per cubic yard of concrete; and designed for the required concrete strength at 1 day, within the range given in column (2), with a water/cement ratio not more than 0·5.	6,000 to 7,500 at 1 day

* Concrete may be provisionally accepted as satisfactory if the 7-day cube strength is at least two-thirds of the required 28-day strength.

(*b*) The mix is to be designed to have a mean strength that exceeds the specified works cube strength by twice the expected standard deviation.

(*c*) Trial mixes have to be made on three separate days.

(*d*) Aggregates are to be properly graded, each size being kept in separate stockpiles, and tests made regularly.

(*e*) Batching of materials is to be by weight, and the water measured.

(*f*) Water content. The water/cement ratio is to be controlled.

(*g*) Tests are to be made regularly for strength in accordance with B.S. 1881, and a continuous statistical analysis of the results is to be made.

Reference should be made to the Code and the Amendment (1965) for the full requirements. As regards requirement (*b*), it is suggested that a check should be made using Table 7.4, to ensure that the requirement is likely to be met. A further provision is made for "standard mixes." Mix proportions by weight are quoted "for producing concrete of the three specified strengths indicated, assuming a standard deviation of 1,000 lb/in^2. Conditions are attached which make it doubtful whether "standard mixes" will prove economically justifiable in preference to nominal mixes except for a limited amount of concrete where high quality is essential; reference should be made to the Code for details.

An important change in C.P. 114, 1957, from the previous Code is the provision for the testing of three cubes (clause 208, *c*, iii) at each age. The appropriate strength requirement is considered to be satisfied if all three cubes exceed the specified strength, or if the average strength of the three cubes is not less than the specified cube strength and the difference between the greatest and least strengths is not more than 20 per cent of that average.

It is likely, however, that this arrangement will be dropped in future Codes, on the grounds that it is better to have single cubes from a greater number of batches, compared with cubes in threes from one third the number of batches.

Water/cement ratio

An inspection of Fig. 7.1 indicates that for any compressive strength at 28 days there is a comparable water/cement ratio; for instance, continuing with the example given on p. 112, with ordinary Portland cement (scale *A*), a compressive strength of 5,250 lb/in^2 requires a water/cement ratio of 0·50, providing the surface index is between about 0·5 and 0·65. For indices below or above these limits, the strength falls away and it is, therefore, preferable to grade the aggregate to remain within them. The fall in strength for the lower surface indices arises from the lack of fines and the greater difficulty in compacting to remove air voids. As may be seen from the shaded areas, the fall is more marked for lean mixes than rich mixes in which the cement helps to fill the voids. If a variation in surface index is expected, it is therefore preferable that the index should exceed 0·65, rather than that it should fall much below 0·5, since the strength drops more slowly for the higher indices than for lower values and there is less chance of segregation.

The curves in Fig. 7.1 are plotted for ordinary Portland cement having a mortar cube strength in the region of 7,000 lb/in^2 at 28 days. For most mix design purposes, it is sufficient to assume that British cements will attain this strength. If, however, evidence is available on the mortar cube strength of the cement to be used, an adjustment should be made. The increase or decrease in concrete strength may be taken as roughly proportional to the percentage increase or decrease in the mortar cube strength. Suggested adjustments for different types of cement, in the absence of other data, are given in Table 7.7. Note that the average compressive strength required is divided by the factor given, before using Scale *A*, Fig. 7.1. Scale *B* given for rapid-hardening cement for convenience bears the relationship to Scale *A* indicated in Table 7.7.

TABLE 7.7 *Adjustment to Compressive Strength of Concrete for Type of Cement*

Type of cement	Factor by which average compressive strength required should be divided in order to use Scale *A*, Fig. 7.1
Ordinary Portland cement	1
Rapid-hardening Portland cement	1·2
Sulphate-resisting Portland cement	1
Supersulphated cement	1·0 to 1·25
High-alumina cement	2
Low-heat Portland cement	0·6 to 0·7

This Table gives an average value. Concrete made from different batches of cement will vary from this value in proportion to the mortar cube strengths.

Note that for high-strength concrete (i.e. over about 6,000 lb/in^2 as a minimum requirement) requiring a mix rich in cement, the grading and shape of the aggregate become less important. The sand content may be reduced with advantage. In mixes as rich as 1 : 2½ or 1 : 3 of cement : total aggregate, the strength of the cement itself becomes of major importance.

The curves in Fig. 7.1 are plotted for an irregular flint gravel aggregate found in the London area. It has been found, however, that hard crushed limestone, some granites, and other aggregates give rather higher strengths than irregular flint gravel, while some rounded and polished aggregates give slightly lower strengths. The advantage in strength for a given water/cement ratio indicated for certain aggregates is largely counterbalanced in practice because, being crushed aggregates, they are angular and they require a higher water/cement ratio to give the necessary workability than irregular or rounded aggregates.

Suggested correction factors for use in determining the water/cement ratio necessary to give a required strength are given in Table 7.8. In case of doubt, it is a good plan to design two mixes, one on the basis of the curves given in Fig. 7.1, and the other on whatever increase or decrease is thought likely from Table 7.8, using the more conservative value where a range is given. The two mixes can later be checked by comparative trial mixes.

TABLE 7.8 *Adjustments for Different Types of Aggregate*

Aggregate		Adjustment to 28 day crushing strength before determination of water/cement ratio by use of Fig. 7.1, scale *A*. Multiply by coefficient below
Coarse	Fine	
River gravel (irregular)	River sand (irregular)	nil
River gravel (rounded)	River sand (irregular)	1·05
River gravel crushed (angular)	River sand (irregular)	nil
River gravel (irregular)	River sand crushed (angular)	nil
Crushed whinstone or granite (angular)	River sand (irregular)	0·95
Crushed limestone (sub-angular)	River sand (irregular)	0·85–0·95
Crushed limestone (sub-angular)	Crushed limestone (sub-angular)	0·80–0·90
Crushed whinstone or granite (angular)	Crushed whinstone or granite	0·85–0·95*

* Crushed whinstone fines sometimes reduce strength rather than causing an increase as indicated here.

Workability and mix proportions

The workability required may be assessed from Table 7.9, for the conditions of placing and compaction anticipated. If concrete will be required for several different purposes from the one mixer, it is wise, in the interests of simplicity, to design the mix for the most workable concrete necessary. Frequent changes of mix on the site are confusing and often time-consuming, so that any possible saving in cement is largely nullified.

The relationship established in Chapter 6, for the assessment of the compacting factor is:

$$\text{c.f.} = 0{\cdot}74\left\{\frac{10(w/c - w_v)}{f_S f_A (A_v - 2)} + 0{\cdot}67\right\} \qquad (1)$$

where w/c is the water/cement ratio by weight

w_v is the water/cement ratio necessary to bring the cement to a fluid consistency, which for most purposes may be taken as 0·25

f_S is the surface index

f_A is the angularity index

A_v is the aggregate/cement ratio based on solid volumes. (See p. 100 for definition)

$$A_v = \frac{3{\cdot}15 A_w}{G_v C_w} \qquad (2)$$

where A_w = total weight of dry aggregate

C_w = weight of cement in same units

G_v = average bulk specific gravity, on a saturated surface dry basis, of the aggregate.

The water/cement ratio has been determined from Fig. 7.1, and the required compacting factor has been decided from Table 7.9. The next

TABLE 7.9 *Uses of Concrete of Different Degrees of Workability*

Degree of workability	Slump in inches	Compacting factor	Use for which concrete is suitable
Very low to low	0 to 1	0·80–0·87	Vibrated concrete in roads or other large sections, where powerful vibrating equipment can be used. Vibrated piles, precast beams and railway sleepers, and similar work where a high strength is required, say 6,000 lb/in² or over at 28 days.
Low to medium	1 to 2	0·87–0·93	Roads with ordinary forms of vibrating and finishing machines, and manually operated compactors and finishers of the vibrated beam type. Mass concrete foundations without vibration. Simple reinforced sections with vibration.
Medium to high	2 to 4	0·93–0·95	Hand-tamped roads with slump 2–3 in. For normal reinforced work without vibration and heavily reinforced sections with vibration.
High	4 to 7	Over 0·95	For sections with congested reinforcement. Other work where placing is difficult. Not normally suitable for vibration.

steps are to determine the mix proportions for the particular aggregate to be used, and to decide the proportions of coarse and fine aggregates. By rearranging equation (1) and assuming $w_v = 0{\cdot}25$, the value $f_S f_A(A_v - 2)$ can be determined:

$$f_S f_A(A_v - 2) = \frac{7{\cdot}4(w/c - 0{\cdot}25)}{\text{c.f.} - 0{\cdot}50} \tag{3}$$

For convenience this relationship is given graphically in a slightly different form in Fig. 7.2.

The value of f_A is determined by the type of aggregate. See p. 53 and Table 6.3.

A suitable surface index is chosen, being based on a knowledge of the reliability of the aggregate supply, an inspection of the quarry or pit and sieve analyses. The more uniform the supply the lower the surface index acceptable, within the following limits.

Values suggested for the aggregate (or combined aggregates) used in a concrete mix are as follows:

1. $1\frac{1}{2}$ in. max. size. Aggregates drawn from a pit or quarry efficiently maintained, $f_S = 0{\cdot}55$.
2. $\frac{3}{4}$ in. max. size. Aggregates from a similar source to that indicated under 1, $f_S = 0{\cdot}60$.
3. $\frac{3}{8}$ in. max. size. Aggregates of reasonably uniform grading $f_S = 0{\cdot}70$.

These values should be regarded as for guidance, and may be reduced by up to 0·05 if reliance can be placed on the grading of the aggregate, particularly that of the sand, being more than usually uniform. If aggregate is drawn from several sources or is likely to be variable, it may

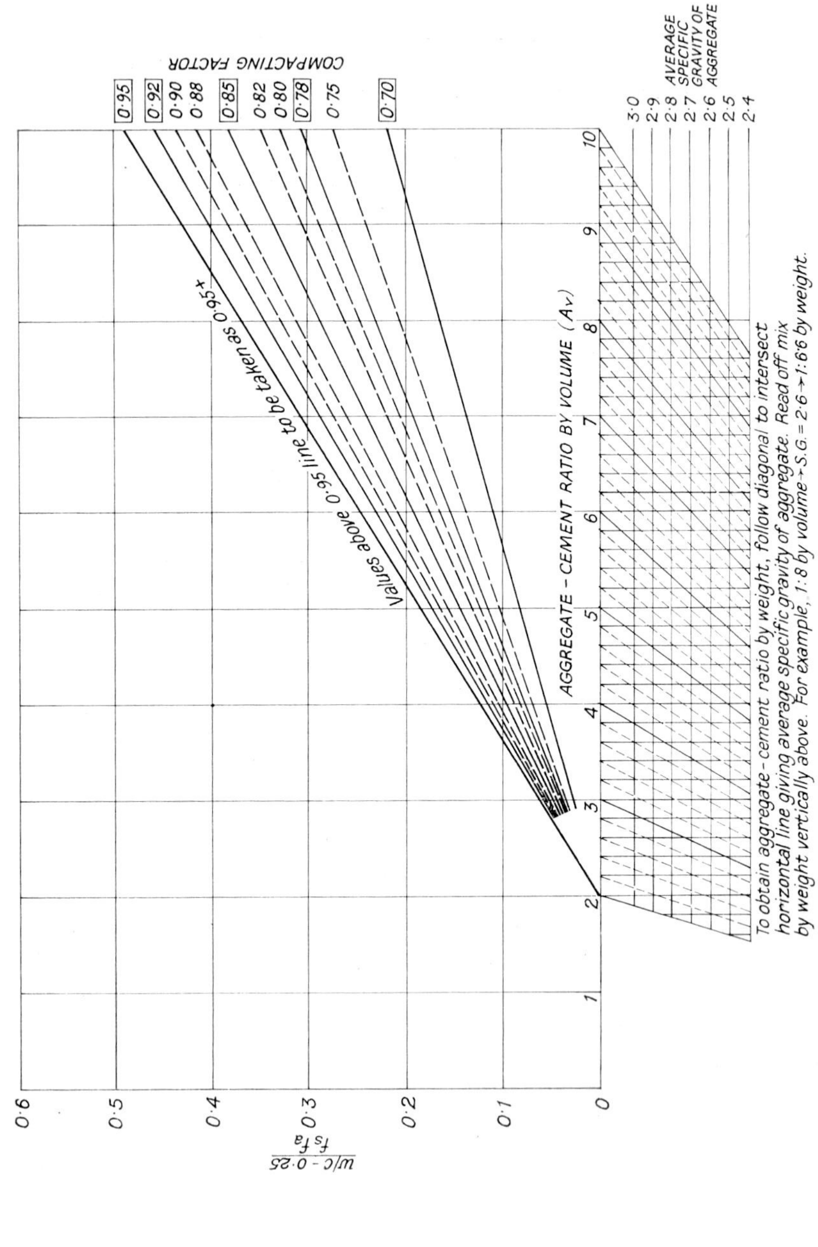

Fig. 7.2. Relationship between compacting factor, water/cement ratio, surface and angularity indices, and mix proportions

be advisable to increase the above values. The aim should be to produce a design of mix which suits the average grading to be expected, but which will still provide sufficient fines when the aggregates reach their coarser limits.

If possible, samples of aggregate representing the range in grading expected should be subjected to sieve analysis. The average grading of these aggregates should be determined, and hence their surface indices. The next step is to determine the proportions in which the aggregates need to be proportioned to give the chosen combined surface index. An example to illustrate the method of doing this is given below.

Having determined the proportions of coarse to fine aggregate for average gradings, the surface indices should be determined for the extremes of grading, using the same proportions in all cases. If the lowest value for f_S falls below 0·50 for $\frac{3}{4}$ in. aggregate or 0·45 for $1\frac{1}{2}$ in. aggregate, the proportion of fines should be increased, that is, assuming that the aggregates at the extremes of grading cannot be rejected.

If the higher limit for f_S exceeds 0·9, the mix may become too cohesive for a rotating drum mixer to handle. A more satisfactory upper limit is 0·8.

Assuming then that a value for surface index has been decided, and also a value for the angularity index, $f_S f_A$ can be calculated, and hence A_v; see equation (3).

Figure 7.2 is plotted in such a way that the aggregate/cement ratio by weight can be determined directly, providing the average specific gravity of the aggregate has been calculated. See below and equation (2) above.

Average specific gravity

As explained in Chapter 3, the specific gravity of aggregates can vary with the particle size and if an accurate value is to be obtained it is necessary to make a determination for each size used in the mix. For the purposes of equation (2) the average specific gravity of the combined aggregate may then be found as follows:

$$G_w = \frac{A_w G_A + B_w G_B}{A_w + B_w}$$

where G_w denotes average bulk specific gravity
A_w denotes parts by weight of coarse aggregate
B_w denotes parts by weight of fine aggregate
G_A denotes bulk specific gravity of coarse aggregate
G_B denotes bulk specific gravity of fine aggregate.

Grading and surface index calculations

The proportions of two aggregates of known individual surface indices required to give a predetermined combined surface index can readily be determined as follows:

Assume the use of 1 part aggregate with surface index y

Then parts aggregate a with surface index x

$$= \frac{z - y}{x - z} \qquad (4)$$

where z denotes surface index for combined aggregate.

A useful example to illustrate these points may be given by considering aggregates complying with the grading requirements of B.S. 882:1965. The grading requirements of this Specification are summarized in Tables 3.4 and 3.5, from which it will be seen that certain tolerances are allowed in the grading of the coarse aggregate, while the sand is required to be within one of four zones.

On the site it is not unusual to find that a sand, even from the same pit, varies considerably in grading, and it is reasonable to assume that it may vary as between consecutive deliveries anywhere within a zone. As an example, the surface index may vary from 0·93 to 1·19 for sands lying within Zone 2. Note that the sands giving these values for f_S are not at the coarser and finer limits, as indicated by the left- and right-hand figures in the respective column.

Variations in the grading of the coarse aggregate have comparatively little effect on the surface index, and may usually be ignored.

On the site it is usually only practicable to cope with one mix which must prove satisfactory, both as regards strength and workability, when using any sand within the zone.

It has been suggested under the heading Workability that a suitable surface index for the combined aggregate, $\frac{3}{4}$ in. max size is 0·60.

Having decided upon $f_S = 0{\cdot}60$ as suitable, the next step is to determine the proportions of coarse aggregate to the average of the sands (in this example the average for sands varying within Zone 2), which will give a surface index of 0·60.

Assume that for the coarse aggregate, $\frac{3}{4}$ in.–$\frac{3}{16}$ in., the value of $f_S = 0{\cdot}3$. The value of f_S for sand lying in the middle of Zone 2 is about 1·10.

Then, using formula (4) above, the proportions required to give a combined value of $f_S = 0{\cdot}60$ are:

$$a = \frac{0{\cdot}60 - 0{\cdot}30}{1{\cdot}10 - 0{\cdot}60} = \frac{0{\cdot}3}{0{\cdot}5} = \frac{1}{1{\cdot}67}$$

i.e. proportions required are 1:1·67 of average sand:coarse aggregate.

Now with these same proportions determine the combined surface index using the sand in Zone 2 having a surface index of 1·19:

$$f_S = \frac{1{\cdot}67 + 0{\cdot}3 + 1 \times 1{\cdot}19}{2{\cdot}67} = 0{\cdot}633$$

And similarly for the coarse grading having a surface index of 0·93:

$$f_S = \frac{1{\cdot}67 \times 0{\cdot}3 + 1 \times 0{\cdot}93}{2{\cdot}67} = 0{\cdot}536$$

Both values are within the acceptance limits suggested on p. 117.

Combining with Angularity

The angularity index will have been assessed roughly from the examples given in Table 6.3, or more exactly, and preferably, by tests as described on p. 53.

For this example, assume $f_A = 1{\cdot}9$

Then $f_S f_A = 0{\cdot}6 \times 1{\cdot}9 = 1{\cdot}14$ (average)

with a range from $f_S f_A = 0{\cdot}633 \times 1{\cdot}9 = 1{\cdot}20$

to $f_S f_A = 0{\cdot}536 \times 1{\cdot}9 = 1{\cdot}02$

Deciding mix proportions by volume

In the example given on p. 114, the water/cement ratio required to give an average compressive strength at 28 days of 5,250 lb/in² was 0·5.

Substituting this value, and that of $f_S f_A$ in equation (3) it is found that for a compacting factor c.f. = 0·85, $(A_v - 2) = 4{\cdot}6$, and $A_v = 6{\cdot}6$.

Alternatively, first calculate:

$$\frac{w/c - 0{\cdot}25}{f_S f_A} = \frac{0{\cdot}5 - 0{\cdot}25}{1{\cdot}14} = 0{\cdot}22$$

Then using Fig. 7.2, draw a line across from 0·22 on the vertical scale to intersect the diagonal line for compacting factor 0·85. From the intersection draw a line vertically downwards and the value of A_v is given directly. In this example it is 6·6 and taking the ratio of sand:coarse aggregate as 1:1·67 the mix proportions by solid volumes are 1:2·47:4·13. These volumes do not include the voids present when the materials are shovelled into a measuring hopper.

It is now necessary to assess the influence of the variation in grading of the sand within Zone 2:

(1) assuming the water/cement ratio is maintained constant

(2) assuming the workability is maintained constant and the water/cement ratio is varied. This is the usual occurrence on site, since a constant workability is important during placing.

Case (1)

$$\frac{w/c - 0{\cdot}25}{f_S f_A} \text{ ranges from } \frac{0{\cdot}25}{1{\cdot}20} \text{ to } \frac{0{\cdot}25}{1{\cdot}02}, \text{ i.e. from } 0{\cdot}21 \text{ to } 0{\cdot}245$$

With $A_v = 6{\cdot}6$, Fig. 7.2 indicates a range in c.f. from 0·83 to 0·89.

Case (2)

$$\frac{w/c - 0{\cdot}25}{f_S f_A} \text{ must remain constant} = 0{\cdot}22$$

$$\therefore \quad w/c - 0{\cdot}25 = f_S f_A \times 0{\cdot}22, \text{ i.e.} = 1{\cdot}20 \times 0{\cdot}22 \text{ to } 1{\cdot}02 \times 0{\cdot}22$$
$$= 0{\cdot}26 \text{ to } 0{\cdot}22$$

$$\therefore \quad w/c \text{ ranges from } 0{\cdot}25 + 0{\cdot}26 \text{ to } 0{\cdot}25 + 0{\cdot}22$$
$$\text{i.e. from } 0{\cdot}51 \text{ to } 0{\cdot}47$$

The resulting range in strength for this variation in w/c ratio can be assessed from Fig. 7.1.

If either the range in workability or in strength is considered excessive for the purpose for which the concrete is required, a less variable sand must be found. The variations indicated in the example are probably as large as are desired for good quality concrete.

Mix proportions by weight

The mix proportions by weight are calculated from the value A_v by suitable substitution in formula (2), or by the use of the conversion grid at the bottom of Fig. 7.2. The method of using this is given below the diagram. For the design example quoted above, where $A_v = 6{\cdot}6$, the mix by weight is 1:5·35 for aggregate with average specific gravity = 2·55. With a sand:coarse aggregate ratio of 1:1·67, the mix becomes 1:2:3·35 by weight, assuming the sand and coarse aggregate to have the same specific gravity.

A strength requirement of 3750 lb/in² at 28 days, would necessitate the use of the following mixes to accord with C.P. 114:1957 and amendment No. 1.

	By volume	Equivalent mix by weight
(1) Nominal mix	1:1½:3	1:1·7:3 (approx)
(2) Standard mix	—	1:1·56:3·35

The equivalent mix from A.C.I. tables is approx. 1:1·9:3·3 by weight. It will be noted that this is very similar to the proportions given by the surface index method. C.P. 114 sand contents tend to be a little lower in nominal and standard mixes.

Mix design charts

Mix design charts based on the surface index method are given in Figs. 7.3, 7.4 and 7.5, for rounded, irregular and angular aggregates of

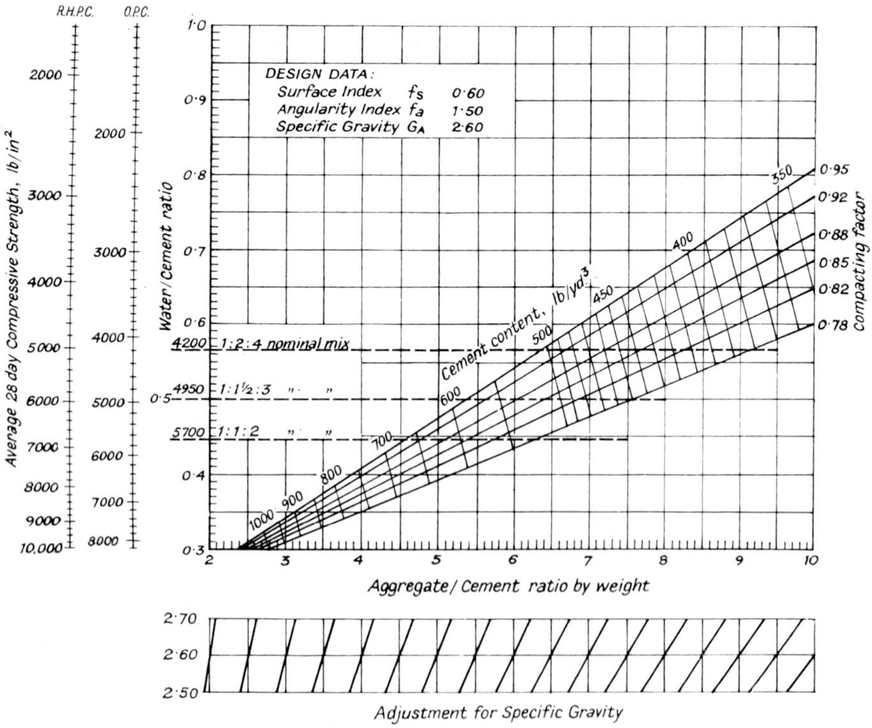

Fig. 7.3. Relationship between a/c ratio, w/c ratio, cement content and specific gravity for concrete mix designs for estimating purposes
¾ in. rounded gravel

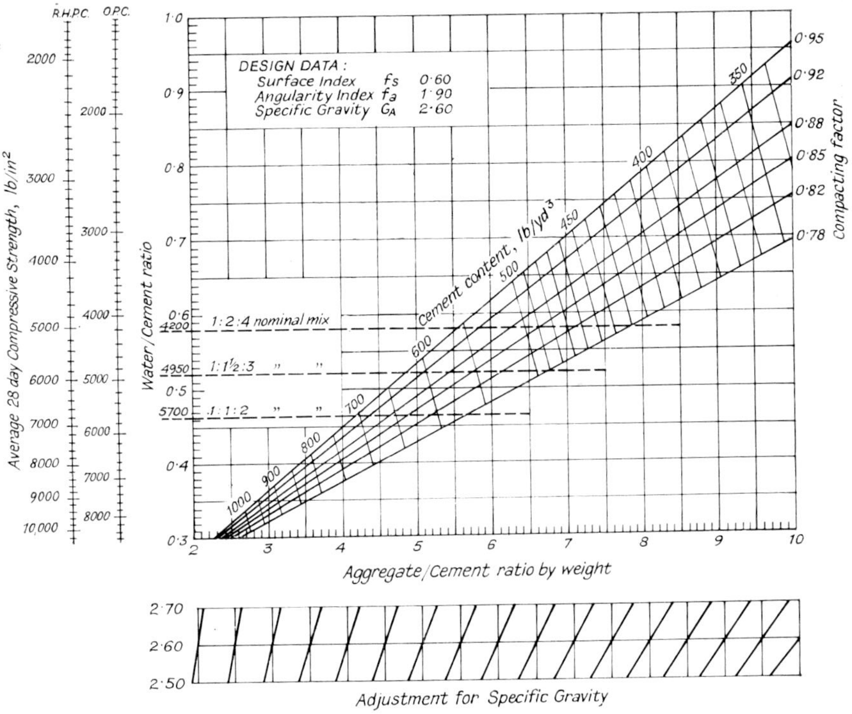

Fig. 7.4. Relationship between *a/c* ratio, *w/c* ratio, cement content and specific gravity for concrete mix designs for estimating purposes
¾ in. irregular gravel

the commonly used ¾ in. maximum size. The charts may be used to choose mix proportions and to determine cement contents without knowledge of the actual gradings of the coarse and fine aggregates. When these are known the proportion of coarse to fine aggregate can be adjusted to give a surface index of 0·6 as described on p. 120. Similar charts may be constructed for other aggregate sizes and values of the surface index for the combined aggregate.

As an example of the use of the charts, assume the aggregate is irregular in shape, that a minimum 28 day compressive strength of 3000 lb/in² is specified, and that the average works compressive strength requirement has been assessed at 4200 lb/in². Follow the dotted line across the chart (Fig. 7.4), until it meets the sloping line for compacting factor, say for example, 0·92. Now drop to the aggregate/cement ratio scale, and in this example the value is about 5·8. The cross lines give the cement content. In this example the value is about 540 lb/yd³. A convenient method of adjusting the mix proportions to allow for differences in the average specific gravity of the aggregate is given at the bottom of the chart. In the example given the aggregate/cement ratio would become 6·0 for a specific gravity of 2·70 or 5·6 for a specific gravity of 2·50.

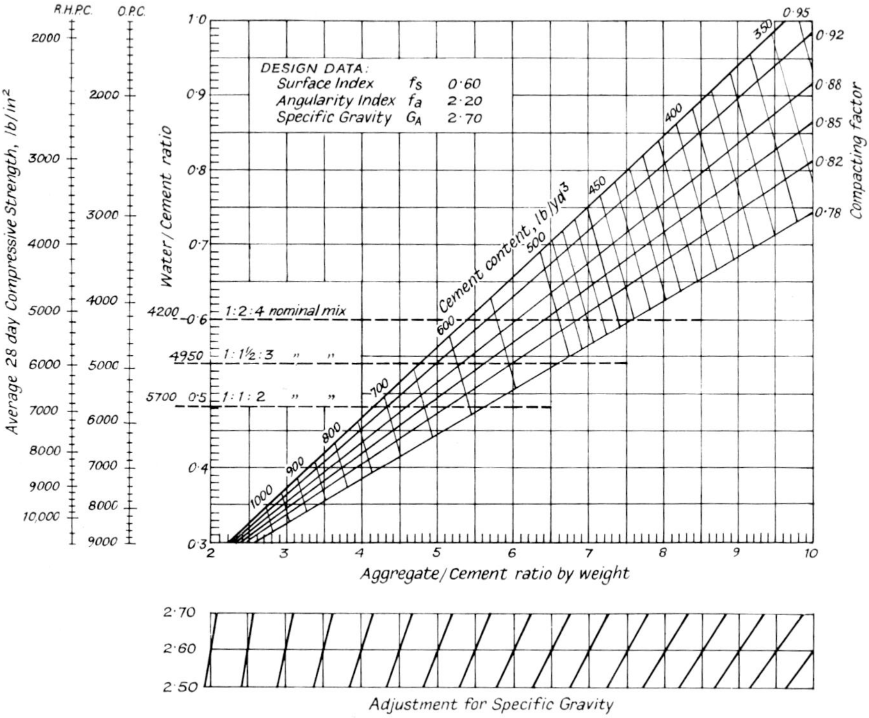

Fig. 7.5. Relationship between *a/c* ratio, *w/c* ratio, cement content and specific gravity for concrete mix designs for estimating purposes
¾ in. angular stone

Cement content of concrete

It is possible to design concrete mixes to satisfy strength requirements, but which will be permeable and therefore not suitable for situations where durability and watertightness are important. Specifications giving only a strength requirement are therefore not satisfactory, and a minimum cement content should be given. The following minima are suggested:

Concrete having a minimum strength requirement at 28 days of	Recommended minimum cement content
1,500 lb/in²	350 lb/yd³
3,000 lb/in²	500 lb/yd³
3,750 lb/in²	550 lb/yd³
4,500 lb/in²	650 lb/yd³

Except in the case of concrete having a minimum strength of 1,500 lb/in² at 28 days, these values apply to all reinforced structural concrete. In order to provide reasonable protection to reinforcement, and satisfactory durability for concrete in protected or reasonably sheltered positions, the recommended minimum cement content is 500 lb/yd³.

Chemical attack, severe exposure or other circumstances, should be taken into account in deciding the minimum cement content, irrespective of the strength requirement for structural purposes. See also Chapter 24 on Durability.

The cement content per cubic yard of concrete may be determined quite accurately if the mix proportions by weight, the water/cement ratio, and the specific gravity of the aggregates are known. The method is often known as the absolute volume method but a more accurately descriptive term used throughout this book is the solid volume method and consists of estimates of the volume occupied by each of the constituents. The volume of concrete is then assumed to be equal to the sum of these volumes and the cement content per unit volume of concrete can be calculated. An example of the method of calculation is given in Table 7.10.

TABLE 7.10 *Weights of Material Required to Produce 1 Cubic Yard of Concrete*

	Mix proportions by weight (lb) M	Specific gravity G	Solid volume $\frac{M}{62{\cdot}5G}$ ft³	Weight of material per cubic yard of concrete $= \frac{27M}{\sum\left(\frac{M}{62{\cdot}5G}\right)}$ lb
Coarse aggregate	390	2·58	2·42	2,102
Fine aggregate	210	2·63	1·28	1,132
Cement	100	3·15	0·51	540
Water	50	1.0	0·80	270
	$\sum\left(\frac{M}{62{\cdot}5G}\right) =$		5·01	

The cement content can conveniently be estimated from the formula:

$$C = \frac{1{,}685}{0{\cdot}317 + R/G + \phi} \tag{4}$$

where C denotes cement content in lb/yd³
R denotes ratio of aggregate : cement by weight
G denotes average bulk specific gravity of aggregates calculated by the method given in the formula under the heading Average Specific Gravity, p. 119
ϕ denotes water/cement ratio by weight.

An indication of cement content can also be obtained from the charts, Figs. 7.3, 7.4 and 7.5.

The solid volume method assumes that there are no losses of water from the concrete during transport and placing, and that the concrete is fully compacted.

Full compaction implies that there are no air voids, a condition never completely achieved in practice. The error is, however, small since the amount of air entrapped should, with good compaction, only amount to 1 to 2 per cent of the total volume.

In air-entrained concrete the total air content rises to about 4 to 7 per cent of the volume of the compacted concrete, and it becomes necessary to make an adjustment.

This may be done by increasing $\sum\left(\frac{M}{62 \cdot 5G}\right)$ in Table 7.10 by the percentage of air entrapped.

The weights of material per cubic yard of concrete calculated by this method do not allow for loss of materials on site, or loss of water during transit or placing.

In assessing overall quantities of materials required on site, an allowance should be made for loss of concrete either at the mixer or during transport and placing. These losses may be considerable, and due allowance should be made depending on site conditions, and the effectiveness of any measures taken to minimise losses.

Conversion from proportions by weight to proportions by volume

The conversion from weight proportions to volume proportions and vice versa involves assumptions for the weights per cubic foot of the fine and coarse aggregates. The values for the aggregates may be obtained by experiment, but some variation must be expected.

Once the proportions by weight have been established, equivalent proportions of aggregate by volume may be calculated as in the following example:

	Cement	Fine aggregate	Coarse aggregate	Water
(1) Weight of material: lb	112	224	381	62 lb
(2) By volume dry to 112 lb cement	112	$\frac{224}{100} = 2 \cdot 24$ ft^3	$\frac{381}{90} = 4 \cdot 23$ ft^3	62 lb
(3) Add for bulking		$+ 20\% = 0 \cdot 45$ ft^3		
(4) Proportions of damp aggregate to 112 lb cement	112	$2 \cdot 69$ ft^3	$4 \cdot 23$ ft^3	62 lb

Line (4) gives the final proportions for use in the field under the conditions stated.

REFERENCES

7.1 *Design of Concrete Mixes*, Road Note No. 4, Road Research Laboratory, 1947. H.M.S.O.

7.2 MURDOCK, L. J., The control of concrete quality, *Proc. Inst. Civ. Engrs*, Part 1, July 1953.

7.3 NEVILLE, A. M., The relation between standard deviation and mean strength of concrete cubes, *Mag. Concr. Res.*, **11**, No. 32, July 1959. Cement and Concrete Association.

7.4 *The Variation of Works Test Cubes*, Report TRB/287, Jan. 1958. Cement and Concrete Association.

7.5 C.P. 114:1957, *The Structural Use of Reinforced Concrete in Buildings* (see also Amendment 1965). British Standards Institution.

CHAPTER 8

Proportioning Aggregate, Cement and Water

SITE control of concrete begins with an inspection of the aggregates as they are delivered. Methods of testing have already been described in Chapter 3, but a few remarks on rapid methods of checking aggregates as they are delivered, may usefully be given. Sieve analyses can be made on coarse aggregates without previously drying them, and the results obtained are accurate enough to indicate serious errors. It is advisable to obtain the sample from different parts of the load after it has been deposited, as the grading may not be the same at the bottom of the lorry as at the top.

Sands can be sieved quickly if they are washed through the sieves with running water. The sand separated on each sieve is weighed in the sieve after allowing a few minutes for the water to drain away. While this method tends to give high values for the smaller particle sizes it does provide an indication of serious changes in grading. It is preferable, however, to have the sand in a dry condition before sieving and it is usually possible to arrange a calor gas ring or primus stove and metal trays for rapid drying of the sand. The same equipment is also useful for rapid moisture content determinations for fine and coarse aggregates.

Clay lumps can be picked out by a visual inspection. Silt and clay are found by shaking up the aggregates with water and allowing them to settle. After a little experience with the particular aggregate in use, it is possible to differentiate between clean and dirty samples by the muddiness of the water immediately after the larger particles have settled.

A first indication of the cleanliness of a sand can be obtained by picking up a little sand and rubbing it between the palms of the hands. If the palms remain clean the sand may be considered clean. If the palms are stained then it may be an indication that excessive amounts of clay and silt are present and that a more accurate determination of the amount is required. Details of tests suitable for use on site are given in Chapter 25.

In regions where dust or sand is carried in the wind, it may be necessary to erect protective screens to prevent contamination. Some form of shelter is also of advantage in hot desert regions to protect the aggregate from the heat of the sun's rays, so reducing the temperature of the concrete at the time of mixing as much as possible.

The deposition of the aggregates should be carefully regulated to prevent different sizes from getting in the same heap, and to make certain that different sizes are separated sufficiently to preclude any intermingling.

Occasionally the method of handling the aggregates enables stockpiles of the different aggregate sizes to be formed some distance apart. Generally the stockpiles are close together and it is therefore essential to provide some form of partition to separate the different sizes. Suitable partitions can be made by driving old rails into the ground and laying railway sleepers or other timber between them, by building brick walls, or by making rough concrete walls. Where there is more than one size of coarse aggregate it is advisable to display prominent notices, giving the sizes in bold figures, such that the drivers of delivery lorries can see at a glance where each size should be tipped.

If the mixer is to stay in one position for any length of time, the area over which the aggregate is deposited should be covered with a layer of concrete 3 to 4 in. thick. Otherwise it is necessary to regard the first 6 inches or so of aggregate as wasted in the mud.

The concrete should be laid to fall away from the mixer in order to facilitate free drainage of water from the aggregate. Variations in the moisture content of coarse aggregate as delivered are generally not sufficient to have much effect on the control of the water/cement ratio compared to the effect of the variations which commonly occur in the moisture content of the fine aggregate, i.e. the sand. Therefore only in special circumstances need the coarse aggregate be stockpiled for some hours to allow for drainage before use. Given reasonable drainage it can be expected that the moisture content of fine aggregate will be reasonably uniform throughout the heap, after having been deposited for at least 14 hours, except for the bottom foot or so, which will remain saturated. To allow for drainage, it is desirable to have two separate stockpiles of fine aggregate, so that fresh deliveries can be tipped into one while the other stockpile is being drawn on. If two stockpiles are provided and if care is taken not to draw from the bottom of the heap a reasonably uniform water/cement ratio can be maintained in the concrete produced, without undue difficulty. Duplicate stockpiles of fine aggregate are, however, seldom practicable or economical where only moderate quantities of concrete are required. This is one reason why it is generally more difficult to achieve good control on small works than on large.

Methods of handling aggregates to prevent segregation have already been discussed. The stockpiles at the back of the mixer should be so arranged that the aggregates are deposited and fed to the mixer economically and in the correct manner. When coarse aggregate is delivered graded from the maximum size to $\frac{3}{16}$ in. it sometimes happens that the aggregate producer fills the lorry with, say, two grabs of $\frac{3}{4}$ in. to $\frac{3}{8}$ in. and one grab of $\frac{3}{8}$ in. to $\frac{3}{16}$ in. The aggregate as delivered is unmixed and on small jobs some difficulty may be experienced in obtaining a uniform mix. The best solution is to insist that the aggregate producer mixes the sizes efficiently at the pit. If, however, it is necessary to mix the aggregate on the site, this should be done by taking portions from different parts of the heap.

In some cases it is possible to obtain from suppliers coarse and fine aggregates already combined. These premixed aggregates are seldom satisfactory for good quality concrete because of segregation during transportation and tipping.

Some of the points mentioned above are illustrated diagrammatically in Fig. 8.1. The diagrams are mainly concerned with the smaller works; for larger works the aggregate would probably be lifted by dragline, grab, or conveyor system to overhead storage hoppers discharging directly into weigh batchers. The slope of hopper bottoms should not be less than 50° to the horizontal if a free flow of aggregate is to be maintained. Plant

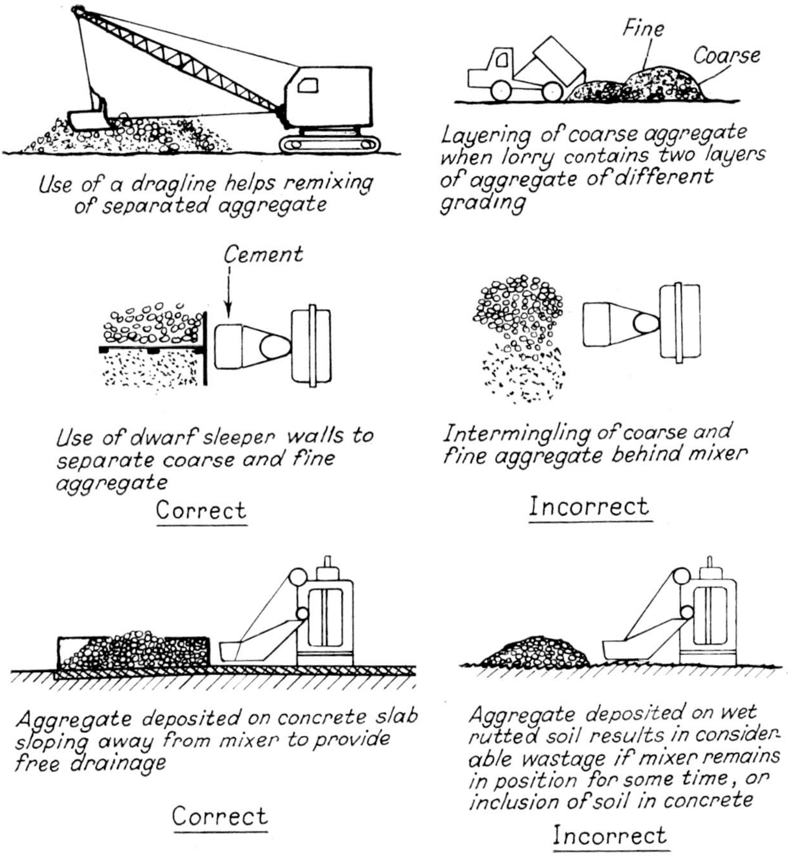

Fig. 8.1. Handling of aggregate at the mixer

arrangements in which storage hoppers feed weigh batchers are shown in Figs. 8.2 and 8.3.

The possibility of contamination of the aggregate behind the mixer by extraneous matter should not be overlooked. As an example, serious and extensive spalling of concrete at the entrance to a coalmine was traced to the deposition of spent carbide on the edge of the heap of aggregate, by the miners going off duty. Leaves, particularly pine needles, can delay the setting of cement, and cause local weakness. A broken bag of plaster and inclusion of a small quantity in the mix can cause a serious expansive failure. In food factories, aggregate may become contaminated with sugar, very small quantities of which are likely to delay or prevent setting of the cement.

Methods of proportioning concrete mixes depend, to some extent, on the quantity of concrete required. On large works it is now generally accepted practice for all to be batched by weight. For small contracts batching is still more commonly done by volume.

Fig. 8.2. Batching and mixing set-up incorporating overhead storage hoppers (filled by grab or dragline), weigh batching bucket and Hi-lift mixer. If bulk cement is used a suitable cement silo with its own weighing bucket can be positioned at one side so as to feed directly into the mixer hopper

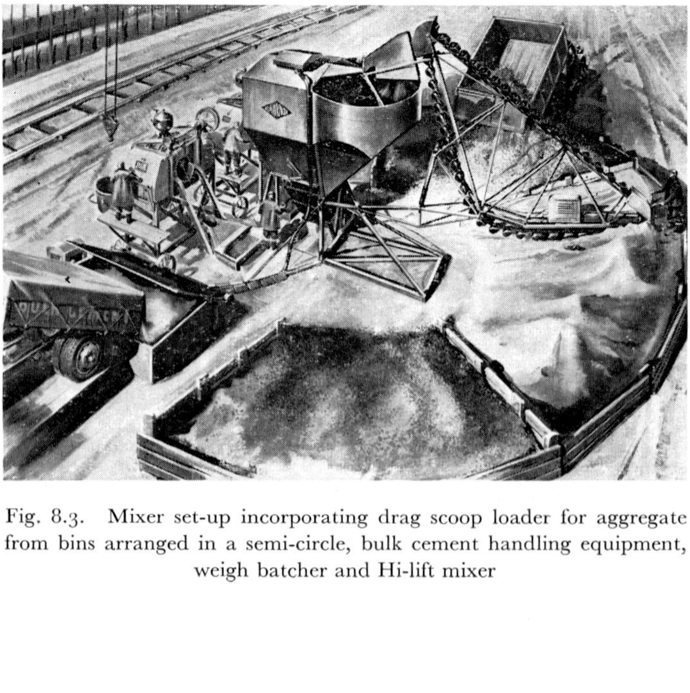

Fig. 8.3. Mixer set-up incorporating drag scoop loader for aggregate from bins arranged in a semi-circle, bulk cement handling equipment, weigh batcher and Hi-lift mixer

Volume batching of aggregate

Any method of measurement in which a gauge-box or hopper is filled to a certain level depends on the closeness with which the material packs; thus, if the material packs closely with few air voids, the solid volume of material is greater than if the material is packed loosely. In the volume batching of aggregate, there are, therefore, two possible sources of error: (1) variations in the solid volume of aggregate contained in a specified measured volume, and (2) errors in the measured volume. The first variation can be minimized although it is difficult to eliminate it, but errors in measurement can be reduced by careful attention to the type of measure.

Although variations occur in volume batching this method is probably as good as weigh batching on small works where supervision is not good, and where proper maintenance of weigh batching equipment is not possible. Volume batching need not be greatly inferior to weigh batching; experience shows that the standard of supervision on the site is of far more importance than the type of equipment used.

Occasionally the practice of batching by filling the mixer skip with so many shovelfuls of each aggregate is adopted. This is bad practice for the following reasons:

(1) No two men load the shovel with the same quantity of material. The difference may be considerable.

(2) The number of shovelfuls required is large, and the men rarely count them except under supervision.

(3) The quantities are difficult to check except by getting each man to load a measure of known volume.

Batching directly by dragline, digger and grab buckets is unsatisfactory, except perhaps for large quantities of weak concrete for mass filling.

The shape of the measuring-box or hopper is important and the best results are obtained if the plan area is relatively small. Flat boxes with a large plan area should be avoided since any error in the height of material in the box causes a comparatively large error in volume. Wheelbarrows are not good measures for the same reason. Properly made gauge-boxes filled by hand and tipped into the mixer are satisfactory, but unless the supervision is strict they are unlikely to be used.

Several methods of marking the skip itself may be adopted, such as band marks, angles fitted round the sides, or bolts running across at the correct level, but none of these is entirely satisfactory. The last named is probably the best method since a definite guide is given when levelling off. A further arrangement sometimes used consists of vertical division plates mounted on lengths of screwed rod at predetermined positions. The coarse aggregate, the fine aggregate, and the cement are filled into the respective divisions of the hopper and the division plates are then pulled out before raising the skip. All these skip marking methods present problems when frequent changes in mix proportions have to be made.

Honeycombing resulting from the inclusion of one or more under-sanded batches is shown in Fig. 8.4. The concrete for this prestressed caisson, being constructed for harbour works in France, was volume batched.

Fig. 8.4. Honeycombing in prestressed concrete construction, caused by the inclusion of one or more under-sanded batches. Note also the evidence of the sloping pour planes, that the concrete was placed at one point and allowed to flow along the shuttering

Batching of aggregates by weight

Batching by weight eliminates errors due to variations in the proportion of voids contained in a specified volume, a point of special importance in connection with the batching of sand. Measurement by weight is therefore logical, and provided the weighing machine retains its accuracy on the site, errors in proportioning should be negligible. Unfortunately, weighing machines need careful maintenance and regular calibration if reasonable accuracy is to be maintained. This is not always easy to arrange on small or isolated sites where volume measurement by deep measuring boxes or hoppers is likely to prove as satisfactory.

Wherever good and regular maintenance is practicable, weigh batching gives greater uniformity between successive batches of concrete.

Different types of weigh batching plant are available for use on large and small contracts. On large sites the weigh bucket may be of the type indicated in Fig. 8.5. This is fed from a large overhead storage hopper and discharges by gravity straight into the mixer. The weigh bucket shown weighs two sizes of aggregate in succession to provide a batch without discharging. Weighing is done through a lever-arm system and two interlinked beams and jockey weights. The required quantity of, say, coarse aggregate is weighed, having only the lower beam in operation. After balancing, by turning the smaller lever to the left of the beam, the two beams are interlinked, and the fine aggregate is added until they both balance. The final balance is indicated by the pointer on the scale to the right of the beams. Discharge is through the swivel gate at the bottom.

On smaller sites the weigh bucket may be of a different type situated at ground level and discharged into the skip of the mixer, in which it is then elevated to the drum.

Fig. 8.5. Large aggregate weigh bucket, in which two sizes of aggregate can be weighed consecutively before discharge

A common type, shown in Fig. 8.6, consists of two weighing buckets, each connected through a system of levers to spring-loaded dials which indicate the load. Discharge is through a door in the bottom of the hopper released by pulling a lever. The two weighing buckets are mounted on a central spindle about which they rotate. Thus one can be loaded while the other is being discharged into the mixer skip.

Hydraulic weighing mechanisms, in which the load is taken on a sealed capsule connected to a pressure gauge, have recently come into general use. They are used for separate weighing buckets and for integral weigh-batchers. The latter term is used when the mechanism is an integral part of the mixer, the aggregates being weighed in the dry hopper of the mixer (Fig. 8.7).

A useful auxiliary which is now fitted to most of the smaller mixers, is the hand scraper for feeding aggregates into the loading hopper, shown in Fig. 8.8. This saves manpower but it needs to be used with care to avoid batching more than the quantity required.

Aggregate weighing machines require regular attention if they are to maintain their accuracy. Check calibrations should always be made by adding weights in the hopper equal to the full weight of aggregate in the batch. Errors may occur for a number of reasons such as:

(1) The knife-edges and other moving parts become clogged or blunt, causing stiffness of movement.

(2) The dashpot damping device on beam and jockey weight types may become detached or be filled with the wrong type of oil or water.

Fig. 8.6. Swing weigh batcher in use for the production of relatively small quantities of concrete. The cement is measured into the mixer hopper by the bagful, or from a bulk cement silo with its own weighing device

Fig. 8.7. A 30 ton overhead mounted cement silo which automatically weighs and delivers cement direct into a mixer having a loading hopper fitted with an integral weighing mechanism for batching the aggregates

Fig. 8.8. View of a hand scraper fitted to a 12 ft^3 mixer for loading aggregate into the mixer hopper

(3) Again on the beam and jockey weight types, the beam gets out of balance due to movement of the counter-balances, either through ignorance or by loosening of the locking device.

(4) Levers or rods may be bent and bind on some part of the framework.

(5) With dial types the interior mechanism may become worn, or the springs may become weak or broken. If the weighing bucket is not locked in position during transit, sufficient wear of these parts can occur during a single journey to cause substantial errors.

(6) Perspex covers warp and may impede the free movement of pointers causing quite considerable errors.

(7) If the weigh batcher has two weighing buckets, and both are in use, the one for sand and the other for gravel, confusion may easily arise as to which is to be filled with sand and gravel, resulting in the relative proportions being reversed. It is therefore essential to label the buckets clearly to indicate the material with which they should be loaded.

(8) Spectators sometimes add their weight to the bucket.

(9) Spillage of aggregate sometimes accumulates and gives local support to the weigh bucket; this can easily occur with most integral weigh batchers as they seldom have much clearance above the ground.

The method of delivering the materials into the weigh-hopper has some influence on the accuracy of the batching. Delivery from overhead hoppers with servo-operated gates is likely to be more accurate than by other methods.

Batching of cement

Cement is batched either by the bag (normally 112 lb in Great Britain; 94 lb in America) as weighed at the cement works or, for bulk cement, by weighing machines installed on the site.

Whichever method of batching is used, precautions should be taken to see that cement is not lost as the loading skip empties it into the mixer. Tests on exposed sites with strong winds blowing have shown considerable losses of cement during this operation.

Volume batching of cement should only be permitted in special circumstances; it is subject to serious errors because Portland cement may weigh from 75 to 95 lb/ft^3 depending upon its fineness and the method of filling the measure. Other errors may result from failure to fill the measure accurately and from warping or denting of the measure.

Batching by bags of cement

Provided whole bags of cement are used, the only error lies in a variation in the weight in each bag as supplied by the cement works. This is not likely to be serious, although variations from 100 lb to 120 lb in single bags have been recorded. Under the ASTM specifications, individual sacks may be as much as 5 per cent below the standard weight of 94 lb, provided that the average weight of 50 sacks taken at random is not less than 94 lb. For important work an occasional check is justified.

Whenever possible, the size of the batch of concrete should be adjusted so that full bags of cement may be used. Splitting of bags is liable to cause substantial error and for accurate gauging the part bags should be weighed.

Weighing cement on site

Where it is practicable there are several advantages to be gained from using bulk cement stored in a silo and weighing it on site:

(1) Bulk cement is cheaper than bagged cement.

(2) A gang of men is not needed to off-load the cement, as is the case with bags of cement.

(3) The trouble of ensuring adequate protection of the cement during storage is avoided.

(4) The mixer can always be used to capacity without the risk of the errors which may result from splitting bags.

With regard to the last advantage, it is not uncommon for mixers to be used well below their capacity, or to be overloaded, because the amount of cement for a batch to the mixer's full capacity does not coincide with a whole number of bags (see Table 9.1, p. 146).

However, the weighing of cement usually provides more difficulty than that of aggregate, mainly because it is difficult to obtain a uniform flow, and compressed-air assistance is often needed to overcome arching in the valve opening. When it is appreciated that the normal rate of flow in a large weighing machine is about 50 lb per second, it will be realised that any sudden rush, when the quantity in the weigh bucket is near the correct weight, may result in some excess which will be reflected either in increased workability or in increased strength for the particular batch. Another difficulty which arises in weighing cement is the tendency to build-up in the weighing hopper, particularly when the atmosphere is damp. If cement and aggregate are weighed together, the dampness from the aggregate causes a serious build-up of cement mortar in the weighing hopper, which measurements have shown to amount to 1 cwt or more in a matter of ten batches, even when some shaking is applied to the hopper to hasten discharge.

Weighing machines operating in conjunction with bulk supplies of cement are often used on both large and small sites. They may be of the spring-balance type or the beam and jockey weight type; or they may be of the hydraulic capsule type (referred to on p. 133).

Cement in bulk is delivered in pressurized lorries which blow it into lightweight portable silos of 10 to 30 tons capacity (illustrated in Fig. 8.7). These silos generally incorporate a mechanism for weighing the correct quantity for a batch and dispensing it into the mixer hopper.

Cement weighing machines should receive regular attention to the following points:

(1) Building up of cement in the weigh hopper and discharge chute, which may result in the use of too little cement.

(2) The cement should flow freely and continuously from the storage hopper, compressed air being blown in at the valve, if necessary. If the cement does not flow freely it is difficult for the operator or the weighing mechanism to cut off at the correct quantity.

Vibration of the silo is more likely to result in compaction of the cement and blockage, than to cause a free flow, hence the use of compressed air.

(3) The knife-edges and all other moving parts should be kept completely free from coatings of cement in order to avoid friction and stiffness. The leather apron which prevents the cement from escaping from the dispenser during filling should be kept clean, since any accumulation will be included in the weighing.

(4) If a dashpot damping device is fitted, this should be filled with the correct oil. Its connections should be inspected occasionally to make sure it does not become loose or detached.

(5) Levers and rods should be inspected occasionally to ensure that they have not been so bent as to bind on the framework.

(6) Care should be taken to see that spectators do not lean on the weigh bucket, especially when calibrating.

(7) The balance of the beam should be checked frequently, in case the counter-balances become moved, or cement adds to the weight of the beam itself.

Cement weighing mechanisms should always be checked by adding weights to the dispenser to the full amount required in the batch. The necessity for this is demonstrated by the curves given in Fig. 8.9, which are plotted from calibrations of a cement weigh batcher over a period of several months. If weights to a maximum of only 300 lb had been used, it would have been assumed that the scales were weighing light, whereas above about 500 lb the quantity of cement delivered was in excess of that indicated on the dial. The necessity for regular checking is apparent from the extent of the errors.

Cement weigh buckets generally discharge directly into the mixer hopper, but in large plants the cement storage bin is often situated to one side of the plant to save height, and means must be provided to convey the cement horizontally to the weigh bucket. A worm feed is often used for this purpose; but a more efficient system is provided by the air slide,

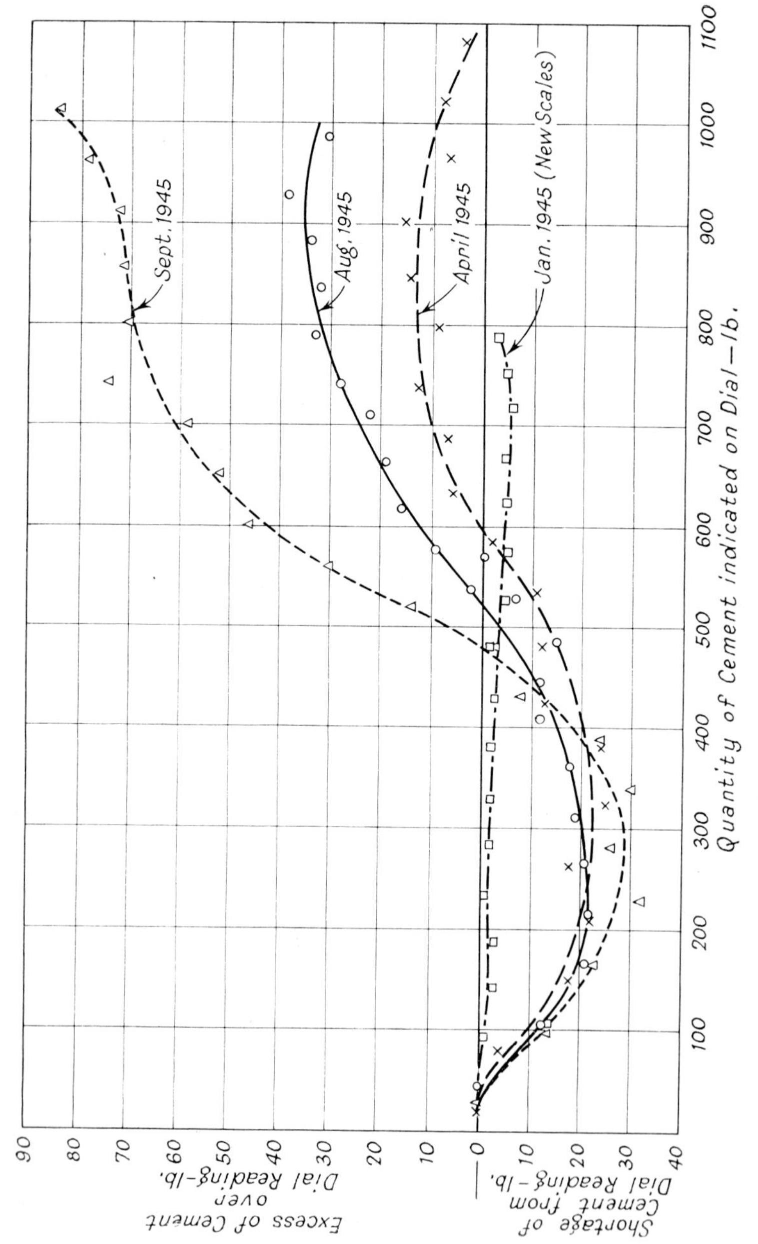

Fig. 8.9. Deterioration in accuracy of cement scales

which is a small duct with a mesh or canvas floor through which low pressure air is introduced, causing the cement to flow freely, even with a very small slope.

Automatic batching plants

Some large and small plants are available with automatic weigh-batching. This generally means that the operator has only to press one or two buttons to put into motion the weighing of all the different materials, the flow of each being cut off when the correct weight is reached. In their most advanced forms, automatic plants are electronically operated on a punched card system; this type of plant is particularly suitable for the production of ready-mixed concrete, in which very frequent changes in mix proportions have to be made to meet the varying requirements of different customers.

With most large plants, particularly those made in the U.S.A., recorders can be fitted which note automatically and graphically the weight of each material delivered to each batch; that is, they are meant to record the actual weight as distinct from the intended weight. In practice their accuracy generally leaves something to be desired, and some plant manufacturers prefer not to introduce them unless the specification insists on their use.

Moisture meters for fine aggregate

Electrical apparatus for measuring automatically the moisture content of sand is sometimes fitted to large batching plants. The accuracy of such apparatus is somewhat limited and it is desirable to check it at least daily by measuring the moisture content of the sand by more conventional means. Nevertheless, the apparatus provides a very useful indication of changes in moisture content and gives the plant operator, indirectly, a constant indication of adjustments needed to the quantity of added water required. It can contribute to good control.

Water supply

Purity of water

Water to be used for making concrete must be reasonably free from deleterious materials such as silt, clay, organic matter and organic acids, alkalies, and other salts. B.S. 3148:1959[8.1] gives methods of testing water and in the appendix are notes on the suitability of water for making concrete.

Silt and clay can be removed by allowing them to settle out in suitable basins or tanks, but the other materials, being soluble, are difficult to eliminate. No specific limit to the quantity of dissolved salts can be given, but unless clear water has a noticeably brackish or saline taste it may be safely used. In any case of doubt it is advisable to send a sample of the water to a laboratory for analysis and comparison with water known to be satisfactory. Care should be taken when sampling water, that a clean bottle is used, and that the bottle is washed out with the water to be submitted, before finally taking the sample.

The experience and recommendations of a large number of investigators and authorities on the use of sea water are conflicting, some reporting adverse results while others report no harmful effects.(4.5) Whenever possible the use of sea water for making concrete should be avoided, since it is liable to cause unsightly efflorescence as moisture movements carry soluble salts to the surface. The chlorides in seawater are liable to cause serious corrosion of reinforcement, and unless measures are taken to ensure that the concrete will be dense and impermeable, the use of seawater in reinforced concrete is better avoided.

Drinking water from the mains supply is safe and convenient.

Water supply on the site

Water is required both for mixing the concrete and for curing. When it is considered that each batch of concrete from a 10 ft^3 mixer may require 8 gallons of water, and 80 batches a day may be made, giving a total of 640 gallons a day per mixer, it will be realized that the quantities required for mixing alone are considerable. More important, the 8 gallons of water in each batch must be replaced in the time taken for mixing, say 1 minute, in order that the measuring cylinder is full, ready for the next batch. The size of hose-piping connecting the mixer with the source of supply must, therefore, be of sufficient size, having regard to its length and the water pressure, to cope with this intermittent high rate of demand.

The question of an adequate water supply is commonly neglected on concrete constructional works, and results in loss of efficiency in operation of mixing machines. Not only is time lost waiting for the water to refill, but the measurement of the quantity of water discharged into the mixer becomes inaccurate. Most systems of measurement rely on complete filling of the tank, and the mixer driver, when in a hurry, is apt to omit to do so. An adjustment is then made to the quantity of water indicated on the scale, and when a longer period occurs between batches, the mix becomes too wet.

An idea of the water supply requirements on a large contract may be judged from the fact that the quantities used may be equivalent to the normal supply in a small town.

In the hot dry regions of the world, the mixing water has often to be brought from a considerable distance, and if the pipeline or water tanker is exposed for long in the sun, it can become too hot to touch. The use of such hot water, together with hot aggregates and cement increases the rapid evaporation of water from the concrete during transport, and reduces the setting time, so that placing may become difficult. In the hot season, it is therefore necessary to bury the water pipeline, and to cover water tankers.

Water control at the mixer

With given proportions of aggregate and cement, the water/cement ratio is determined by the workability required for satisfactory placing. With the workability controlled by close observation at the mixer, so that any variation is small, it may be taken that any changes in the water/cement ratio will also be small and the strength of the concrete will be reasonably uniform. Further control may be obtained by making tests on the moisture content of the aggregate, particularly the sand, at frequent intervals, and adjusting the quantity of water added at the mixer accordingly. Except where large stockpiles of aggregate are maintained, however, the variations in moisture content of the sand from one load to another, and from one part to another of a small stockpile, make the value of tests for moisture content doubtful. The cost of making such tests continually is, in any case, excessive when small quantities of concrete only are to be produced, and under these conditions it is suggested that the following measures be adopted in order to secure concrete of the required quality.

The first few batches of concrete should be produced under the supervision of an experienced engineer or inspector, the aggregate and sand being carefully weighed and checked for surface and absorbed moisture content. When the aggregate is saturated the absorbed moisture may be taken as constant and should not be included in the water/cement ratio. The surface moisture will be variable and should be deducted from the theoretical quantity of water required when determining the water to be added at the mixer. If the aggregate is dry, it will be necessary to determine the quantity of water absorbed in bringing the aggregate to a saturated surface dry condition, and to add this additional water to the theoretical quantity required. In the tropics it is advisable to keep the aggregate wet, if an ample supply of clean water is available.

The workability should be checked by means of a slump test, or compacting factor apparatus, and cubes made for crushing strength tests. With most concrete mixers an appreciable quantity of fine material from the first few batches remains in the drum. Therefore only after producing two or three correctly proportioned batches will the concrete have settled down to a uniform consistency. The appearance of the mix should be particularly noted and should be observed by the mixer driver and the placing ganger. This concrete should then be placed and observations made on its suitability for placing in the formwork, on ease of consolidation, and on any tendency to segregate. If the concrete is found to be satisfactory in all respects, the mixer driver and the placing ganger should be instructed that all further concrete produced should have a similar appearance.

Reference is also made to the control of consistency in the field on page 102.

Allowance for surface moisture

Several different methods of determining surface moisture are available, such as the simple drying method using a frying pan and gas ring, and the rather more difficult procedure using a pycnometer. Both methods are fully described in Chapter 25 and Appendices B and C, pp. 387, 390.

Measurement of water at the mixer

Water gauges on concrete mixers may be of three types. Their principal features are given below.

(1) *Horizontal measuring tank.* This type, which has a movable syphon to regulate the discharge, is inherently inaccurate, since any change in the tank inclination causes errors in measurement. Both filling and discharge are effected through the three-way valve, and it is essential that

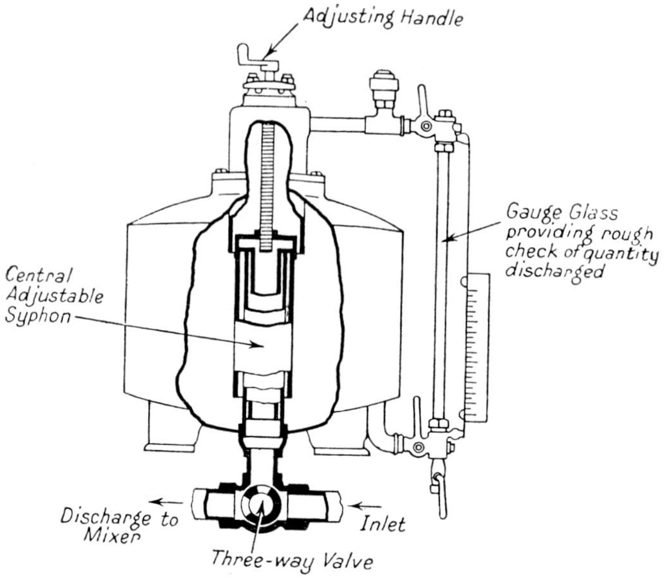

(a)

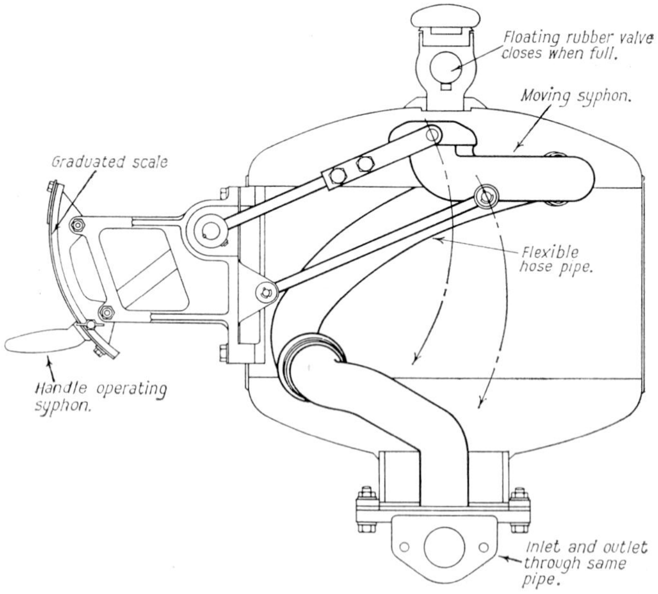

(b)

Fig. 8.10. Different arrangement of two vertical water measuring tanks no leakage shall occur at this point. The tank is filled completely after each discharge, air being expelled through a rubber valve, which closes under water pressure. The gauge glass indicates whether the tank is completely full. The tank should be calibrated after repairs, as the setting is often affected. This type of tank is going out of use.

(2) *Vertical tanks* were developed to overcome inaccuracies due to changes in inclination. Examples are shown in Fig. 8.10.

The tank shown in Fig. 8.10 (*a*) is set by winding a small handle on top and counting the turns, a procedure which is not entirely satisfactory, since it is not possible to check the quantity delivered without turning back to zero and counting the number of turns. A large number of turns are required and errors in counting often occur. The other tank, Fig. 8.10 (*b*), overcomes this difficulty by the provision of a scale at the end of the lever which actuates the central moving syphon. A flexible hose allows vertical movement of the syphon. This is more satisfactory than a bellows for this purpose, being less liable to develop leaks.

(3) *Water meters* fitted in the main water supply line to the mixer. For use on batch mixers, the meters are provided with a reset hand to indicate batch readings, and a register of the total water passed. The latter is particularly useful in determining the average water content of a group of concrete mixes.

Water meters are affected by dirt, scale, and sediment more than measuring tanks, and are therefore less suitable for portable mixers. They are most satisfactory for stationary mixers and central mixing plants fed direct from the mains water supply.

Before setting the water gauge on a mixer for the first time, it is essential to check the calibration. Errors of as much as 2 gallons in 10 have been found with some types of water gauge. Regular attention should be given to the cleanliness of the tank, and to the fitting of any valves or syphoning arrangement.

American machinery is likely to be calibrated in U.S. gallons, equivalent to 8·33 lb of water, compared with 10 lb for the Imperial gallon. This point should also be remembered when reading American literature.

REFERENCE

8.1 B.S. 3148:1959, *Tests for Water for Making Concrete*. British Standards Institution.

CHAPTER 9

Mixing and Transporting Concrete

THE mixing of concrete is almost invariably carried out by machine, but a very brief reference to hand mixing is necessary to cover those occasions, which occur on many jobs, when hand mixing must be done.

Hand mixing should be done thoroughly. A clean surface should be selected for mixing. If this is not available, a wooden platform with close joints to prevent loss of mortar should be obtained. It is usually specified that the concrete shall be mixed three times dry and three times wet, turning over from one spot to another. The colour and uniformity of the concrete is, perhaps, the best gauge of the efficiency of the mixing.

In order to compensate for the lower strength which usually is obtained with hand mixing, it is advisable to allow an extra 10 per cent of cement above that normally required.

Machine mixing

Many types of machine are available for mixing concrete, some having a revolving or stationary pan, with paddles and sometimes rollers to mix the materials, others having tilting or revolving drums. The continuous mixer, although in a class on its own, relies for the actual mixing process on a revolving drum in one type and on a revolving shaft with "tines," in a stationary trough, in another type.

Very little is known about the relative mixing efficiencies of the various types of mixer, but what evidence there is appears to suggest that pan mixers with a revolving star of blades are the most efficient; they are especially suitable for stiff and lean mixes, which present difficulties with most other types of mixer, mainly due to sticking of the mortar in the drum. They are, however, somewhat cumbersome and difficult to maintain in good condition.

Uniformity of mixing is difficult to measure, but there is evidence that considerable variations can occur with most types of mixers in the proportions of the constituent materials within the same batch; for instance, variations of as much as 20 per cent in the proportion of coarse aggregate have been measured in a fair proportion of batches mixed.[9.1] In conventional rotating-drum mixers this variation is due principally to the lack of interchange of materials from one end of the drum to the other. Some lack of uniformity of concrete mixed in pan mixers appears to be due to the throwing of stones from one part of the batch to another by the high-speed action of the mixing star.

Mixing time and output

Some research has been made both in America and in this country[9.1] on the time required to mix concrete intimately and uniformly. It appears that the number of revolutions of the mixing drum or pan is more important than the mixing time, and in general no more than

20 revolutions are necessary for adequate mixing. For most types of mixer up to the 1¼ cubic yard size operating at their correct speed, this represents a mixing time of a little over 1 minute. Normally a mixing time of between 1 minute and 1½ minutes may be regarded as adequate. Mixing for longer periods does not appear to give any useful improvement in the efficiency of mixing. With some high speed pan mixers, it is claimed that 35 seconds is adequate mixing time for most types of concrete.

The minimum mixing time generally specified for Bureau of Reclamation work in America,[3.4] the time to start after all ingredients, except the last of the water, are in the mixer, is as follows:

Capacity of mixer yd^3	Time of mixing minutes
2 or less	1½
3	2
4	2½
6	3

In considering the potential output of a particular mixer, it is the time of the total mixing cycle, not just the time of mixing, which, with the batch size, determines the output. For instance, with a mixer which requires 1½ minutes mixing time, the design or condition of the loading hopper may be such that the materials flow only slowly into the mixer. The loading time may be 30 seconds or more; and the discharge of the mixed concrete may be slow, taking up to 1 minute, particularly if the mix is dry. Adding the time to raise the loading hopper and allowing for minor delays, the total cycle may well be 4 minutes or more, so that only some 15 batches an hour, or less, can be produced. In comparison, 35–40 batches per hour are possible with some of the more efficient mixers.

The output per hour from any mixer is dependent, to a large extent, on the efficiency of organization, both in charging the mixer and in transporting away the mixed concrete.

Mixer size

Concrete mixers complying with the requirements of B.S. 1305:1967 "Batch Type Concrete Mixers" are designated for size by a number representing the nominal batch quantity in cubic feet, together with letter(s) indicating the type of mixer. The nominal batch quantity is the volume in cubic feet of compacted concrete which can be mixed satisfactorily in one batch. The letters used are:

Tilting drum type	T
Non-tilting drum type	NT
Reversing drum type	R
Forced action type	P

However, the input of the separate ingredients in cubic feet has often been given as a prefix, for instance, the capacity of a 14/10 mixer is 14 ft^3 of unmixed materials, or 10 ft^3 of mixed concrete. The prefix of input capacity is very approximate and should be disregarded when all-in aggregate is used.

Table 9.1 *Capacity of Concrete Mixers*

Nominal mix proportions by volume	Amount of materials to produce a batch of concrete without overloading the mixer — Size of mixer: 7 ft³				10 ft³				12 ft³				14 ft³			
	Bags of cement	Cubic feet of aggregate: Fine	Coarse	% of mixer capacity	Bags of cement	Cubic feet of aggregate: Fine	Coarse	% of mixer capacity	Bags of cement	Cubic feet of aggregate: Fine	Coarse	% of mixer capacity	Bags of cement	Cubic feet of aggregate: Fine	Coarse	% of mixer capacity
1:1½:3	1	2	4	73	1	2	4	52*	2	4	8	81	3	6	12	100
1:2:3	1	2⅔	4	80	1	2⅔	4	57	2	5⅓	8	89	2	5⅓	8	73
1:2:4	1	2⅔	5⅓	93	1	2⅔	5⅓	67	1	2⅔	5⅓	52*	2	5⅓	10⅔	85
1:2½:5	—	—	—	—	1	3⅓	6⅔	81	1	3⅓	6⅔	63	1	3⅓	6⅔	52
1:3:5	—	—	—	—	1	4	6⅔	86	1	4	6⅔	67	1	4	6⅔	55
1:3:6	—	—	—	—	1	4	8	95	1	4	8	74	1	4	8	60
1:4:8	—	—	—	—	—	—	—	—	1	5⅓	10⅔	96	1	5⅓	10⅔	79

* Use 2 bags of cement and cut aggregate slightly. Allowance for bulking of sand not included.

Table 9.1 gives the capacity of four of the more popular sizes of mixer assuming full bags of cement are adopted. It will be noted that the size of mixer providing the most efficient usage depends on the nominal mix proportions. Any allowance for the bulking of sand may be ignored in assessing the batch quantity, since the bulking does not affect the quantity of wet concrete on which the mixer capacity is based.

As indicated in Chapter 8, one of the advantages of using bulk cement is that the quantity of cement batched can be chosen, according to the mix proportions, so that the mixer is used to capacity.

General principles in the use of concrete mixers

There are five main types of concrete mixer in use in the United Kingdom, and while their method of operation varies considerably, the following general principles are usually applicable:

(1) It is an advantage to feed the cement, sand and coarse aggregate into the mixer simultaneously and in such a manner that the flow of each extends over the same period. The concrete produced is more uniform than that obtained when the ingredients are introduced one after the other, or in unmatched quantities.

With most mixers it will be found that filling the loading hopper in more or less horizontal layers of gravel, cement and sand, possibly with another layer of a different size of gravel on top, provides the best conditions.[9.1] Placing of the gravel at the bottom of the loading hopper is more likely to result in a self-cleaning process, so avoiding the formation of a hardened layer of mortar at the bottom of the hopper. The piling-up of any one size of material in the throat of the loading skip should be avoided. With large mixers (4 yd^3), the sequence of materials into the mixer is of particular importance.

(2) The water should enter the mixer at the same time and over the same period as the other materials. With many mixers this is not possible, since the rate of flow is limited, and the time taken for the full quantity to enter the mixer may be up to $\frac{1}{2}$ minute. In such cases it is advisable to start the flow of water a little in advance of the other ingredients.

If all the water is added before or after the other ingredients, the batch of concrete is liable to vary from a wet portion to a dry portion. The direction of inflow of water also influences the uniformity and may need adjustment.

(3) Mixing should continue until the concrete is of uniform consistency and colour. With drum mixers it is an advantage if the mixer driver can see into the drum at both ends, so that he can watch the consistency, and make any necessary minor correction in the water added.

(4) The mixer should not be loaded beyond its rated capacity. Overloading results in spillage of materials, and less satisfactory mixing, in addition to imposing undue strain on the mechanical parts.

(5) The mixer should be set up accurately so that the axis of rotation of the mixer drum, except in the case of the tilting-drum type, is horizontal. Inaccurate setting may result in poor mixing, or in the case of continuous mixers, too fast or too slow a passage through the mixer drum. The horizontal type of water measuring tank is also affected by errors in levelling.

(6) For satisfactory performance the mixer should be capable of producing uniform concrete throughout the batch, so that the quantity of the largest size of aggregate at the end of discharge does not differ by more than 20 per cent from that at the beginning of discharge. In order to minimize the risk of stone pockets, due to excess of stone in any part of the batch, it is advisable to discharge the whole batch into a suitable container, rather than to discharge in small separate quantities into wheelbarrows. If the container is suitably designed, the stone will become spread through the batch.

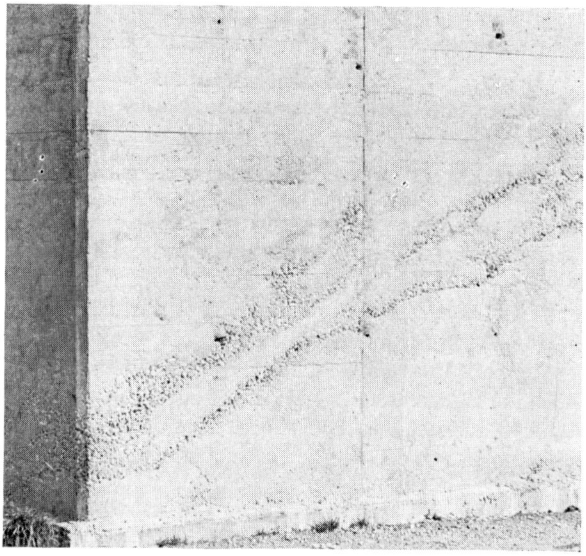

Fig. 9.1. Honeycombing caused by uneven mixing coupled with unsatisfactory placing indicated by the sloping pour planes

Honeycombing, resulting from uneven distribution of stones and sand in batches of concrete, is seen in Fig. 9.1.

(7) The mixer should be run at the correct speed as stated by the manufacturer. The speed should be checked regularly.

The effect of the speed of the drum on the efficiency of mixing is illustrated in Fig. 9.2.[9.1] The lower the "efficiency number," the greater the efficiency of mixing.

(8) Some cement mortar from the first batch of concrete mixed is left on the blades and round the drum. In order to avoid difficulties in placing due to shortage of fines an extra 10 per cent each of cement and sand should be added for the first batch.

(9) Considerable amounts of concrete adhering to the blades or the inner surface of the drum reduce the efficiency of the mixing. Regular cleaning at the end of each spell of mixing is necessary to prevent concrete building up, especially if stiff mixes are in use.

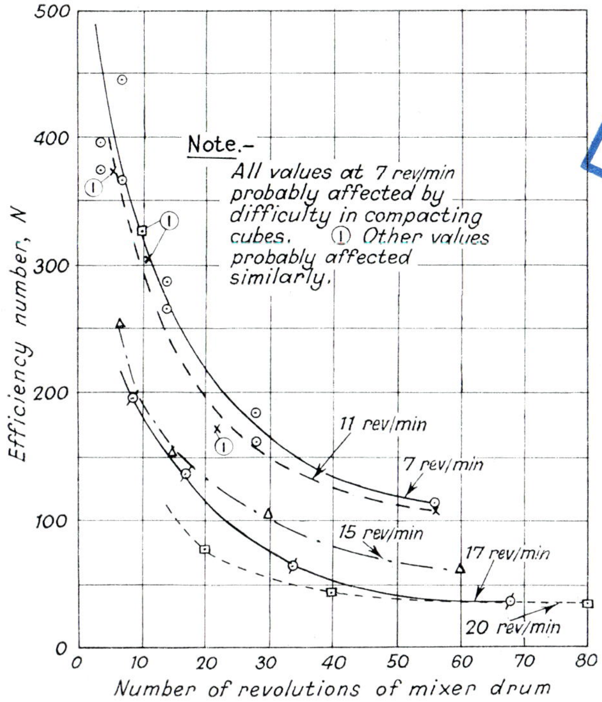

Fig. 9.2. Relation between efficiency number, drum revolutions and mixing speed

(10) Badly worn and bent blades reduce efficiency, and should be replaced. Wear of the inlet and discharge chutes eventually results in loss of materials, and should be remedied by suitable repairs.

(11) Adherence of cement is reduced by rubbing grease or oil over the mixer after cleaning. Layers of cement are particularly apt to build up in the nose of the loading hopper and should be chipped off at frequent intervals.

Non-tilting drum mixers

Non-tilting drum mixers normally have a single drum rotating about a horizontal axis. The drum is mounted on rollers and is turned either by a rack-and-pinion drive or by a chain passing round the drum and a cog-wheel driven from the engine. The blades inside the drum are arranged somewhat differently by the various makers of mixing machines, but the general tendency is for the blades to work the concrete towards the discharge end of the mixer, in order to provide a rapid rate of discharge. Examples of blade arrangements are given in Fig. 9.3.

With many rotating drum mixers it is important for uniform mixing that the water delivery pipe be so positioned that the water is directed to the rear part of the drum, i.e. the side into which the dry materials are fed.

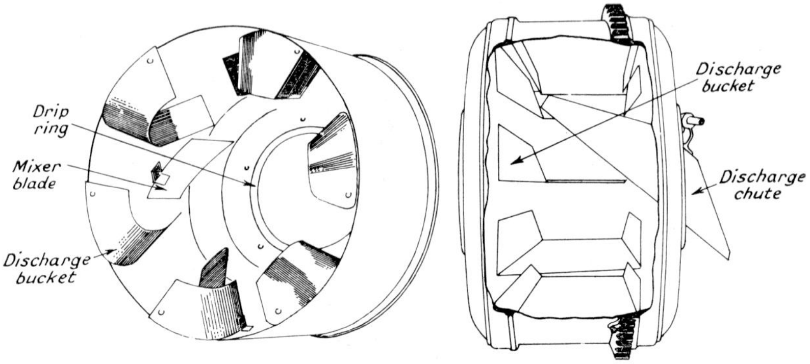

Fig. 9.3. Interiors of two non-tilting drum mixers showing bucket arrangements

An example of a mixer designed to produce 10 ft^3 of mixed concrete per batch is given in Fig. 9.4. The drum has ten similar blades arranged around the periphery. Discharge is effected by a manually operated chute.

A Paver mixer of rated capacity 28 ft^3 of mixed concrete is shown in Fig. 9.5. This mixer is intended to run under it own power along the side of the road, and is provided with a bucket and boom to distribute the concrete. This mixer is similar in principle to the smaller mixers already described, but the controls are operated mechanically instead of manually and may be largely automatic. The water measuring tank is of the vertical type, the inlet and outlet valves being operated by the raising and lowering of the loading hopper. The concrete is discharged

Fig. 9.4. Non-tilting drum type concrete mixer
10 ft^3 capacity, and provided with a vertical water measuring tank

Fig. 9.5. Non-tilting drum paver mixer showing control platform

Fig. 9.6. A compact central mixing plant

The cement storage silo discharges into a separate weigh hopper directly over the pan mixer. The aggregates are discharged into a weigh hopper which travels up the inclined rails and discharges into the mixer. The small drag line operates from the end of the boom above the aggregate stock piles. On being discharged from the mixer the concrete falls into a wet hopper.

into the travelling bucket by a mechanically operated chute. A batch-meter is fitted to control the mixing time, and to record the number of batches of concrete produced.

Paver mixers may have two drums in series, arranged so that the second drum can be refilled while the first drum is completing the mixing of the previous batch. By this means some 60 batches an hour can be produced with a mixing time of $1\frac{1}{2}$ minutes for each batch.

Non-tilting drum mixers are available in sizes 7, 10, 12, 14 and 27 ft^3 nominal batch capacities.

Split drum mixers

Split-drum mixers have been used for some years in Belgium and a number are now in use in the United Kingdom. Normally of $1\frac{1}{2}$ yd^3 capacity, their distinctive feature is that the drum separates into two halves along a vertical plane, allowing the mixed concrete to be discharged rapidly.

Tilting-drum mixers

Small tilting-drum mixers of the type shown in Fig. 9.7 are commonly used for all types of building work and in Great Britain are generally

Fig. 9.7. Concrete being discharged from a tilting-drum mixer

available in sizes of 3, $3\frac{1}{2}$, 5 and 7 ft^3 nominal batch capacity. Those of capacity up to 5 ft^3 of mixed concrete are often loaded by shovelling straight into the drum, but this method is unsatisfactory for high-quality concrete. A scoop suspended from a spring balance in such a manner that it can be easily tipped into the mouth of the mixer is a much more satisfactory method.

Medium size tilting mixers are provided with a loading skip similar to that used for non-tilting drum mixers.

The large sizes (Fig. 9.8) may be 1 yd^3 to 10 yd^3. They are common overseas in large scale concrete production plants; in the United

Fig. 9.8. Large size tilting-drum mixer

Kingdom they are more likely to be seen in ready-mixed plants than on sites. Their suitability for large plants is to some extent due to the way in which three or four of them can be arranged radially, such that each in turn can be fed from the same chute bringing materials from overhead hoppers, while each discharges in turn into the same wet hopper. Also tilting drum mixers are the most suitable type for concrete with large sized aggregate, e.g., 6 in.; and because of their rapid discharge they are suitable for low workability concrete.

Reversing drum mixers

Drum mixers which discharge the mixed concrete by reversing their direction of rotation have been used on the Continent for many years, and in the smaller sizes, such as that shown in Fig. 9.9, they have recently been gaining popularity in the United Kingdom. They give a clean discharge of all grades of concrete.

Forced action mixers

The several types of forced action, i.e. pan, mixer which are made all have the advantage, particularly over conventional non-tilting drum

Fig. 9.9. Reversing drum mixer of 12 ft³ capacity with built in batch weigher and powered scraper shovel. The arrangement of the blades in the drum is such that on reversing the direction of rotation the mix is discharged. Note the rubber drum cone extension for discharging cleanly into a rollover crane skip

Fig. 9.10. Action of the mixing star in a "Cumflow" forced action or pan mixer

mixers, of being able to mix stiff and lean concrete efficiently. For this reason they are widely used for precast concrete manufacture. One common type of pan mixer, with a rotating pan, is fitted with a mixing star of cast-steel paddles (Fig. 9.10), mounted eccentrically to the pan; the star revolves either in the same direction as the pan or in the counter direction, but at greater speed.

In the past one of the disadvantages of pan mixers has been their comparatively slow rate of discharge. One type which is satisfactory in this and other respects is shown in Fig. 9.11. This, in common with most modern large-pan mixers, has a stationary pan.

Fig. 9.11. $\frac{3}{4}$ yd^3 high-speed forced action or pan mixer. The electrical drive for the paddles is housed in the centre of the stationary pan. Steel cover plates removed to show the interior of the pan

The pan mixer is not an easily transportable mixer, and wear is fairly heavy, so that its use is mainly confined to precast concrete factories and large civil engineering works.

A small version of this type of mixer having a rated capacity of $1\frac{1}{2}$ ft^3 of mixed concrete has proved very useful for trial mixes in the laboratory, and for the preparation of small quantities of coloured facing for architectural cast stone. Even with this mixer, however, it is found that some segregation of the larger sizes of aggregate occurs, and for trial mixes it is necessary finally to turn the concrete over by hand to ensure uniformity throughout the batch.

Forced action mixers are available in sizes of 7, 12, 14, 20, 27 and 54 ft^3 nominal batch capacity.

Continuous mixers

Continuous mixers are no longer being manufactured in the United Kingdom, largely due to the erroneous impression that they were incapable of producing reasonably uniform concrete. Most jobs requiring at least moderate quantities of concrete can be organized so that the steady discharge of concrete can be handled and placed without interruption. Such construction work as dams, foundations, retaining walls, and mass concrete filling under roads is well suited to this type of machine and they have been found particularly useful for the production of dry-lean concrete for road construction.

One type of continuous mixer developed on the Continent is fed by a continuous weigh-batching system. A similar plant with automatic controls is installed in at least one ready-mixed plant in the United Kingdom.

Ready-mixed concrete

Ready-mixed concrete is produced in a central plant, and should have the advantages of strict control which this provides. Either the concrete is mixed before being filled into the truck which then only agitates it sufficiently to keep it plastic, or the batched materials are deposited in the truck and mixing is commenced just before reaching the job or on arrival. An example of a truck mixer is given in Fig. 9.12.

Fig. 9.12. Delivery of ready-mixed concrete from a 6 yd^3 truck mixer

The use of ready-mixed concrete is increasing, particularly when the concrete is required on a congested site, or in small quantities at intervals.

Truck or transit mixers, which can deposit the concrete directly into the forms, have definite advantages in hot climates, and the cost of this form of mixer may be more than mitigated by the easing of other items in the placing procedure.

Transport

The transport of concrete from the mixing plant to the point at which it is to be placed must comply with three main requirements, namely:

(1) Transport must be rapid so that the concrete does not dry out or lose its workability or plasticity during the time which elapses between mixing and placing.

(2) Segregation must be reduced to a minimum in order to avoid non-uniform concrete. For the same reason any loss of fine material or cement and water should be prevented.

(3) The transport should be organized so that during the placing of any particular lift or section, delays will not result in the formation of pour planes, cold joints, or construction joints.

Dry concrete may prove difficult to transport, since it may compact so hard that it has to be hacked from the containers. It is also likely to stick in concrete buckets, skips and transfer hoppers. Providing the container is well designed there is only one remedy and that is to make the mix more workable.

Some of the mortar from the first one or two batches of concrete transported, nearly always sticks to the container. It has been suggested previously in dealing with concrete mixers that an extra 10 per cent of cement and sand should be added to the first batch produced; this should be sufficient to compensate for any loss of mortar in transit as well as in the mixer.

Various methods are available for transporting concrete, of which the following are the most important:

(1) Wheelbarrows, handcarts, light trucks and skips on rails.
(2) Dumpers.
(3) Lorries.
(4) Steel skips and buckets.
(5) Chutes.
(6) Belt conveyors.
(7) Mono-rail transporters.
(8) Concrete pumps and pneumatic placers, with pipe lines.

A number of general points concerning the handling and transport of mixed concrete are illustrated in Fig. 9.13.

Wheelbarrows, handcarts and light trucks. Wheelbarrows and handcarts may be used over short distances. It is advisable, however, to discharge the whole of the contents of the mixer into an intermediate hopper, from which the concrete can be drawn by gravity into the barrows, in order to overcome variations in the consistency of the concrete as discharged from the mixer, to avoid segregation in long chutes from the mixer, and to minimise delays.

Light trucks or skips on rails may be used for longer deliveries, depositing into wheelbarrows or handcarts which are hoisted to the level of placing. Wheelbarrows are well suited for wheeling along the scaffolding on building work.

With any of the above methods jolting and vibration should be avoided as far as possible, for instance barrow runs should be made as smooth as possible. For longer deliveries, particularly in hot, dry atmospheres, it is advisable to cover the concrete with wet sacks or a tarpaulin. The high rate of evaporation from a heap of concrete is not always appreciated.

Dumpers. Dumpers are useful for transporting many constructional materials and are particularly useful for transporting concrete for small scale roadworks and for foundations and ground floor slabs on housing

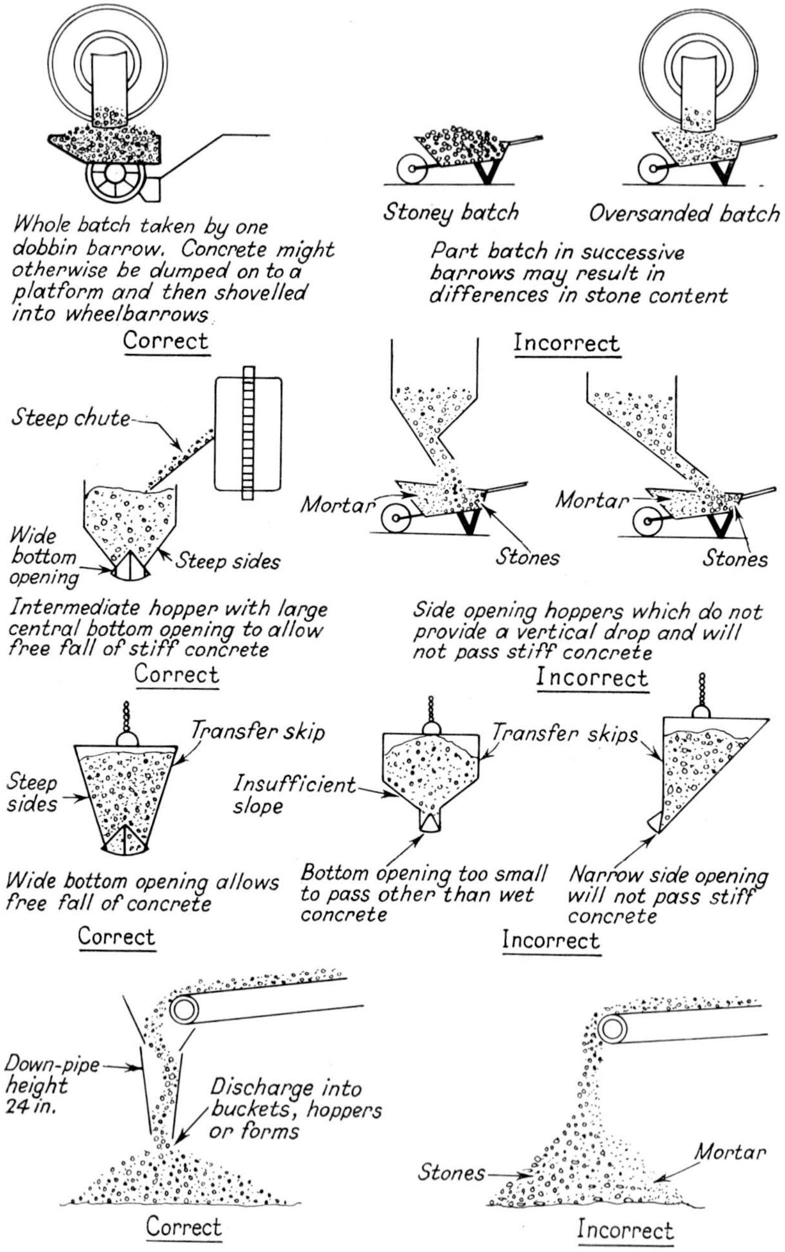

Fig. 9.13. Methods of handling mixed concrete

sites. Segregation or compaction frequently occurs when dumpers are used and care in transportation is particularly necessary.

Lorries. Lorries are commonly used in conjunction with central mixing plants for airfield and other work. They are loaded with three or four cubic yards of concrete from a storage hopper of the type shown in Fig. 9.6. Reference is made to the use of side-tipping lorries for the transport of concrete in road construction in Chapter 19.

High-lift lorries have been developed for circumstances where it is possible and convenient to tip from a height of about 4 ft.

With all types of lorry there is some risk of segregation of wet mixes and of compaction of dry mixes. In general, the slump needs to be between $\frac{1}{2}$ in. and 3 in. if difficulties are to be avoided.

Evaporation of water from the concrete in hot, dry regions during transport can be quite serious, and the only alternative to increasing the water/cement ratio at the mixer, is to provide some cover to the lorry or other transporting medium, unless the distance is very short. Truck or transit mixers, which can deposit the concrete directly into the forms, have definite advantages in this type of climate, and the high cost of this form of mixer may be more than mitigated by the easing of other items in the placing procedure.

Steel skips and buckets. Well-designed steel skips and buckets, properly operated, afford a most satisfactory means of handling and placing concrete.

Transfer skips for use with the smaller mixers are usually made to hold about $\frac{1}{2}$ yd^3. Two types are in use, a wedge-shaped skip with bottom opening, and a lay back skip which is allowed to fall over on its back for filling from a low level mixer discharge. The latter type has the advantage that the impact of concrete discharged into it does not result in compacted concrete immediately over the bottom opening, with resultant difficulty in discharging. With either of these types of skip it is essential that the bottom opening is large enough to allow the easy discharge of reasonably stiff concrete mixes. Any changes of direction or restrictions should be avoided, since these cause arching and difficulty in discharge, and the inevitable result on the site is that more water is added at the mixer. A suitable design of wedge-shaped skip is illustrated in Fig. 9.13.

Steel dump buckets for the transport of concrete are usually square or circular in plan, and hold about 4 cubic yards. Other sizes may of course be adopted to suit the scale of the work being done and buckets up to 8 cubic yards in capacity were used in the construction of the Shasta and Boulder Dams in the U.S.A. Discharge may be through manually or mechanically, usually compressed air, operated gates at the bottom of the bucket. Care is necessary to prevent excessive impact and possible damage to the formwork by releasing the concrete from too great a height. Some detail of a 4 yd^3 bucket is shown in Fig. 9.14.

Buckets are ordinarily transported by truck, derrick and cableway. An example of the use of a bucket slung from a derrick, together with final distribution by conveyor, is given in Fig. 9.15. This photograph was taken during the construction of the Haweswater Dam, Cumberland. Some types of hoist have a bucket which is elevated and discharged

Fig. 9.14. 4 yd^3 dump bucket, with compressed air operated discharge mechanism

Fig. 9.15. Concrete-placing plant, incorporating crane skip, feed hopper and distributing conveyor, discharging mixed concrete on Haweswater Dam

into a hopper at the level of placing. From the latter it is distributed by wheelbarrow or possibly by chutes.

Chutes. Long open chutes tend to cause segregation and drying of the concrete and are one of the most objectionable methods of transporting concrete. If chutes are used they should be designed so that the concrete is to some extent remixed at the lower end by passing down through a funnel-shaped pipe or drop chute.

Alternatively, they should discharge into a storage hopper, from which the concrete should be transported to the point of placing by wheelbarrows or other means.

The concrete mix should be workable and cohesive, if it is to flow satisfactorily down a chute.

Belt conveyors. Belt conveyors have very limited application in concrete construction. Occasionally a fixed conveyor between two points is useful, which in some conditions needs to be enclosed in a protective cover throughout its length. One type of conveyor consists of a series of short belts with swivel connections such that the discharge end can be moved about readily if a flat running surface is available for its wheeled support.

Belt conveyors tend to cause segregation on steep inclines; at transfer points; as the belt passes over the rollers; and as the concrete is discharged from the belt. A method of providing a certain amount of remixing may sometimes be necessary at the discharge end.

Mono-rail transporter. The Mono-rail transporter shown in Fig. 9.16, is valuable on sites where relatively small quantities of concrete have to be transported over a considerable area congested with other work proceeding, and crossed by trenches and other obstacles. With suitable support the rail can be inclined so that concrete can, for instance, be

Fig. 9.16. Mono-rail transporter being used for the placing of foundation bases and other work on a building site

conveyed to the upper floor of a two storey structure. The Mono-rail skip is started on its journey by the mixer driver and is automatically stopped at the required location by a trip device.

Concrete pumps and pipelines. Pumping through steel pipelines is one of the most satisfactory methods, though somewhat expensive, of transporting concrete, particularly in congested areas. Concrete may be pumped through a straight horizontal pipe to a distance of 1,000 ft,

Fig. 9.17. Concrete pump
Mobile versions of the same type are sometimes used.

but curves, lifts, and harsh concrete materially reduce the maximum pumping distance. An allowance of 40 ft should be deducted from the maximum horizontal pumping distance for every right-angle bend, and 8 ft for every 1 ft of head. Good workable concrete can be raised to a height of 100–120 ft within a short distance of the pump.

An illustration of the most common type of concrete pump is given in Fig. 9.17. The mixed concrete is fed into the conical hopper above the pump, from which it passes through an inlet valve into the pump cylinder during the retreating or suction part of the piston stroke. As the piston moves forward, the inlet valve closes and an outlet valve opens to allow the concrete to pass into the pipeline. In starting the pumping operations a quantity of mortar equivalent to the concrete with the coarse aggregate omitted, is fed into the hopper. This mortar is necessary to lubricate the pipeline, and about 1 yd^3 is sufficient for 1,000 ft of pipe.

The most usual size of pipeline is 6 in., although 7 in. and 4 in. pipelines are sometimes used. The output from a conventional 6 in. pump varies between 8 and 20 yd^3 per hour, depending on conditions. These figures do not allow for time for overhauls, site inefficiencies, etc., which may reduce output by 25 per cent.

A new type of concrete pump developed on the Continent provides an output of up to 30 yd^3 of concrete per hour. The pumping unit is operated by high pressure water which is pumped into the cylinder behind the piston ram on the working stroke and sucked from it on the return stroke. Experience so far indicates that less maintenance costs

Fig. 9.18. Concrete being discharged from pipeline

are involved than with the conventional type of pump. However, there are risks of water leakage past the piston when wear occurs.

Concrete for pumping must be well graded without any tendency toward segregation.

Mixes richer than about 1:6 cement:total aggregate can usually be pumped without much difficulty. Leaner mixes require greater care both as regards the grading of the aggregate and the accurate proportioning of the mix. Weigh batched concrete at 1:9, cement:total aggregate, has been pumped successfully over about half the distances mentioned above.

A slump of $1\frac{1}{2}$ in. to 3 in. but not more than 3 in. appears most suitable. Segregated or harsh lean concrete is usually the cause of stoppages in the pipeline. Too dry or too wet a consistency, and interruptions which allow the concrete to stiffen are other causes of trouble. Concrete as delivered by pump is generally of uniform quality and practically free from segregation by reason of these requirements. In Fig. 9.18 the concrete is shown being discharged and spread to form a concrete floor.

If the concrete discharged from the pipeline should appear wetter than the concrete fed into the pump hopper, it is possible that water used in the pump to keep the pistons free from grit and so reduce wear, may be leaking past the glands. Good maintenance arrangements are essential when concrete pumps are in use.

Cleaning of the pump and pipe line at the end of concreting operations is very important, and for this purpose a plug or "go-devil" is inserted at the pump end, and forced through the pipe line with compressed air. A bundle of sacks or hessian forms a satisfactory plug. As the plug may be ejected from the end of the pipe line with considerable force, the pump operator should make sure that all persons are behind the mouth of the pipe during the cleaning out operation.

Pneumatic placers. Another means of conveying concrete through steel pipe lines is by means of compressed air, using a pneumatic placer such as that shown in Fig. 9.19. The important distinction between

Fig. 9.19. A mobile pneumatic concrete placer of $6\frac{1}{2}$ ft^3 capacity having an output of 4 to 6 yd^3 per hour. For each batch of concrete air pressure is built up until the concrete is propelled along the pipe line

pumping and pneumatic placing is that, while in the former case a continuous plug of concrete moves slowly along the pipe line, in the latter case separate batches of concrete are "shot" at considerable velocity along the pipe line. In tunnel lining work, the discharge end of the pipe line is best kept just buried in the previously placed fresh concrete between the shutter and tunnel walls. In almost all other circumstances a trap is fitted on the end of the pipe line.

Pneumatic placers vary in size from 4 ft^3 up to about $1\frac{1}{2}$ yd^3, the larger sizes seldom being used except in tunnel work. Placers involve much less capital outlay than pumps, but require considerable quantities of compressed air, a costly commodity, particularly when working at the limit of their range, which is of the same order as that of pumps. Although in some circumstances either a pneumatic placer or concrete pump may equally well be used, their different characteristics render each most suited to somewhat different circumstances. For instance, the pump is likely to be more suitable where considerable quantities of concrete need to be delivered at a steady rate into one area of the site. Pneumatic placers are satisfactory for intermittent conveyance of modest quantities, in which conditions pumps are likely to prove impracticable.

With pneumatic placers highly workable concrete of 4 in. slump or more is necessary, and with all mixes there is a tendency for some segregation to take place, though with suitable mixes this need not occur to a detrimental extent.

Small pneumatic placers, fitted with a suitable nozzle, can be used for the pneumatic spraying of mortar or concrete with small aggregate, a process often used in the repair of old concrete, and sometimes in new construction.

Cableways. It is primarily in the construction of concrete dams that cableways are used for transporting concrete in skips or buckets, the latter being of up to 8 yd^3 capacity or more on large projects.

Transit-mixers and transit-cars on rail tracks. In tunnel lining construction the concrete is generally transported along a rail track from the batching plant outside the tunnel by means of transit cars of 2 to 8 yd^3 capacity. If used as transit mixers, the batched materials are mixed with water during the journey to the point of placing; though the capacity of the horizontal axis drum is then less than if it is used for concrete already mixed.

REFERENCE

9.1 MURDOCK, L. J., The efficiency of concrete mixing plant, *J. Inst. Civ. Engrs*, **31**, Nov. 1948.

CHAPTER 10

Preparation and Joints

Preparation for placing

THE preparations necessary before placing commences are dependent on the type of construction and the position in which the concrete is to be deposited. In roads and foundations it is necessary to prepare the soil surface or sub-base, while in structural work it is essential to clean construction joints thoroughly in order to obtain a good bond between the fresh and the hardened concrete.

Foundations

Methods of preparing foundation surfaces depend on the type of subsoil, and on the degree of bond required if the concrete is to be placed in contact with rock.

Clay, loam, hoggin and similar materials which have been consolidated by rolling or punning, are commonly used as a foundation for concrete. The surface should be clear of debris, mud and pools of water. In dry weather the sub-base should be thoroughly moistened. During winter, when contamination of the concrete by the earth is difficult to avoid, it is an advantage to lay a few inches of weak concrete and to allow it to set before placing the dense concrete.

Porous materials such as sand, gravel, and ash, should be thoroughly consolidated and covered with waterproof paper, polythene sheeting, or similar material to prevent absorption of water from the concrete.

When it is necessary to obtain a good bond on to a rock foundation, as for instance in the construction of dams, the rock surface should be well roughened, and thoroughly cleaned of all loose rock, dirt, soft pockets and seams, and other foreign material. Cleaning is accomplished by air and water jets, sand blasting, etc. and thorough washing with water and stiff brooms. Immediately before placing, a layer of mortar $\frac{3}{4}$ in. thick and of similar mix and consistency to that contained in the concrete to follow, should be thoroughly worked with stiff brooms into all irregularities in the surface.

Construction joints

A construction joint is a concrete surface which has hardened because of a limitation or delay in construction so that fresh concrete cannot be integrally incorporated with it.

At construction joints no allowance is usually made in the design of the structure for subsequent movement to occur.

Under laboratory conditions it has been shown that fresh concrete can be bonded to hardened concrete in such a way that the strength of the joint can be as good as the strength of the concrete itself.[10.1] Laboratory tests, however, do not necessarily bear any relationship to work done on site under completely different conditions, where the nature of joint preparation and the efficiency of the completed joint is likely to be affected by:

(*a*) The size of the joint.
(*b*) The presence of reinforcement and water bars.
(*c*) Whether the joint is horizontal or vertical.
(*d*) The degree of restraint or freedom of the members on each side of the joint.
(*e*) Workmanship.

(*a*) *Size of the joint.* The specified joint preparation is more likely to be achieved on sections up to about 12 in. square, such as in columns and beams, than with sections of walls, where in the case of both vertical and horizontal joints the section may be 12 in. by 20 ft or even greater. Placing of the fresh concrete at horizontal joints in columns and walls where the height of pour may be up to 12–20 ft calls for the exercise of great care in both placing and compaction.

(*b*) *The presence of reinforcement and water bars.* The preparation of hardened concrete is made more difficult by the presence of projecting reinforcement, and particularly when water bars are present, especially if the preparation is a manual one. Furthermore, the placing of the fresh concrete is likely to make it difficult for full and adequate compaction to be achieved because of the possible restriction of the reinforcement on the flow of concrete. The degree of compaction of the fresh concrete around water bars has to be of a higher standard and quality than may be acceptable for joints without water bars, in order to prevent the formation of paths of weakness along which the water or contained liquid could flow.

(*c*) *Horizontal and vertical joints.* The placing of fresh concrete at a vertical joint requires a different technique from that at a horizontal joint. For example, some specifications require a thin coat of sand/cement to be applied to the surface of the prepared hardened concrete before the fresh concrete is placed; this can, under some circumstances, be achieved with horizontal joints but is not practical for vertical joints having cross-sectional areas exceeding, say, 12 in. by 24 in. high. In addition, the placing of mortars on hardened concretes on site is made more difficult and sometimes impossible by the fact that the formwork has to be erected initially. Placing mortar along a horizontal joint at the bottom of shutters only 6 ft high is almost an impossibility, and for vertical joints must undoubtedly be so.

(*d*) *The degree of restraint.* The effectiveness of a construction joint is affected by the tendency of the jointed parts of concrete to move relative to each other. Stresses, usually tensile, at a joint are likely to arise due to temperature changes during the setting and hardening of the concrete and to subsequent thermal or shrinkage movements. The magnitude of such stresses will be a function of the restraint, or lack of restraint, of the jointed parts. For example, if the two portions of a beam are jointed with the ends of the beam free to move on sliding supports, then it is unlikely that stresses due to thermal or shrinkage movements of any significant magnitude will be realised across the joint. If, on the other hand, the ends of the beams are fixed in position, tensile stresses will be set up at the joint, as well as in the concrete member.

There may also be a differential form of restraint in the case of a wall panel which is cast adjacent to and on top of hardened concrete. At the horizontal joint at the bottom, because of the frictional resistance, there will be less tendency for this concrete to shrink compared with the top surface. At vertical joints the vertical shrinkage of the fresh concrete is likely to be greater than that of the older hardened concrete and shear stresses may result.

(*e*) *Workmanship.* In the laboratory the preparation of the joint and the subsequent placing of concrete are likely to have been carried out by a technician working under close supervision. He is likely to have been interested in the work. Construction joints on site will inevitably be prepared by unskilled labour, usually disinterested, and with relatively far less supervision. The quality of workmanship and the care which is exercised in joint preparation and the subsequent placing of the concrete will affect the efficiency of the joint.

The form of treatment necessary at a construction joint will also be dependent on:

(*f*) Whether or not the finished appearance of the joint is important. Joints which are to be masked or hidden by finishings may not require the same attention to preparation as would be required in a joint which will be exposed.

(*g*) Whether a satisfactory bond between the concrete is required. There seems little point in attempting to achieve a high degree of bond in compression members, provided that the joint is capable of transmitting the compressive forces. Where exposed, the durability of the joint has to be taken into account.

The following points are those generally considered to be good practice in the preparation and placing of concrete at construction joints:

(*a*) All foreign material such as laitance, dirt, oil, wood chippings, shavings, and wire tie clippings must be removed.

(*b*) Roughening of the joint face is desirable. Wet sandblasting followed by air and water jetting is usually accepted as the best method of cleaning the surface. Wire brushing is effective if the concrete has only been placed a day or two, but bush-hammering or chiselling may be necessary on older concrete which has had time to become fully hardened. In either case it is necessary to remove all loose particles, preferably by washing if this does not cause damage or serious inconvenience to other work in progress below. An air jet provides a reasonable alternative.

(*c*) The face of the hardened concrete should be kept dry. Research has shown that this contributes towards the strongest joints.

(*d*) Opinions differ as to whether or not a layer of mortar applied to the surface of the hardened concrete is necessary. If used, the mortar layer should be well worked or scrubbed into the old surface and the first layer of the new concrete should be well tamped into the mortar, especially at the corners, to ensure a good contact.

(*e*) The best results are achieved by simplicity of construction and joggled joints are best avoided.

(*f*) Any cement mortar which has leaked under imperfectly fixed vertical stopboards should be chipped off as soon as possible after removing the formwork. If this is not done, it is liable to spall off later.

(*g*) The concrete engineer should make a final inspection just before the concrete is placed to make sure that all foreign substances have been removed, that the forms are correctly positioned and tight, and that the reinforcement is accurately and firmly fixed.

(*h*) Wet and badly graded mixes which readily segregate and "bleed" tend to produce excessive quantities of mortar and laitance at the top of

Fig. 10.1. Segregation resulting in a layer of mortar with very little coarse aggregate at the top of the pour

a lift, as illustrated in Fig. 10.1, and should be avoided. Examination of water-retaining structures has shown that leakage is as likely to occur through the porous layer of laitance at the top of a lift as at the joint itself. The amount of laitance is least when the concrete, especially in the uppermost portion of the lift, is the stiffest that can be compacted properly, and when the surface is worked as little as possible after placing.

Construction joints are normally the weakest part of the structure, and every effort should be made to ensure that a good joint is obtained. Many concrete buildings are disfigured by the careless preparation of construction joints or lift planes for further placing, especially if they are positioned indiscriminantly without regard to other architectural features. Examples of defective and badly placed construction joints are given in Figs. 10.2, 10.3 and 10.4.

Fig. 10.2. Concrete breakwater made with blastfurnace slag aggregate After 60 years' exposure to sea action cracking has occurred along the construction joints.[10.2]

Fig. 10.3. Severe blemishes at a construction joint caused by leakage and movement of the formwork away from the surface of the previous lift. In this case honeycombing at the joint has been caused by the leakage of grout

Fig. 10.4. Disfigurement caused by construction joints
The sparrow-pecked finish has not masked the joints.[2.16]

Position of construction joints

In external walls and columns, construction joints should be horizontal and arranged at levels to blend with the general architectural appearance. Construction joints can be masked by such devices as the lines of tiles on Twickenham Bridge (Fig. 10.5), and the formation of grooves or projections as in Figs. 10.6 and 10.7. Note that in the right-hand example in Fig. 10.6, the line of the joint is preferably near the top of the groove.

Vertical joints should be formed with properly constructed stop boards, preferably at positions where the length of joint is shortest or where a symmetrical pattern can be arranged. It is bad practice to allow the concrete to take a natural slope as has been done in Figs. 8.4 and 9.1.

It is not possible to give general rules for the position of construction joints covering the whole field of reinforced concrete construction. In columns, beams, or slabs, however, the following rules should be observed:

(1) Columns should be filled to a level, preferably a few inches below the junction of a beam, or alternatively to the bottom of any haunching.

(2) Joints in beams and slabs should be formed at the points of minimum shear. Thus for beams, the joint should be at the centre of span or within the middle third.

Fig. 10.5. Use of lines of red tiles to mask lift planes
The surface is bush-hammered gravel concrete.(2.23)

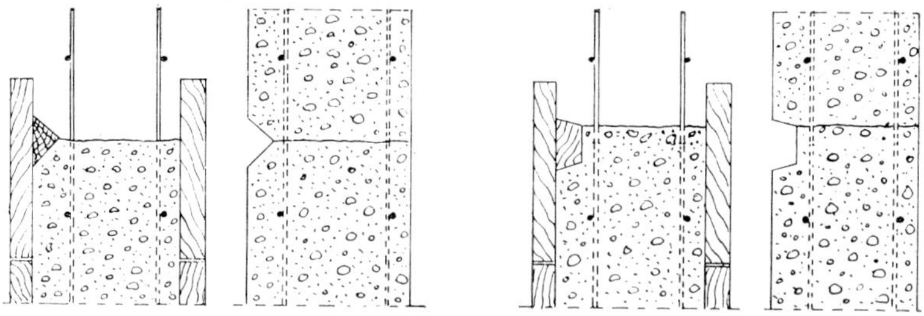

Fig. 10.6. Formation of grooves to mask construction joints

Fig. 10.7. Formwork joints deep set to give a panel effect for the in-situ wall of a church at Oslo

Slabs spanning in one direction and of short span should have the joints at middle span if they are at right angles to the direction of spanning. Joints in the direction in which the slabs are designed to span should be within the middle third. Joints in slabs spanning in two directions should be within the middle third on either span. The two cases are illustrated in Figs. 10.8 and 10.9.

(3) If for any reason a joint has to be made between the slab and the beam, it is then necessary to provide some form of key and to add shear reinforcement to meet any weakness at the joint.

Other problems are likely to arise regarding the positions of joints in the structure, and in any case of doubt reference should be made to the designer concerning these details. In general, joints should not be made in positions subject to large shear forces.

Movement joints in buildings

There is no well-established practice for providing movement joints in reinforced concrete structures. A common length between movement joints in a building is 100 ft, sufficient reinforcement being used intermediately to control cracking. Wall panels are sometimes constructed to fit into keyways in the vertical columns, leaving only the wall beams subject to movement. With the provision of enough temperature reinforcement to prevent the formation of visible cracks, the distances between movement joints have been in some cases extended to 200 or 300 ft.

Movement joints in buildings should completely separate the structure from top to bottom, and should preferably be made by means of a double column or beam. The joint should be waterproofed, the method depending on the water pressure to be resisted, if any. Some examples of movement joints are given in Figs. 10.10 and 10.11.

The thermal expansion of concrete roofs is a common cause of damage to buildings, usually by pushing of parapets and end walls as illustrated in Figs. 10.12, 10.13 and 10.14.

Cracking is also liable to occur, either through the plaster at ceiling level, or of partition walls near ceiling level, particularly near the ends of a long building.

Methods which may be adopted to prevent the defects indicated are:

(1) to allow the roof slab to slide on the supporting walls, with suitable breaking of the plaster finish at the junctions.

(2) the incorporation of suitably designed expansion joints at convenient intervals.

(3) in conjunction with (1) and (2), the use of a roof covering having a high thermal insulation and the whitening of the roof surface in order to reduce temperature movements.

Expansion and dummy joints in roads are described in Chapter 19.

Movement joints in water-retaining structures

All joints in a water-retaining structure, whether it be a dam, a reservoir, a water tower, or any other structure, are a potential source of leakage, and their design and construction require special attention. They have to accommodate repeated movement of the structure resulting from

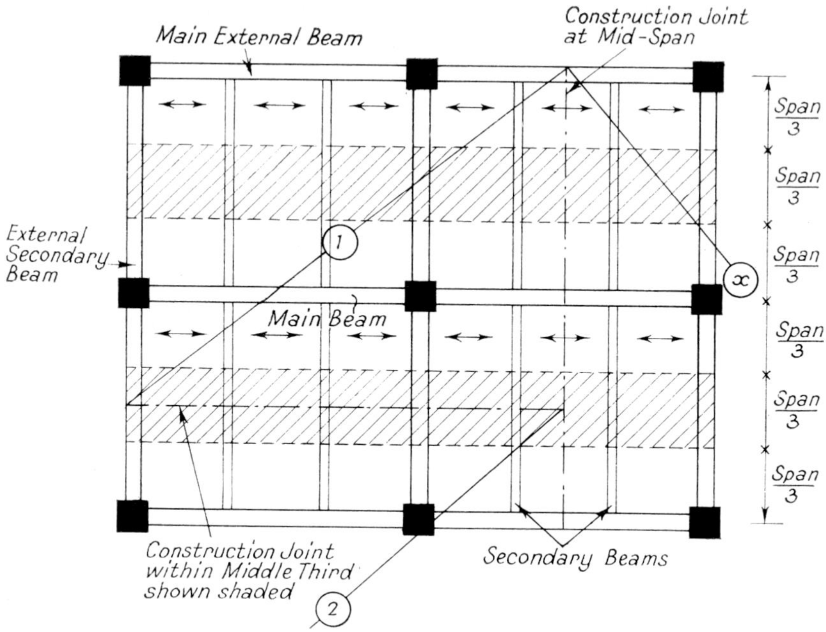

Fig. 10.8. Construction joints in beams and slabs designed to span in one direction

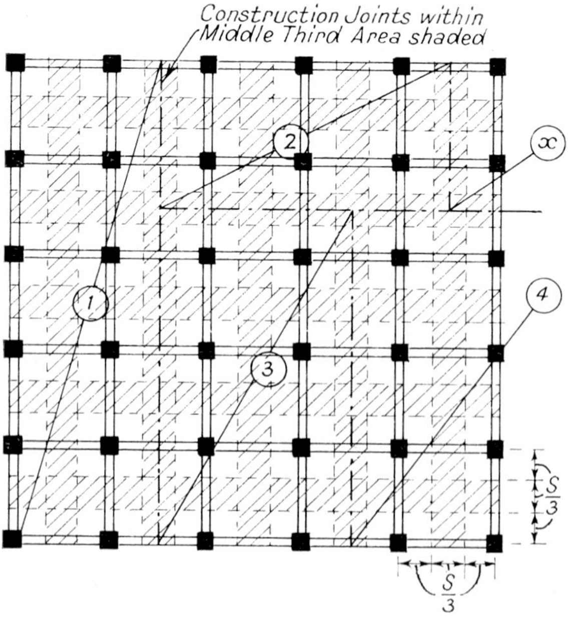

Fig. 10.9. Construction joints in slabs spanning in two directions

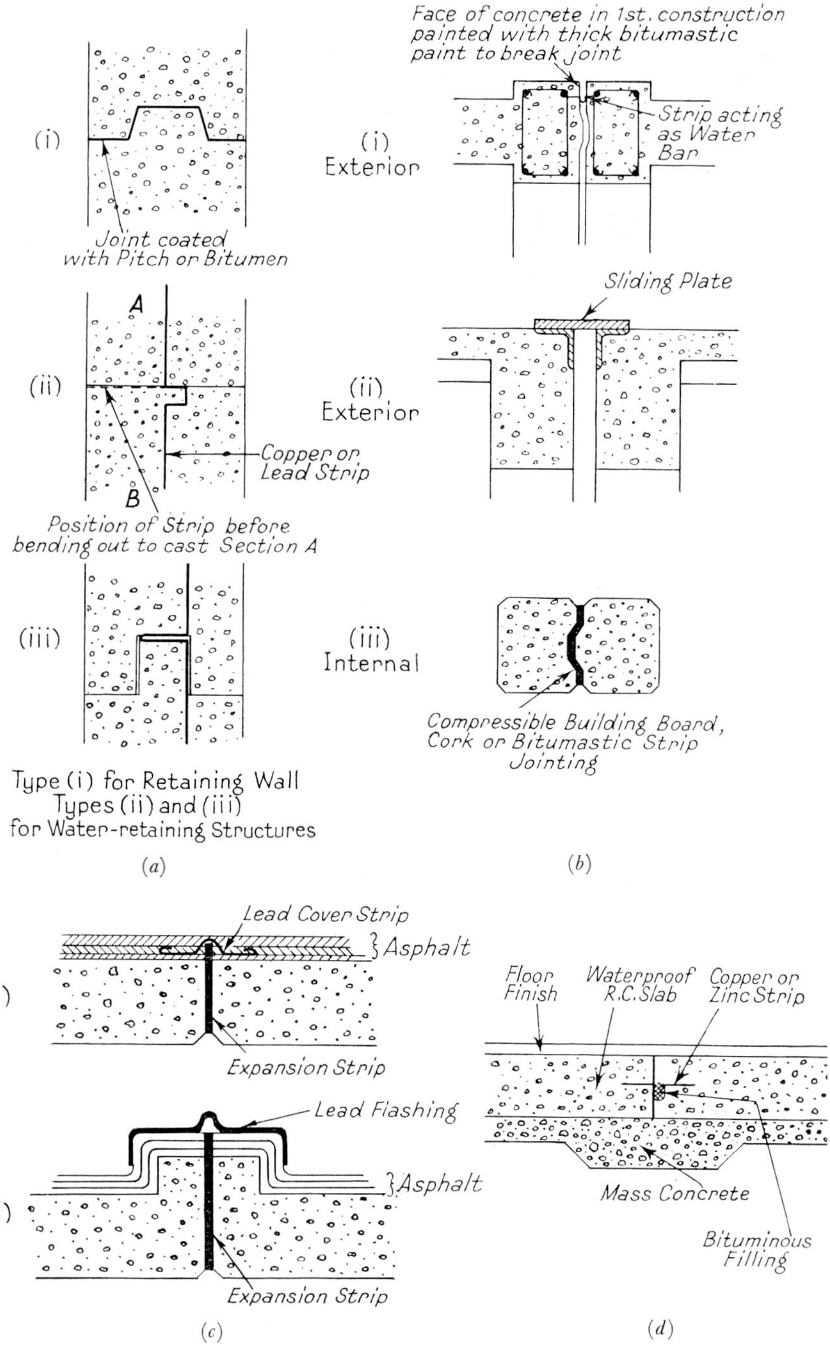

Fig. 10.10. Examples of formation of movement joints
(*a*) Joints in walls, (*b*) Joints at columns, (*c*) Joints in roofs, (*d*) Contraction joint in floor to resist ingress of ground-water under pressure.
P.V.C. or rubber mouldings may be used instead of metal strips—see also Fig. 10.11.

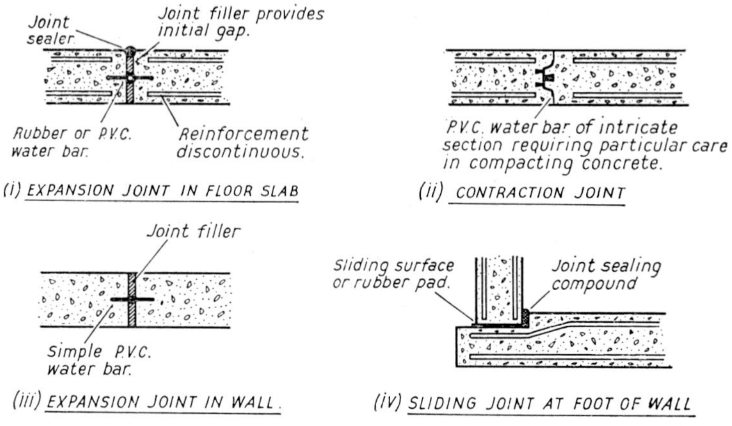

Fig. 10.11. Further examples of movement joints, more particularly suited for water-retaining structures

changes in temperature or moisture conditions, without loss of watertightness. At the same time the joint must provide for the exclusion of grit and debris. This is achieved by the use of a joint sealing compound, which is an impermeable ductile material capable, ideally, of adhering to wet concrete throughout the range of joint movement. This latter requirement is difficult to achieve, so that the use of a sealing compound alone should not normally be relied upon to give watertightness.

In order to achieve watertightness water bars of the type shown in Fig. 10.11 are nearly always incorporated at movement joints and frequently at construction joints. Other arrangements and a more detailed description of the design and construction of movement joints for water-retaining structures are given in the Code of Practice for Water-retaining

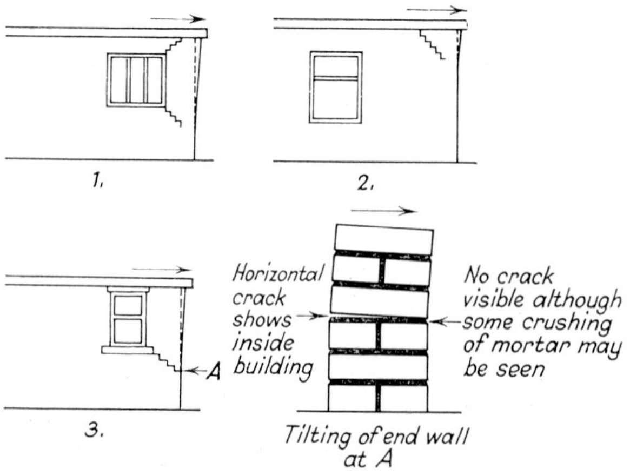

Fig. 10.12. Cracking of end walls of buildings due to the expansion of the reinforced concrete roof

Fig. 10.13. Cracking of brickwork at the end of a long concrete beam due to thermal movements. An expansion joint should have been provided at the end of the beam, and provision made for sliding of the beam on its brickwork seating

Fig. 10.14. Cracking in brickwork due to movement in reinforced concrete slab above

Structures.(10.3) Whichever type of water bar is used, great care is necessary to ensure that it is continuous, with carefully made joints at all junctions. The surrounding concrete should be compacted very thoroughly round the joint. Any failure to achieve good contact with the water-bar may result in leakage, which will be difficult both to locate and to cure.

REFERENCES

10.1 DAVEY, N., *Construction Joints in Concrete*, Building Research Special Report No. 16, 1930. H.M.S.O.

10.2 DAVEY, N., Concrete: Its Appearance and Durability, *J.R.I.B.A.*, January, 1945.

10.3 Civil Engineering Code of Practice, C.P. 2007:1960, *Design and Construction of Reinforced and Prestressed Concrete Structures for the Storage of Water and other Aqueous Liquids*. British Standards Institution.

CHAPTER II

Placing Concrete

Placing concrete

SERIOUS defects in the finished structure may result from the omission of certain precautions in placing concrete in the formwork. There is a temptation in the interests of economy both in equipment and operation to drop the concrete from the point to which it has been transported irrespective of the height above the forms. Such a procedure may lead to displacement of the reinforcement, to damage of the concrete already placed, and damage to the formwork, as well as causing segregation and the entrainment of air bubbles. The free fall of concrete into the forms should be reduced to a minimum unless vertical pipes or suitably designed drop chutes of rubberized or other flexible material are provided.

An example of segregation resulting from the free fall of concrete from a barrow into a large column base is given in Fig. 11.1. The stones segregated out of the mix, and rolled down the cone which formed

Fig. 11.1. Honeycombing on a large column base. Placing was by tipping from the top, a cone being formed in the centre down which the stones rolled to the face

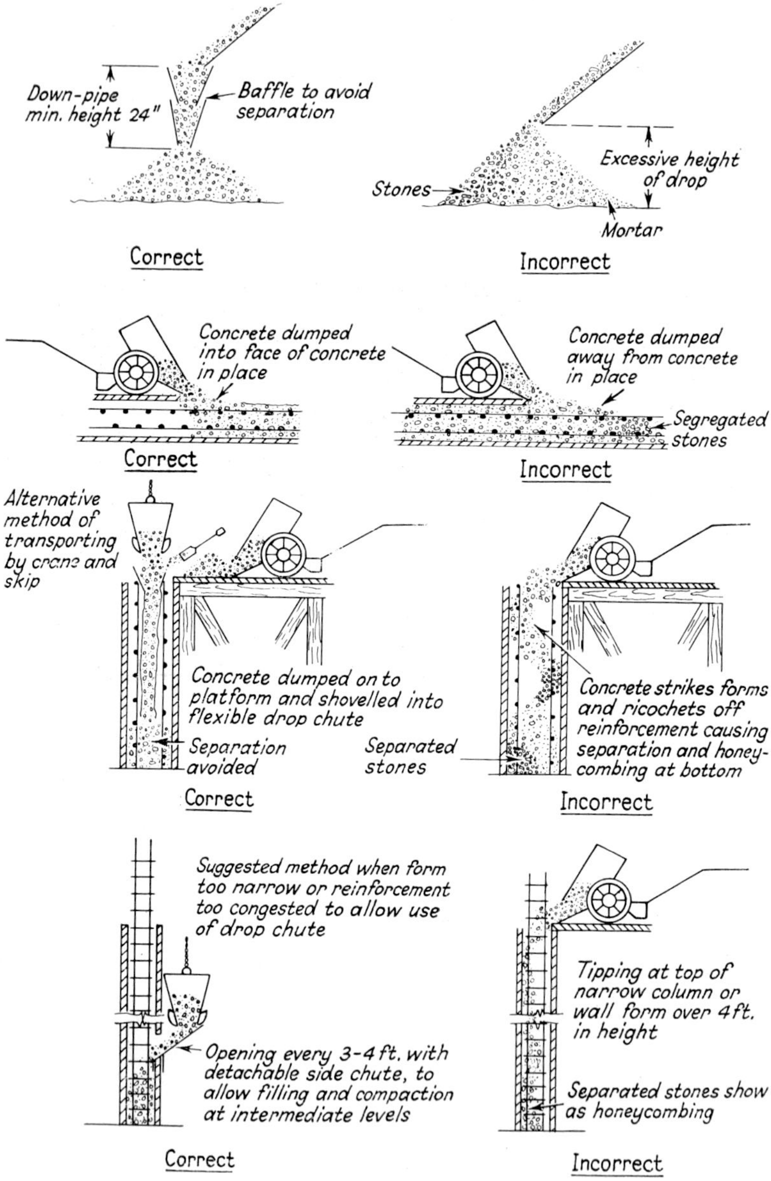

Fig. 11.2. Methods of handling mixed concrete

to the side forms. The cone formation can be traced in the honeycomb pattern.

Suggested details for controlling the fall of concrete are given in Fig. 11.2, which is based partly on the recommended practice published by the American Concrete Institute,(11.1) and partly on the author's own experience.

Among other precautions which should be taken during placing, the following should receive special attention:

(*a*) The concrete should be deposited as near as practicable to its final position and should not be deposited in a large quantity at any point and allowed to flow along the forms. Failure to do so is apt to cause segregation, honeycombing, sloping pour planes and poor compaction. Segregation resulting from the flow of concrete along the forms is shown in Fig. 11.3.

Fig. 11.3. Honeycombing resulting from segregation. Concrete was allowed to flow along the shuttering, passing over a stop end on the left. This is part of a large retaining wall in an exposed whinstone finish, the blemishes shown being repeated at several points along the length

(*b*) Concrete should be deposited in horizontal layers, and each layer should be compacted thoroughly before the succeeding layer is placed. As far as practicable each layer should be placed in one continuous operation, the thickness depending on the size and shape of the section, the consistency of the concrete, spacing of the reinforcement, method of compaction and the necessity that the next layer shall be placed before the previous one has hardened. In reinforced concrete work it is good practice to place the concrete in layers from 6 in. to 12 in. thick. In mass concrete thicker layers from 15 in. to 18 in. thick are commonly used. Several such layers may be placed in succession to form one lift, provided they follow one another quickly enough to avoid "cold joints." With care and well designed shuttering, concrete can be placed to heights of 15 ft or more in one lift, but there must be adequate space within the shuttering and around the reinforcement to allow access for compacting equipment.

(*c*) Concreting should be carried on continuously in order to avoid the appearance of an unsightly lift line on the finished structure.

(*d*) The concrete should be worked throughly into position around the reinforcement and bedded fixtures, and into the corners of the formwork.

(*e*) Full advantage should be taken of the possibility of using drier concrete in the upper portions of the lift in order to reduce and compensate for water gain, thus improving the uniformity in appearance of the concrete.

Placing in thin walls and columns

The space available between the reinforcement in thin walls and columns is sometimes insufficient to allow the insertion of any drop chutes to break the fall of the concrete. Further, visibility is often limited to the top 3 or 4 ft. of the form, so that compaction of concrete from the top of a form of greater depth would have to be done blindly. Under such circumstances it is advisable either to arrange for one side of the formwork to be made up in panels, say, 4 ft high, which can be erected one after the other, as the concrete is placed, or to leave gaps in the formwork at every 4 ft in height through which placing and compaction can be satisfactorily carried out. With suitable design it is possible to insert panels in these gaps and to wedge them in place, ready for the placing of further concrete in a matter of a few minutes.

Openings

If a window or door opening occurs in a wall to be concreted, a difficulty arises at the top of the opening, in that if concreting is carried on continuously the subsequent settlement of the wet concrete before hardening causes a crack to form at the corners. The risk of cracking is greatest when wet concrete is used. This trouble can largely be overcome by stopping concreting at the level of the top of the opening, and allowing at least four or five hours for settlement and stiffening to take place before continuing to the top of the formwork. If it is convenient to form a construction joint at the level of the top of the opening, the trouble is eliminated altogether.

Placing in wet conditions

Concrete has sometimes to be placed in positions into which water is continually seeping, as in deep trenches. In such cases it is usually best to dry up small areas and to place the mortar coat and quickly follow it with the concrete. In other instances it may be advisable to insert lengths of pipe at suitable points to allow the water to drain away.

Placing should be discontinued when heavy rain causes water to collect in pools, or washes the surface of the fresh concrete, and shelter cannot be adequately provided. It is advisable during wet weather to maintain a slope on any large area of fresh concrete, sufficient to result in self drainage of the rain water. Before continuing placing any puddles should be dried up. In road work, placing during heavy rain causes serious damage to the finished surface.

Placing under water

Concrete may be placed under water either by the use of bottom dump buckets such as those seen in Fig. 11.4, or by means of tremies. If buckets are used they must not be discharged from the bottom until

Fig. 11.4. Underwater concreting by means of bottom dump buckets

contact has been made with the foundation or surface of concrete already placed. A certain amount of washing is liable to occur.

A tremie is a pipe having a funnel-shaped upper end into which the concrete is fed. The bottom or discharge end must be kept continually below the surface of the soft concrete so that there is as little disturbance as possible. Pouring of the concrete proceeds continuously, the pipe being raised slowly as the depth of the concrete deposited increases. If the seal is inadvertently broken at the bottom the succeeding batches should be made richer in order to compensate for cement lost in the water. In general, it is necessary to use a highly workable concrete in order that it can flow laterally from the tremie pipe over the horizontal distance necessary to reach the sides of the forms with a reasonably flat and smooth upper surface. The less disturbance to the concrete once it has flowed into place the better. Puddling of the surface should be avoided.

In underwater work it is often necessary to place the concrete on a rock surface and it is then essential to ensure that the tremie concrete does not leak away from under the forms. An effective method of doing so is to attach a long strip of canvas to the form near its lower edge and weight this down with broken stone in bags. Mud and silt should be removed by pumping.

The success of a tremie operation depends on securing a watertight seal at the bottom of the tremie pipe so that the concrete is not subject to washing at the commencement of a pour. One method of sealing tremie pipes is to insert a "go-devil" consisting of a bundle of hessian, into the mouth of the tremie after it has been suspended in the form in the position in which it is to be used. This "go-devil" is forced down the tremie by feeding concrete into the top of the pipe, and in doing so forces the water and air out at the bottom. By the time the "go-devil" is expelled from the bottom of the pipe the column of concrete following it is descending quickly and rushes out between the bottom of the tremie and the foundation to form a mound round the bottom of the tremie pipe. The flow of concrete is then checked by lowering the pipe to the bottom while more concrete is fed into the top. The pipe is then raised just sufficiently from the bottom to permit a continuous flow of concrete.

The use of tremies presents a problem when frequent removal is necessary as the work proceeds from one point to another, since in the situations in which tremies are used there are usually congested conditions due to bracings, reinforcement, etc. otherwise tremies are preferable to bottom dump buckets.

Rich mixes of about 1 : 1 : 2 to 1 : 1½ : 3 of cement : sand : coarse aggregate are necessary, whether placing underwater by bottom dump buckets or by tremie.

Cement injection into placed aggregate

A method sometimes advocated for concreting consists of placing aggregate of minimum size ¾ in., though often not less than 1½ in., in the formwork, and then injecting cement grout through vertical steel pipes arranged in the aggregate with their lower ends near the bottom of the mass to be concreted. The grout is pumped in and rises steadily through the aggregate. The grout usually contains a plasticiser or air-entraining agent to improve the flow properties, and when mixed in a high-speed mixer is usually described as "colloidal grout." When injected into air voids, the grout flows fairly freely, but under water the difference in specific gravity between the grout and water are insufficient to ensure a horizontal flow of more than a few feet. The angle of slope of the grout surface under water may be expected to be at least 30°.

Such concrete is not always free from honeycombing and the high water/cement ratio necessary to ensure a free flowing grout is not conducive to high strength and impermeability. In underwater concreting the method is also open to objection that mud and silt tend to settle in the interstices of the aggregate before it is grouted.

The method has apparently had some success when used in the dry. A number of failures, including that of a bridge foundation[11.2] in America, have, however, been reported when grouted concrete has been attempted under water.

REFERENCES

11.1 A.C.I. 614–58, *Recommended Practice for Measuring, Mixing and Placing Concrete.* American Concrete Institute Committee 614, A.C.I. Book of Standards, 1965.

11.2 THORNLEY, J. H., Building a foundation through a foundation, *Engineering News-Record*, Aug. 28, 1958.

CHAPTER 12

Compaction of Concrete

THE object of the compaction of concrete is to eradicate air holes and to achieve maximum density. Compaction also ensures an intimate contact between the concrete and the surfaces of reinforcing steel and embedded parts.

In order to obtain maximum density it is necessary to use a mix which is of adequate workability to enable the operator to place it in position without difficulty with the equipment he has available. On the other hand, it is important that the mix shall not be too wet, as this is liable to result in segregation, excessive laitance at the top of the pour, weakness, and a lower density due to the volume occupied by the excess water. An indication of the loss of strength due to the presence of voids is given in Fig. 12.1. These voids include both the volume of entrapped air, and voids formed by the drying out of excess water, so that it is apparent that a compromise must be reached, depending on the conditions of placing, between the provision of adequate workability and the necessity for reducing the water/cement ratio as much as possible as discussed in Chapter 6.

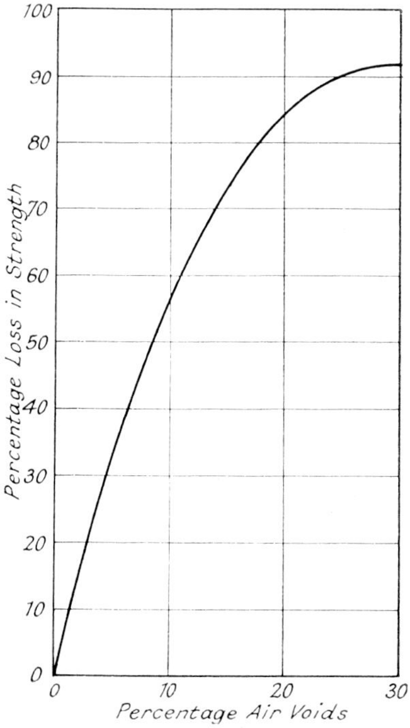

Fig. 12.1. Relationship between loss of strength and air-void space

When considering the method of compaction to be used for any particular concrete structure or member, it is necessary to decide whether to use a workable mix with hand compaction, or a stiffer mix with vibration. In general, it is found that a better surface with less blow holes is obtained when using the more workable mix with hand compaction, but it is necessary to increase the cement content in order to offset the increased water/cement ratio necessary to achieve workability. The cost of this increased cement content is not likely to be of importance in structural members, the cross-sections of which are usually relatively small, as compared with shuttering and other costs.

The cost of handling and placing stiff concrete, together with the cost of vibrating equipment and the heavier formwork necessary when it is used, cannot be overlooked, and it may well be found that on works where the value of the concrete is relatively small, that hand compaction will prove more economical, whereas in mass concrete and roadwork, the saving in cement possible with a stiffer mix, more than offsets the cost of handling, placing and vibrating.

When compacting concrete it is important that the reinforcement should not be disturbed and the formwork should not be damaged or displaced. Care must be taken, however, to ensure that the concrete is worked thoroughly in the neighbourhood of the formwork so that the finished surface will be even, dense, and free from honeycombing and pronounced blowholes.

Hand compaction

Ordinary hand methods of compaction consist of rodding, tamping, and spading with suitable tools. During the construction of floor slabs, the tramping of the workmen also plays an important part. For road work, a heavy timber beam prepared to the correct camber and spanning from form to form is used for compaction. Hand methods of rodding and tamping necessitate the use of a fairly wet mix if the sections are at all narrow and the reinforcement closely spaced. A slump of 4 in. to 7 in. may be necessary if the reinforcement is congested.

Compaction by vibration

Although the use of vibrators has extended into almost every class of concrete work, compaction by hand, properly carried out, gives satisfactory results for many purposes, and because of the simple equipment is often more economical. Vibration makes it possible to use drier mixes, resulting in increased strength and lower drying shrinkage for given mix proportions. Even for heavily reinforced sections the slump should not be more than 4 in. Suggested values for the consistency for various purposes have already been given in Chapter 7. The stiffness of mixes is limited by the characteristics of much of the equipment for mixing, handling, placing and finishing the concrete. Difficulties are often encountered in:

(1) Discharging very stiff concrete from mixers.

(2) Discharging stiff concrete from lorries, chutes and other means of transport.

(3) The lack of adequate vibration at points not immediately in contact with the vibrating equipment, resulting in segregation.

(4) The finishing of horizontal surfaces as, for instance, in road work to a uniform texture free from ridges and stony patches.

As the stiffness of the mix increases, so also does the cost of the various operations from mixing to finishing; the consistency should, therefore, be carefully adjusted to secure a proper balance between the numerous factors pertaining to the particular job.

Fig. 12.2. Pumped concrete being vibrated by poker type vibrators. The concrete is too wet to be vibrated, and would be better compacted by hand

Different types of vibrator require different consistencies of concrete for the most efficient compaction, and the characteristics of the machine available can be the governing factor in determining the consistency of the mix to be produced on the site. A joint sub-committee of the Institutions of Civil and Structural Engineers,[12.1] appointed to consider the effect of vibration on concrete, concluded that the process in practice is little affected by the characteristics of the vibration other than its energy. It follows that the vibratory machines in common use giving 3,000 vibrations per minute and an acceleration of $4g$ or more in the concrete are quite satisfactory.

Concrete can be too wet for effective compaction by vibration, as well as too dry; an example of the use of internal vibrators to compact concrete pumped to the site, and which would have been better compacted by hand, is shown in Fig. 12.2.

Vibrating machines are usually operated by petrol engines, compressed air or electricity, and those suitable for site use are of three main types, namely:

(1) Vibrators, used internally.
(2) Vibrators which are attached to the formwork.
(3) Surface vibrators.

Where their use is practicable, internal vibrators are more effective than the other types. Considerable care is, however, necessary with all types to avoid segregation and non-uniform compaction, and the following precautions should be observed:

(*a*) All forms should be as tight as possible to avoid leakage of mortar and consequent honeycombing. The use of tongued and grooved boarding, or a lining of plywood may be justified to prevent leakage, if the surface finish is important. The formwork should also be rigidly braced to prevent displacement, and all wedges and distance pieces nailed. With steel forms, adequate bracing is equally necessary.

(*b*) In order to avoid air being trapped, the lift should be as shallow as practicable, although not less than 6 in. The danger of trapping air increases with depth, and a 2 ft layer is about the maximum advisable.

(*c*) Vibration should not be prolonged in any one position; such a procedure produces honeycombing and lack of homogeneity.

(*d*) Care should be taken to avoid overworking and thus drawing to the surface an excess of mortar which may later craze, or scale

Compressed air vibrators have certain advantages and disadvantages compared with electric vibrators. The advantages are that the weight handled by the operator is usually less, and there are no electrical troubles. On the other hand, the air cylinders are apt to freeze in cold weather due to the sudden release of air pressure. This can be prevented by the use of dry air from a larger receiver, by trickling an "anti-freeze" agent or thin oil into the air line, or by arranging for the air to pass through a coil of pipe heated by a brazier or fire devil. For comparable service, air compressors are more cumbersome, and expensive than electric generating units, but may be of value for other services.

Internal vibrators

Internal vibrators, sometimes referred to as poker or immersion vibrators, have a higher efficiency than other types, since all the energy is transmitted directly to the concrete. They are also easier to manipulate and, being portable, can be readily used in difficult positions.

Internal vibrators usually consist of blades or tubes with or without rigid handles. In the rigid type of machine, the vibrator blade is connected rigidly to the air turbine or electric motor by a steel tube. The vibration is produced by an eccentric shaft enclosed in the tube and vibrator head. In the flexible type, the eccentric shaft which produces the vibratory effects is within the head or blade. This eccentric shaft is driven through a flexible drive from an electric motor, air turbine, or small petrol engine. The choice depends on site conditions. Petrol engines are not always easy to start, and failures from damage or inadequate maintenance are not infrequent. Electric motors provide a constant speed and are conveniently portable, but unless a reliable supply of electricity is available require a generator with its attendant maintenance costs. Compressed air motors are relatively inefficient, requiring considerable quantities of air, trouble is experienced due to icing in cold weather, and air compressors are expensive to maintain.

Vibrators of this type, driven by electric motors and petrol engines respectively, are shown in use in Figs. 12.3 and 12.4.

Internal vibrators are now available with heads ranging from 1 in. to 4 in. diameter. The 1 in. diameter head has been developed for use in small, heavily reinforced, sections.

Fig. 12.3. Use of electrically driven internal vibrator in wall shuttering

Internal vibrators in common use give 3,000 vibrations or more per minute with an acceleration of at least $4g$ in the concrete and are quite satisfactory. Machines with frequencies of 5,000 to 10,000 vibrations per minute are available, although some types are not entirely satisfactory due to the occurrence of mechanical troubles at the high speeds involved. Higher frequencies are generally not worth while, since they are attained only by loss in mass or amplitude and are not more effective in compacting concrete. Mechanical troubles also increase.

With all internal vibrators it is necessary to keep the vibrating head in the concrete while running in order to keep the bearings cool and avoid breakdowns. Flexible drives should not be bent to a small radius, since this causes excessive wear.

The amount of concrete which can be compacted by a vibrator is

Fig. 12.4. Vibration of a concrete retaining wall by the use of poker type internal vibrators. Concrete was brought to the wall by means of the bridge and deposited on the sheeting in the middle of the picture, from which it was placed in the wall by hand. The metal formwork was arranged in panels

dependent upon the size and horse-power of the appliance and on the workability of the concrete. With $1\frac{1}{2}$ in. slump concrete, a 2 in diameter immersion vibrator will compact about 8 cu yds of concrete an hour.

Internal vibrators should generally be inserted vertically or nearly so, at points 18 in. to 30 in. apart. They should then be withdrawn slowly, at about 3 in. per second.[12.1] German specifications are similar and give a speed of movement for the vibrator of 8 cm per second, at which the best results should be obtained. Vibrators should not be used to push concrete laterally in the forms, as this is likely to lead to segregation. Fig. 12.5 illustrates the way in which separation occurs. The vibrator should be at least 4 ft from the leading edge.

As a general rule the vibrator should be used not nearer than 4 in. to the form surface in order to obtain a uniform appearance. If the vibrator touches the form surface, a sand streak is very apt to occur;

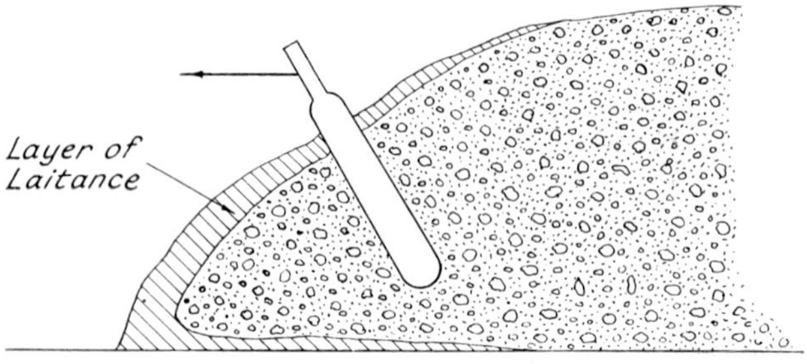

Fig. 12.5. Effect of using internal vibrator to push concrete along form

also the surface of the formwork may be damaged. Immersion for periods of 5 to 15 seconds should be sufficient, but the amount of vibration should be judged by the appearance of cement paste next to the forms, and by the change in sound. Fig. 12.6 shows cement paste working up round a vibrator, which has been held over-long in one position. The vibrator should have been withdrawn as soon as the paste began to show, and reinserted a little farther along. A further indication of sufficient vibration is the cessation in the appearance of air bubbles at the surface

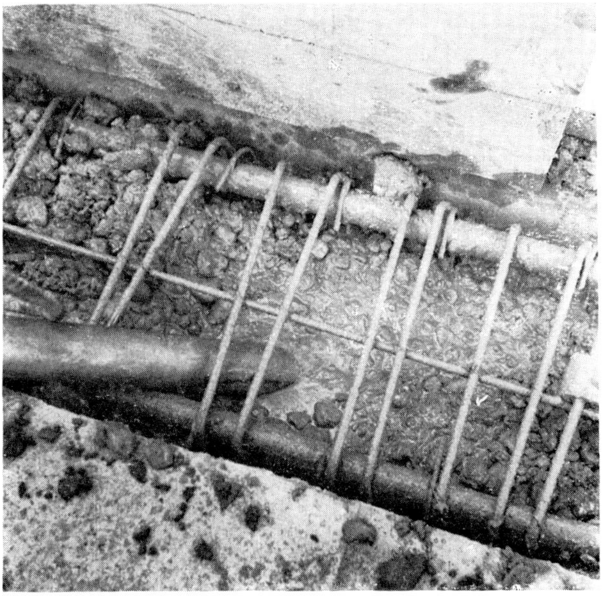

Fig. 12.6. Cement paste beginning to rise round an internal vibrator

as the concrete is being vibrated. The vibrators should be immersed to the full depth of the fresh concrete, and preferably for an inch or two into the previous layer to eliminate any tendency to form a work plane.

Over-vibration should be avoided owing to the danger of segregation and streakiness.

Blowholes on the surface of the concrete are, if anything, more troublesome when vibrators are used than when hand tamping is adopted. No simple procedure is known for eliminating blow-holes formed at the surface, apart from the use of absorptive form linings or vacuum forms. Judicious working of the concrete at the surface of the forms with a thin blade may, however, give some reduction in the amount of pitting. Shutters faced with plywood produce surfaces less pitted than if metal shutters are used. Absorptive form linings have been used to an increasing extent in the U.S.A. as a means of preventing blow-holes and otherwise improving surface texture. A few trials have been made in this country with this type of lining and there appears little doubt that it is of considerable benefit. Unfortunately, absorptive form linings can only be used once or twice and are therefore expensive. The types so far tried have been mostly insulation board and similar porous material, sometimes with a covering of hessian or muslin. Recent work by the Cement and Concrete Association has shown that the number and size of blowholes can be reduced considerably by selecting the appropriate type of mould oil for the type of formwork in use. This is referred to in greater detail in Chapter 16.

Form vibrators

Form vibrators, or external vibrators, are usually rigidly attached to the forms by means of a clamp or vice and cause a shaking motion of the form through which the pulsations are distributed to the concrete. Form vibration is usually adopted when it is impossible to insert an immersion vibrator, such as into heavily reinforced or small and narrow section members, but since a considerable amount of energy is absorbed by the formwork, more power is required in comparison with internal vibrators. The formwork must be sufficiently strong and rigid to resist the oscillatory action, and especially tight to prevent loss of grout. In their general assembly, timber forms may need screwed connections where nails would normally be used.

Several form vibrators may be seen in Fig. 12.7. They consist of electric

Fig. 12.7. Electric vibrators attached to steel formwork. As the height of concrete increases the vibrators will be moved up to the next clips

motors with unbalanced members, pneumatic hammers or electromagnetic pulsators.

When using external vibrators on the forms, it is necessary to place the concrete in layers no more than a few inches deep to provide a uniform build-up. Air holes are more apt to occur with form vibrators than with internal vibrators, particularly near the top of a lift. It is wise to compact the last 2 ft by hand, or possibly by the use of an internal vibrator, if space allows.

Fig. 12.8. Vibrating table for precast concrete units

Portable types of electric and pneumatic hammers are also used to some extent, being held against the forms at selected points. They are of particular value for the compaction of portions of concrete inaccessible for internal vibration.

Vibrating tables are in effect a type of form vibrator, but with the advantage that the whole mould is vibrated uniformly. They may be operated electrically or pneumatically, and are widely used for the manufacture of precast concrete products. A vibrating table of this type is shown in Fig. 12.8. Other types of vibrating table have been developed which compact the concrete by means of a series of jolts rather than by vibration proper. The compaction obtained by this method is quite good.

Surface vibrators

Surface vibrators consist essentially of a flat horizontal plate or pan on which is mounted one or more electric or pneumatic vibrating units of similar type to those used for attachment to the formwork. They are used for compacting concrete in large structures such as dams, retaining walls, and bridge abutments, where the surface area is extensive.

In such instances they usually follow internal vibrators and are used to embed any excess of coarse aggregate and to provide the desired finish.

In general, surface vibrators are not effective beyond a depth of 8–12 in. depending on the type used.

Vacuum process

In the vacuum process of compaction, special moulds are used incorporating a processing mat which allows the suction of water from the concrete by a vacuum pump working through a water separator. The two main advantages of the vacuum process are the ability of the concrete to stand up immediately after processing, and the increase in crushing strength which is obtained. Also the shrinkage of the concrete is reduced and the quality of the surface finish is generally improved. The requirements for the concrete to be used are not very exacting, but sand with a high proportion of dust or fine particles should not be used, as it is apt to seal the surface of the concrete too soon against the removal of water; also very lean mixes are unsuitable. In general, proportions of cement to aggregate of 1:6 or less, are most suitable.

Vacuum processing should be considered as a special process of limited practical and economical application. The depth of concrete which can be processed from one face is limited to about 6 in., and the amount of water withdrawn varies with the distance from the face and to some extent with the mix proportions and aggregate grading. However, the process has been used in a variety of circumstances, including the precasting of floor and wall panels, large diameter pipes, and shell roof sections, and the casting in situ of structural floors, canal linings and the concrete facing to rock-filled dams and bund walls. It is in the last two of these applications, where a steeply sloping surface has to be

Fig. 12.9. The vacuum process being used for the concrete facing to a rock-filled bund wall. The panel at the bottom of the slope is being processed while the concrete is being placed behind the next

concreted, that the vacuum process offers particular advantages. On steep slopes it is difficult to obtain well-compacted concrete with a good finish without forming the upper face. In these circumstances, vacuum forms, in spite of being in themselves much more expensive per unit area than ordinary forms, can show an economy if the size of the job warrants the special equipment and experienced personnel required; this is because they can be struck and re-used easily and quickly. In the example illustrated in Fig. 12.9 each form, 16 ft long and 2 ft wide, was used many times each day. Details of the formwork construction are explained on p. 247.

Workmanship

Skilled workmen are essential to the production of homogeneous and well-compacted concrete. In large measure the correct placing and tamping of the concrete is dependent on their knowledge and experience, and it is particularly important that in teaching them, every effort should be made to obtain their close co-operation.

Vibration by unskilled operators is probably more dangerous than hand tamping, since there is a risk of either under-vibrating or over-vibrating the concrete. This may happen in small areas, as, for instance, if the vibrator is left in one position too long through ignorance or diversion of attention on the part of the operator.

In small road works the period of vibration and the method of handling the compacting beam materially affect both the compaction of the lower part of the slab and the finish obtained on the surface. With the larger vibrating and finishing machines the operator must learn to set the front screed at the proper height in order to leave the correct amount of concrete for the vibrator, and the vibrator must be set at the correct level in order that the finished concrete shall not be high or low. He must also learn at what forward speed the machine gives satisfactory consolidation and a close even finish.

Vibrating equipment generally must be kept clean and free from coatings of set concrete. Frequent servicing and attention to the mechanical parts is also necessary if the machines are to work efficiently.

REFERENCE

12.1 *The Vibration of Concrete*. Report by the Joint Committee of the Institutions of Civil and Structural Engineers, 1956. Further Report of a sub-committee, see *J. Inst. Civ. Eng.*, 1967.

CHAPTER 13

Curing Concrete

THE chemical action which results in the setting and hardening of concrete is dependent on the presence of water, and although there is normally an adequate quantity for full hydration at the time of mixing, it is necessary to ensure that the water is either retained or replenished to enable the chemical action to continue. Loss of water due to evaporation may cause the hydration process to stop, with a consequent reduced strength development. In addition, evaporation can cause early and rapid drying shrinkage, resulting in tensile stresses which are likely to cause cracking unless the concrete has achieved sufficient strength to withstand these stresses. Methods of curing are therefore designed to maintain the concrete in a continuously moist condition over a period of several days or even weeks either by preventing evaporation and the absorption of water by the forms or subgrade, or by repeatedly wetting the surface.

It is usually required that moist curing shall be carried on for a certain number of days from the time of placing, without any definite reference to the time at which curing shall commence or to the efficiency required. As a result, curing is often commenced too late, and sporadic spraying with water at infrequent intervals is all that the concrete receives. Since tests are rarely made on the finished structure, no evidence is available of the loss in strength and watertightness which is caused. Tests[2.22,13.1] have, however, been made in the laboratory on the effect of curing on permeability and compressive strength. The results have shown that the permeability of 1:2:4 concrete, water cured, is high at short periods from mixing, but gradually decreases until, after the age of one month, the changes which occur are comparatively small. The days immediately following mixing are the most important, for at later periods long curing in water is necessary to improve the impermeability of badly-cured concrete. Wet curing gives the best concrete, and the nearer the curing conditions approach this the more impermeable the product. This is shown in Fig. 13.1.[13.1]

Figure 13.2 shows the number of freezing and thawing cycles* required to reduce the weight of specimens of concrete by 25 per cent, compared with the length of moist curing. The curves show that concrete should be prevented from drying out for at least 10 days in order to obtain the maximum resistance to disintegration by frost. The reduction of permeability with length of moist curing follows a similar curve, and the increased frost resistance with longer curing is probably due in part to the decrease in the water absorbed.

The length of the period of moist curing necessary is dependent to a large extent on the type of cement used. Seven days is usually sufficient

* A test in which the specimen of concrete is alternately frozen and thawed, the loss in weight being determined as disintegration proceeds.

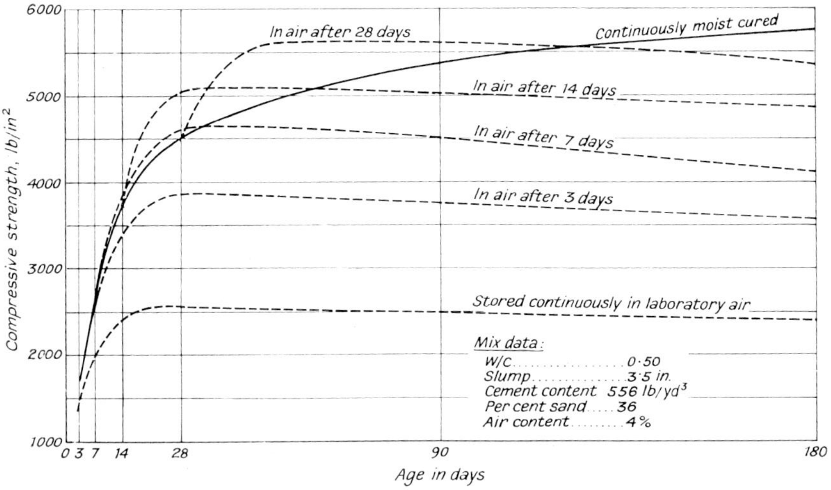

Fig. 13.1. Compressive strength of concrete dried in laboratory air after preliminary moist curing

for normal Portland cement to achieve the required 28 day compressive strength under normal temperature conditions.

Slow-hardening cements such as low-heat cement require 14 days or more of moist curing.

Cements which harden more rapidly require a shorter period, but greater care should be taken. For example, high-alumina content concrete should not be permitted to dry out during the first twenty-four hours,

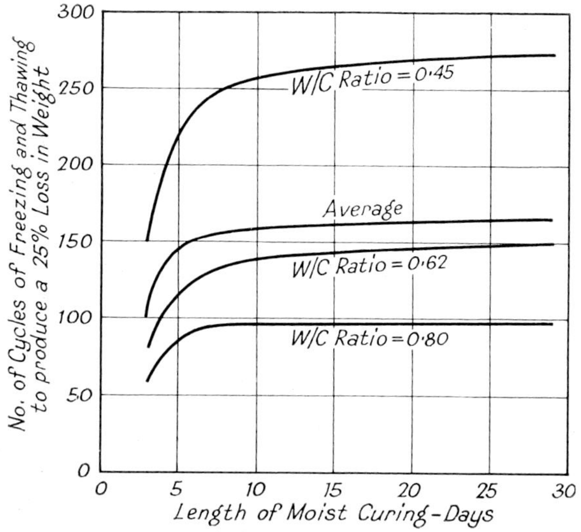

Fig. 13.2. Effect of the length of moist curing on the durability of concrete

and preferably longer. Where the thickness of concrete exceeds about 6 in. it is also necessary to keep the temperature down by a flow of cooling water. Once the temperature of high alumina cement concrete rises above about 27°C (80°F), the strength may be reduced by as much as half.

The actual procedures used for curing concrete vary widely, depending on the type of construction, site conditions, size and shape of the member. However, the treatments commonly used can be broadly divided into two groups, as follows:

(1) Formed concrete such as columns, beams and walls.
(2) Large horizontal surface areas, such as roads and floors.

Formed concrete

Wooden or steel forms generally provide adequate protection against drying out, although it is helpful to spray the outsides of forms with water at intervals after placing, especially during hot weather. Wooden forms which have been treated with a shutter paint or those which are lined with an impermeable coating will be more effective in preventing absorption into the wood. The use of mould oils will prevent absorption by wooden forms, but care should be taken to avoid excessive thicknesses which result in a weak water-repellent surface skin, which is difficult to remove, yet if left on will interfere with the bond of such finishings as plaster.

Exposed surfaces such as the tops of beams and walls should be covered immediately after casting in dry weather.

Good curing to beam, column, and wall vertical surfaces could be achieved by the continuous application of sprayed water. This is seldom practical, particularly in the case of vertical surfaces, because of the nuisance caused by water dripping on to other work below. Small sections such as columns and beams may be cured most satisfactorily by spraying with water and then wrapping in polythene sheeting as soon as the forms are removed. Alternatively, wet hessian or burlap in contact with the surfaces may be suitable, provided that the necessary arrangements are made on the site to ensure that the hessian is kept continuously wet. It should be realised that wet hessian can quickly dry out during warm days, or even on cool days when a strong breeze is blowing; warm dry nights will also cause wet hessian to dry out rapidly and under these circumstances measures must be taken to ensure wetting down during the night.

For some members, curing membranes may be suitable. For concrete surfaces which are to receive further finishing, such as paints or renderings, it is essential to make certain that the curing membrane will have worn off, or can be removed, before applying finishes.

It is not usually practicable to wrap concrete walls in polythene sheeting and recourse will probably have to be made to draping wet hessian or polythene from the top of the wall, taking care that all edges are firmly held down to prevent evaporation. In winter in Britain little curing is usually necessary, since in a cold moist atmosphere the rate of evaporation is slow. If polythene sheeting is not available, wrappings of hessian or straw mats may be used to reduce the amount of watering required and to reduce the dripping of water on to other works proceeding below. Tarpaulins and waterproof paper are also of value for reducing the rate of evaporation of water.

Small concrete sections require more protection than larger masses owing to their comparatively large surface area, in comparison with their volume. The corners and edges are most likely to be weakened by early drying out.

Large horizontal surfaces

Concrete having a large surface area unprotected by formwork requires careful curing from the time of placement, and several methods are available for this purpose. The best curing for large horizontal surfaces is undoubtedly one in which the concrete is kept continuously wet by ponding or spraying. Not only does this give good curing, but it also keeps the temperature down as a result of evaporation. Later thermal shrinkage is therefore reduced.

Ordinary hose-piping connected to a mains supply and with a spray nozzle is commonly used for spraying concrete surfaces. On certain types of structure it may be economical to fit up banks of sprays operated continuously. Water from spray-pipe systems may contain sufficient dissolved iron to cause staining, a trouble which may be avoided by the use of galvanized or alloy piping.

Spraying with water must be carried out judiciously in structural work in order to avoid interference with other work going on below.

Dams may have a large exposed surface area at each lift and curing by means of water and protective coverings is necessary.

Roads and airfields are another example of concrete construction with a large surface area; the curing of them is dealt with in Chapter 19.

Ponding or continuous spraying is not always a practical method and recourse has often to be made to other means.

Spraying with water as soon as possible after placing the concrete and covering with polythene or building paper is perhaps the next most effective method.

Alternatively, damp hessian may be laid directly on the surface as soon as the concrete has hardened sufficiently for the hessian not to cause damage or adhere to the surface. Damp sand, in a 2 in. layer has also been used for curing large horizontal surface areas. Both the wet hessian and damp sand methods are seldom carried out properly and the spraying or wetting at intervals during both day and night is often forgotten. The damp sand method also suffers from the disadvantage of the extra labour required to place the sand and subsequently remove it.

Floor surfacings dry out very quickly owing to their thinness, and this is often the cause of dusty surfacings of poor wear resistance. Adequate measures should therefore be taken to prevent any initial drying out by giving a light fog spray soon after finishing the trowelling operations, followed by a covering of waterproof paper or polythene sheeting, as soon as this can be done without damage to the surface. Curing membranes, which should be applied whilst the concrete is still moist, are generally accepted as satisfactory means of curing. As mentioned previously, however, they may cause a reduction in the bond of subsequent floor finishings.

Curing in hot, dry regions

Protection of open surfaces quite soon after placing is essential if shrinkage cracks, occurring before the concrete has set, and resulting from the evaporation of water, are to be avoided. Such protection is largely a matter of local materials available, and may be straw or reeds, or other vegetable covering, sheet tin, hessian, etc.

Curing with water may be extremely difficult, if not out of the question. Even if supplies are available, they are likely to contain salts and the continual spraying and evaporation leaves concentrations of such salts on or near the surface of the concrete, causing efflorescence, and a possible weakening of the surface layer. It is advisable in such cases to drape hessian or similar covering a few inches away from the concrete surface if at all practicable, and to spray the hessian, without the water reaching the concrete at all.

Sealing compounds

Liquid membrane-forming curing compounds retard the evaporation of concrete from the surface, and an effective membrane, properly applied, will under most conditions retain enough moisture for adequate curing. Retardation of the evaporation does, however, reduce the cooling effect and this is a disadvantage. It is advisable to provide also for moist curing or at least shading in the early stages. ASTM Designation C156-65[13.2] gives methods of test for determining the efficiency of curing compounds.

Three types are available, clear, white or grey-pigmented, and black, of which the first-named causes the least discolouration. The white-pigmented coating is intended to reduce the absorption of heat from the sun's rays, as well as to seal the surface and so reduce evaporation. Tests in the United States have indicated that the white surface closely approaches the effect of shading for maintaining low temperatures in concrete exposed to sunlight.

It is not easy under practical conditions to apply one coat of a sealing compound to form a continuous impervious membrane. Two coats are therefore preferable and should follow one another as quickly as is practicable without causing running or wrinkling of the surface. A coverage not thinner than 300 ft^2 per gallon per coat is necessary. The United States Bureau of Reclamation allows the use of a white-pigmented compound for some of its work, applied in one coat at a maximum coverage of 150 ft^2 per gallon (U.S.). Tests and field observations have shown that this coverage provides the equivalent of 14 days continuous moist curing if the sealing membrane is kept intact for 28 days.

These coverages are recommended for general use; they are less than the values claimed by some manufacturers.

Asphalt and tar cutbacks and emulsions have been applied to paving surfaces on aerodromes to provide some protection to the concrete immediately after finishing and before it has become sufficiently hard to stand the normal forms of curing without marking. The effectiveness of these sealing coats is reduced by their black colour which increases the absorption of heat. The heat absorption may be reduced by the application of a white-pigmented coating. This does not entirely overcome the objection of the black surface, which will again become apparent when the white coating has weathered off. It is advisable in either case to cover the concrete with straw, hessian, or other material, and commence moist curing in the normal manner as soon as possible.

Tar and asphaltic coatings are useful in tunnels and culverts, and on surfaces which will eventually be covered. Heat absorption should be reduced by whitewashing until the concrete is covered.

Application of deliquescent salts

Deliquescent salts have been used in order to keep concrete moist, but they are open to the objection that when the relative humidity of the air drops below a critical point water evaporates instead of being absorbed.

REFERENCES

13.1 PRICE, W. H., Factors influencing concrete strength, *J. Amer. Concr. Inst.*, **47**, Feb. 1951, pp. 417–32.

13.2 ASTM Designation: C156–65, *Standard Method of Test for Water Retention. Efficiency of Liquid Membrane-forming Compounds and Impermeable Sheet Materials for Curing Concrete.* American Society for Testing Materials.

CHAPTER 14

Concreting in Cold Weather

As the temperature of concrete decreases, the rate of hardening and development of strength is slowed down until at a temperature below freezing point the chemical process of hardening ceases altogether. When the temperature rises again hardening is resumed. When concreting during cold weather and when the temperature falls below freezing point, measures should be taken to ensure:

(1) That water in newly placed concrete does not freeze and expand as ice is formed.

(2) That the concrete is protected at early ages, i.e. until it will withstand cycles of freezing and thawing without deterioration.

(3) That the strength development is maintained even though at a lower rate than at higher temperatures.

Unless some form of heat is applied and the concrete can be adequately protected, it is advisable to stop concreting when the temperature reaches about 3°C (38°F) on a falling thermometer. With certain precautions, nevertheless, concrete can be continued at freezing temperatures.

During frosty weather, when the temperature rises above freezing point during the day, it is usually sufficient to use a rapid-hardening cement with or without the addition of calcium chloride, and to protect the concrete by coverings of straw, hessian, or tarpaulins. Timber formwork gives reasonable insulation, except to any exposed surface, against an occasional frost. During continued cold weather further measures are needed and these are described under various headings below.

Temperature of materials

The temperature of the concrete when first made depends on the temperature, specific heat, and weight of the constituent materials. The approximate temperature of the concrete can be calculated from the following formula:

$$T_c = \frac{t_c + At_a + 5Wt_w}{1 + A + 5W}$$

where T_c, t_c, t_a and t_w are the temperatures of the concrete, cement, aggregate and water respectively,

A is the aggregate/cement ratio,

W is the water/cement ratio

As an example, for a mix with a cement/aggregate ratio of 1 to 6 by weight and a water/cement ratio of 0.6 by weight, with the cement and aggregate at a temperature of 2°C (36°F) and a water temperature of 50°C (122°F), the concrete would have a temperature of:

$$\frac{2 + 6 \times 2 + 5 \times 0.6 \times 50}{1 + 6 + 5 \times 0.6} = 16°\text{C } (61°\text{F}) \text{ for calculations in centrigade units.}$$

When using a cold concrete mixer and cold transporting plant, the heat transfer in the first few batches will lower the temperature of the concrete considerably below the calculated figure.

Water will hold five times the amount of heat held by aggregate or cement, and it provides the most convenient way of introducing heat to a concrete mix.

Cement

Provided that the cement, either in bags or silos, is kept free from moisture it will not be affected by frost. Cement in bags should be stored under cover in a dry building. Where this is not possible it should be stacked out of contact with the ground and completely covered.

Cements which produce considerable quantities of heat are an advantage in cold weather. An example is high-alumina cement, which with suitable insulation can be placed in temperatures well below freezing point. No attempt should be made to heat the cement.

Aggregates

For air temperatures above 0°C (32°F), when aggregates are free from ice and frozen lumps, the desired temperature of the concrete can usually be obtained by heating only the mixing water. For air temperatures below 0°C or when the stockpiles contain frozens lumps, ice, or snow, it is necessary to heat the aggregates.

Sufficient aggregate for each day's pouring should be thawed out as far in advance of batching as possible so as to attain a uniform moisture content and temperature; 24 hours is considered the minimum time for this.

Stockpiles of aggregate are somewhat difficult to heat satisfactorily. Several methods are available, such as steam coils, hot-water coils, steam jets, fire flues, and the erection of covers over the stockpiles inside which braziers are used to keep the air warm. The adoption of any particular method will depend on the relative economy, having regard to the quantity of aggregate to be heated and plant arrangements. Live steam is generally satisfactory if adequate drainage is provided in the stockpile to keep the moisture content uniform. It has the advantage that the steam percolates through the stockpile and heats the aggregate fairly uniformly.

Steam heating of aggregates is common, and should probably be adopted where the volume of concrete placed daily exceeds 20 yd^3. Large boilers are only necessary on very large contracts or under the most extreme conditions. The small quick-firing type of boiler is adequate for the majority of sites and conditions. This type of boiler raises steam in less than half an hour and consequently does not require attention during the night, when it can be drained and shut down. The heated aggregate should of course be protected during the night by tarpaulins.

The sand bay should always be kept nearest to the supply end of the steam as this ensures more even distribution of the steam to the remaining bays. A coarse aggregate bay nearest to the boiler means that the bulk of the steam pressure is lost through it before reaching other bays. The exposed surfaces of the aggregates should be covered with tarpaulins or polythene sheets to maintain a uniform distribution of heat and to prevent the formation of ice crusts. It is not possible to generalise as to how much steam will be required, since this will be dependent on factors such as the amount of aggregate, the temperature required, and the efficiency of the tarpaulins. However, as an example, a steam boiler working at 25 lb/in^2 pressure producing 3,000 lb of steam (300 gallons of water being used per day) is sufficient to heat 180 yd^3 of coarse aggregate, 100 yd^3 of sand and 100 gallons of water to produce a concrete at the mixer having a temperature 17°C (30°F) above the air temperature. Frequent use should be made of thermometers to check the aggregate temperatures.

For smaller contracts where the cost of installing a boiler is not warranted the aggregates should be covered by tents over the stockpiles and the enclosure heated by braziers or low output space heaters until the aggregate is completely unfrozen. For small heaps of aggregate, thawing may be accelerated by turning over the heap at frequent intervals. If this is done fumes from the heaters must be vented outside the enclosure to avoid the danger of carbon monoxide poisoning. This method is unlikely to heat the aggregates evenly to a temperature much above 45°F and should only be regarded as a means of keeping the aggregate in a frost-free condition.

During frosty weather, all aggregate stockpiles should be covered when not in use; this is particularly important overnight. Tarpaulins or polythene sheets are useful for temperature only a few degrees below freezing, but it is recommended that insulated blankets are used, since these are more efficient.

Mixing water

It is generally found that the cheapest and easiest way of preheating concrete is to heat the mixing water. The required temperature of the water should be between 50°C–65°C (120°F–150°F), and care should be exercised to ensure that the temperature of the water does not exceed 70°C (160°F).

Mixing water should not be heated above 60°C–70°C, depending on the type of cement, since overheating may produce quick setting or reduce the slump sufficiently to make placing difficult. No trouble was experienced at Boulder Dam when using low-heat cement and water heated to 90°C (194°F). In all cases when the water is heated above about 35°C (95°F) it is advisable to mix the water and aggregate first before adding the cement.

Where steam is available the mixing water may be heated by means of a pipe, about 1½ in. diameter, run from the boiler to an insulated water supply near the mixer. This water tank is additional to the measuring tank on the mixer, and should be positioned above the mixer on scaffolding. It has been found satisfactory to have a 100 gallon tank for use with a 12NT mixer. This steam pipe, which is controlled by a valve placed near the mixer, terminates a few inches above the bottom of the tank. Steam can be periodically injected in sufficient quantities to heat the water as required. Alternatively, the water may be heated by a steam coil in the bottom of the supply tank, but this method does not permit close control over the water temperature.

The heated water is fed into the measuring tank on the mixer, which should also be lagged externally with foamed polystyrene, fibreglass or similar insulant to avoid heat losses. Constant checks on the temperature of the water in the tanks are necessary, particularly if the mixer is stopped for longer than the normal batch cycle time when the water temperature may rise unless the steam supply is reduced. Whenever the mixing water temperature exceeds 65°C (150°F) some cold water should be added or some run off to waste until the temperature is below 65°C. When steam is not available, coal or diesel fired hot water boilers or 100 gallon tanks, well lagged, placed on brick supports with calor gas heaters beneath, may be used for providing the hot water. Electric immersion type heaters have been used but these are expensive to run for long periods, although the cost can be reduced by using off-peak electricity during the night.

Batching and mixing

Mention has already been made of placing an additional water tank for heating above the normal mixer measuring tank. The whole of the mixing plant should be kept under cover as much as possible. Sometimes it will be possible for the whole set-up, including aggregates, to be placed in a part of a building; where this is not possible the plant should be placed in as sheltered a place as possible, such as the lee of a wall. Windbreaks should be used to give protection against high winds and driving rain or snow.

The temperature of the concrete leaving the mixer should not be lower than 10°C (50°F), as there is an unavoidable loss of heat during transport and placing (see below). On the other hand, the temperature should not be unduly high, since the thermal shrinkage is increased by any increase in temperature drop. It is rarely necessary to raise the temperature to more than about 20°C (68°F). As a general rule the temperature of the concrete as mixed should be 3°C–8°C higher than the required temperature after placing; the allowance is dependent on the ambient temperature.[14.1]

The concrete can be warmed by heating the water and the aggregate or by applying heat in the mixer. The simplest method is to heat the water as described above. Methods which rely on the application of heat in the concrete mixer, as for instance by a torch or other flame, are not entirely reliable, both from the point of view of heating efficiency and variability in moisture loss.

Placing

The concrete should be placed as soon as possible after mixing since heat is lost quickly, and even when the work proceeds quite efficiently the drop in temperature from mixing to placing may be as much as 3°C to 8°C. Under normal conditions a loss of 3°C can take place during transportation alone, but when excessive hauls are unavoidable the loss may be as much as 5°C.

The required temperature of the concrete after placing will be dependent on the mass of concrete. For thin sections the temperature of the concrete after placing and for the first 72 hours should be between 10°C and 13°C

(50°F and 55°F). For mass concrete the temperature should be between 4°C and 7°C (40°F and 45°F).

Open chutes and conveyors should be avoided since a good deal of heat is lost and they also present difficulties owing to the possible formation of ice between runs.

The principal danger during low temperatures is the possibility of the water in the newly placed concrete freezing. Frozen concrete may be mistaken for concrete which has set normally; it is not unknown for the shutters to have been removed from such concrete, and collapse not to have occurred until a subsequent thaw. Such a happening can, of course, have serious consequences. To avoid this, the temperature of the concrete when deposited in the forms should be in no case less than 4°C (39°F), and the temperature of the concrete should be kept above this until it has hardened.

Equipment

Skips, wheelbarrows, chutes, and other concreting equipment should be protected from frost when not in use and outside working hours. Any ice that has formed in them should be removed before concreting starts by the use of hot water or, if available, a steam lance from a boiler.

Calcium chloride

The setting and hardening of concrete can be accelerated by adding calcium chloride to the mix. Although the use of calcium chloride does not in itself give protection against freezing, it has the following effects:

1. It increases the rate at which heat is evolved from the cement during the early periods of hardening and as a result of this heat evolution the rate of strength development is increased, especially at early ages.

2. It lowers the freezing point of the wet mix, although this is relatively unimportant because the normal additions of calcium chloride only lower the freezing point of the wet concrete by about 1°C (2°F–3°F).

The quantity used should be 2 per cent of the weight of cement and this must not be exceeded. The effect of additions of 2 per cent of calcium chloride on the compressive strength of concrete cured at different temperatures has already been shown in Fig. 5.1, p. 82. Excess of calcium chloride has adverse effects on concrete, and increases the risk of rusting of reinforcement. Calcium chloride can be added either as a solution in water or more conveniently by adding the material dry, in the form of flakes, direct to the mix. For the latter method a quart measure of calcium chloride in flake form weighs about 2 lb and is suitable for use with a one bag mix. The flakes should be placed in the mixer with the aggregates and water before the cement is added. The use of flake calcium chloride in solution form is not recommended since errors can arise in preparing the solution and difficulties may be experienced in keeping the solution unfrozen.

Calcium chloride may be used with either ordinary or rapid-hardening Portland cement, but its use is not recommended with sulphate-resisting cement or super-sulphated cement.

Calcium chloride should not be used in mixes for prestressed concrete.

Protection of concrete after placing

Protection of concrete after placing may take the form of insulation of the surface itself or the erection of temporary covers with internal heating. The choice of method in any particular circumstances must depend on relative costs in relation to the rate of progress required and the amount of work to be done. Thin reinforced concrete members normally require more protection over a longer period than larger masses. The corners and edges are particularly vulnerable. Mass concrete, nevertheless, requires protection during the early stages of setting until the quantity of heat evolved is sufficient to offset the heat losses. Under normal circumstances this is not fulfilled until 1–3 days after placing.

Strip and pad foundations below the surface of the ground can usually be given sufficient protection by covering with a straw or fibreglass quilt to conserve the heat which will be generated within the mass of the concrete. It cannot be overstressed that concrete should not be placed on frozen ground.

Ground floor slabs placed on unfrozen ground or hardcore and covered with a protected insulating layer of fibreglass or straw mattress will normally be sufficient to conserve the heat generated in the concrete when the daytime temperature is above freezing point. Severe weather conditions should be countered by covering the slabs with low tents of polythene sheeting or tarpaulins and placing heaters inside the area in order to maintain the temperature of the concrete above 4°C (39°F). Care should be taken to avoid early drying out of the concrete.

Suspended concrete floors may be kept warm after laying by erecting covers on frames over and down the sides of the slab, with a propane gas hot-air blower discharging warm air into the air space thus formed. Coke braziers, paraffin heaters or propane heaters below the slab will keep the underside of the concrete warm. Under less severe conditions, heat applied from the underside of the shuttering with the top surface insulated will generally be sufficient. The top surface insulation should consist of fibreglass or two 1 in. thicknesses of woven straw matting between two layers of polythene. An alternative is a 6 in. thick layer of loose straw. Electrically heated insulated mats are now available and have been found especially useful for keeping slab concretes at the required curing temperatures.

Tops of columns, beams and walls should be covered with an insulating material immediately after completion of placing. Side and wall shutters should be completely draped with straw matting and tarpaulins or polythene sheeting. Tarpaulins stretched across from wallhead to wallhead with paraffin or propane heaters within the cells will help to maintain the temperature at the required level, but precautions must be taken against the possibility of carbon monoxide poisoning. Provided that sufficient heat is applied for a further two days to maintain the concrete temperature above 4°C (39°F) vertical shuttering may be struck within 24 hours.

Floor toppings require special protection during cold weather owing to their thinness. It is preferable to enclose the space above and below the floor and to maintain a temperature of 4°C or over for the first few days after laying. Heaters should be well insulated from the floor topping by a thick layer of sand, and the area round them prevented from drying out prematurely.

Formwork

Before concreting it is essential that all shutters and reinforcement are free from ice. Where the formwork and reinforcement are assembled the day previous to concreting, sheeting over and applying heat from braziers or hot-air heaters will be sufficient to keep the formwork and reinforcement free from ice so that concreting may commence first thing the following day.

Where steam is available an additional pipe from the boiler may be run up the building with valve connections at each floor level. A steam lance may then be used for heating formwork and reinforcement.

Provided that precautions have been taken to keep them warm, shutters to vertical surfaces may be struck after 24 hours if heat is to be continued, but whenever possible it is recommended that 48 hours be permitted to elapse before vertical shuttering is struck. It may be desirable to retain external shuttering for a longer period than is strictly necessary, so that it can provide the required heat insulation for internal work, including floor construction. For multi-storey work the external forms can readily be left in position for as long as the pouring cycle permits before they are moved up to the next floor. The external forms, together with the floor shutters erected immediately after striking the internal forms, provide a good means of "tenting" the newly poured walls.

Electrically heated formwork is now available and where practicable affords a convenient and effective method of increasing the rate of hardening of the concrete. The cost should be considered in relation to any expected savings.

Striking of formwork

The minimum time for striking can best be determined by trial on the job, since it is influenced by factors such as the type and amount of cement, curing temperature, size of member, and the type of formwork. The strength of test specimens cured under similar conditions to the structural members provides a conservative basis for determining the safe time for removal of formwork.

The strength of a typical 1:6 by weight concrete mix kept at 2°C (36°F) for seven days is less than half the strength of similar concrete maintained continuously at 18°C (65°F), and remains lower for several weeks. Therefore, even at temperatures above freezing point but below say 10°C (50°F), care must be taken not to strike shuttering too soon nor to load the concrete before it has sufficient strength to carry such loads.

Test cubes

Test cubes cast in cold weather require careful protection. Either they can be used as a guide to the early strength of the concrete in order to provide information for assessing the earliest time for safe removal of the forms, or they can be used as a check on the concrete quality. In the former case they should be placed in contact with the forms and given the same protection as the mass of the concrete, with a little additional insulation to compensate for the fact that they are cast in steel moulds and have a greater relative surface area from which heat losses can occur. If required as a check on quality, they should be cured in accordance with B.S. 1881 : 1968.

REFERENCE

14.1 A.C.I. 604–56, *Recommended Practice for Winter Concreting*, American Concrete Institute Standards, 1965.

CHAPTER 15

Surface Finishing of Concrete

MANY concrete structures have a satisfactory appearance, but unfortunately others are deplorable from the start, and yet others were quite pleasing when new but have weathered badly. There is, therefore, considerable justification for the view variously expressed that concrete is not pleasing to the eye. Many of the unsightly features of concrete structures result from contraction cracks, carelessly constructed and badly placed construction joints, patching of damaged or honeycombed areas, poor formwork, and lack of adequate cover to the reinforcement. All are faults which can be prevented.

If an efficient structure, which is also pleasing in appearance, is to be produced, co-operation between the architect, the engineer and the contractor is essential. A close liaison is especially necessary in arranging such details as the positions of construction joints, the construction of the shuttering and formwork, and the method of finishing.

Considerable care needs to be given to the many details involved in the production of concrete if a uniform colour and a surface free from blemishes and honeycombing is to be achieved. The aggregates should be well graded and of one quality. Fine sands and rich mixes assist in reducing blowholes and colour variations, but can lead to crazing if the water content is excessive. The proportioning, water content, mixing, and placing should be constantly watched to ensure that such blemishes as segregation, honeycombing, blowholes, sand streaks and colour changes do not occur. A fast rate of placing in excess of 6 ft per hour is advantageous in eliminating air bubbles[15.1] from vibrated concrete. Immersion vibrators are to be preferred because air bubbles then tend to travel inwards towards the source of vibration. Construction joints should receive particular attention. Bleeding should be prevented as far as practicable, as the concentration of fine particles of cement with a low water content gives a darker colour.

Much depends on adherence to good practice, described in previous Chapters, and on the use of suitable release agents, see p. 249.

The design of the formwork should be such that it can easily be removed and used repeatedly if required; if removal is difficult or is done carelessly, the use of crowbars and other tools may mar the appearance of the concrete, especially at the edges and corners. Sharp arrises and internal corners should be chamfered whenever possible.

A committee appointed by the Institution of Civil Engineers have drafted a guide to specifying concrete[15.2] which includes the following classification of finishes for formed surfaces.

Class F.1. This finish is for surfaces against which backfill or further concrete will be placed. Formwork shall consist of sawn boards, sheet metal or any other suitable material which will prevent the loss of grout when the concrete is vibrated.

Class F.2. This finish is for surfaces which are permanently exposed to view but where the highest standard of finish is not required. Forms to provide a Class F.2 finish shall be faced with wrought and thicknessed boards with square edges arranged in a uniform pattern. Alternatively, plywood or metal panels may be used if they are free from defects likely to detract from the general appearance of the finished surface. Joints between boards and panels shall be horizontal and vertical unless otherwise directed.

This finish shall be such as to require no general filling of surface pitting, but fins, surface discolouration and other minor defects shall be remedied by methods approved by the Engineer.

Class F.3. This finish is for surfaces prominently exposed to view where good appearance and alignment are of special importance. To achieve this finish, which shall be free of boardmarks, the formwork shall be faced with plywood or equivalent material in large sheets. The sheets shall be arranged in an approved, uniform pattern. Where-ever possible, joints between sheets shall be arranged to coincide with architectural features, cills, window heads, or changes in direction of the surface. All joints between panels shall be vertical and horizontal unless otherwise directed. Suitable joints shall be provided between sheets to maintain accurate alignment in the plane of the sheets. Unfaced wrought boarding or standard steel panels will not be permitted for Class F.3 finish.

Rough finished surface textures are gaining favour. Such patterns in bold relief, as for example that shown in Fig. 15.1, are particularly suited to concrete, and weather better than smooth surfaces on which defects become all too evident. In sooty towns the effect of light and shade is retained, whereas the colour and texture of finer finishes are lost.

Fig. 15.1. The elephant and rhinoceros pavilion, London Zoo, showing the attractive finish obtained by casting the external walls in-situ with vertical ribs subsequently hacked to expose the aggregate

Other types of special finish may be summarized under the following headings:

(1) Board-marked finishes, where emphasis is placed on the pattern produced by the boards, and the rough finished surface texture.
(2) Finishes from special linings.
(3) Tooled finishes.
(4) Scrubbed and exposed aggregate finishes.
(5) Grit-blasted finishes.
(6) Finishes produced by grinding.
(7) Precast slabs.

Board-marked finishes

The formwork pattern is of more importance than any other single factor to the production of a monolithic concrete structure having a pleasing appearance. Well-designed formwork can provide an excellent finish without further treatment, as may be seen in the examples given in Figs. 15.2 and 15.3. Even rough boarding of uneven thickness can give an attractive general appearance with careful planning. A finish from rough boarding masks defects and horizontal construction joints, and weathers well in polluted atmospheres. It is necessary either to have tongued and grooved joints, or to plane the edges of the boards, so as to obtain a close fit which will obviate loss of cement mortar with consequent honeycombing.

A more uniform board-marked finish is obtained by using planed boards. As the finish becomes smoother greater care is necessary to preserve a balanced pattern, and to avoid surface defects. The board

Fig. 15.2. Effective use of board-mark finish on the soffit, mushroom headed columns, and slab edge to walkways at the Royal Festival Hall, London

Fig. 15.3. Exceptionally high standard of boardmarked concrete finish to mushroom headed columns

pattern may be either subdued by removing all projecting fins with chisels, carborundum stones, or other approved means, or it may be accentuated by the use of tongued and grooved boarding with chamfered edges.

The following points should receive attention:

(1) The formwork should be tight and in true alignment.

(2) The boards should be evenly matched in width. Narrow boards usually give better results owing to less warping and the finished appearance is more pleasing (Fig. 15.4). Wrought boards are more easily stripped from the concrete and more easily cleaned.

(3) Old and new boards should not be mixed, since old boards are often more absorbent and have a rougher surface. Variations in texture, therefore, occur.

Finishes from special linings

Smooth finishes are normally produced by lining the formwork with metal sheets, plywood, or compressed fibre boards treated with oil or shellac to prevent the concrete adhering, see p. 249 dealing with mould release agents. These sheets should be arranged in panels so that the joints form a pleasing pattern on the building. The useful life of plywood is dependent on the type used, the best being that which has a waterproof binder. The care used in handling during stripping and in storage on

Fig. 15.4. Use of narrow boards provides pleasing finish to church in Finland. The roof is a copper covered prestressed concrete parabolic shell

the site also materially affects the number of times it can be used. Hardboards also require care in handling and can normally be used only three or four times before fraying at the edges occurs. Chipboards deteriorate rapidly, due to swelling of the individual chips, and are much weaker in bending than timber, particularly when wet.

Highly absorptive insulation boards about ½ in. thick have been used as lining in order to reduce the number of blowholes on the surface. Tests have indicated that when concrete was placed against absorptive form linings, practically all pitting and voids were eliminated, whether compaction was by hand or by vibration. However, differences in absorption can cause colour variations. Muslin and hessian are sometimes used as a covering to the insulating board. The use of such linings is expensive due to the limited re-use possible.

Generally, non-absorbent linings give the most uniform colour to the concrete. Barrier paints and lacquers may be used to reduce the permeability of linings, a suitable mould oil then being used to reduce blowholes and ensure good release.

Smooth finishes are best avoided externally, since uneven weathering, small defects in the concrete and all joints (whether in the shuttering or in the concrete) are plainly visible. Internally, treatment of smooth finishes with alkali-resistant paints can give satisfactory results.

Other form linings which have been used to produce special finishes include crêpe rubber and waterproofed serrated cardboard. The rubber sheeting can be used several times, but the cardboard can normally be used only once. A difficulty arises with these materials in that they are difficult to fix without creasing and it is also necessary to arrange them in small panels with suitably placed joints, in order to avoid unsightly joint-lines.

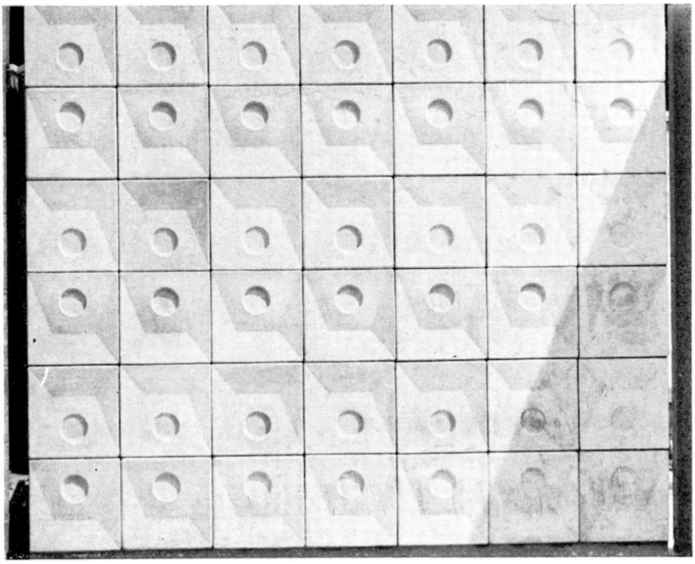

Fig. 15.5. Profiled patterned blocks with an "egg-shell" finish obtained by casting against thermoplastic formwork linings

Awkwardly shaped mouldings with a smooth surface finish have been satisfactorily produced using moulded resin-bonded fibreglass.

An example of a wall built with concrete blocks cast against thermoplastic formwork linings is shown in Fig. 15.5.

Tooled finishes

A number of methods of producing tooled finishes are available, and with the use of suitable aggregates a very pleasing appearance can be obtained which retains its character even after a number of years of weathering. Bush-hammering is the most popular type of tooled finish, and electrical and compressed air equipment is available for this work. It is advisable when bush-hammering to leave a narrow margin at the edges and corners of the area untreated in order to avoid the danger of spalling. Bush-hammering is done with either a disc type of hammer head or a roller type as shown in Fig. 15.6.

Figure 15.7 shows the effect of heavy bush-hammering on ballast concrete. Other forms of tooled finish include sparrow-pecking (Fig. 10.4), which is done with a pointed hammer, comb chisels used to give various types of pattern, and other effects produced by special types of hammer-head. It should be noted that construction joints, honeycombing, and segregation, cannot be masked by tooling and it is very necessary that every care should be taken in placing the concrete.

An example of an attempt to mask a badly made joint with honeycombing at the bottom of the upper lift is shown in Fig. 15.8. The patch in the bush-hammered section is clearly seen. A further example of the failure of bush-hammering to hide defects in construction is seen in

Fig. 15.6. Electrically driven roller type bush-hammer tool in use on a column

Fig. 15.7. Point-tooled white concrete finish

Fig. 15.8. Failure of bush-hammering to mask a honeycombed patch of concrete just above a construction joint. Honeycombing was similar to that shown on the left and was patched with mortar before bush-hammering

Fig. 15.9. Not only do joints show, but there are sundry mortar patches, and the joints between the steel shutter panels are still apparent. Evidence of lipping where the shuttering was not tight against concrete previously placed, is seen at window head and upper sill levels. A contraction crack has occurred below the lower right hand window.

Since construction joints always show in tooled work, it is advisable to arrange a panelled effect, emphasising the joints by methods such as those shown in Figs. 10.5, 10.6 and 10.7, or by leaving untooled margins or recesses at the joints.

Attempts to adjust the line and level after the bulging or slipping of formwork always result in differences in the tooled texture, a bad example being seen in Fig. 15.10.

Patches always show through tooling sooner or later.

Scrubbed and exposed aggregate finishes

Removal of the cement skin may be effected while the concrete is green by means of a stiff fibre brush or a wire brush. The earlier the scrubbing is done the deeper is the aggregate exposed. Plenty of water,

Fig. 15.9. Bush-hammered gravel concrete wall to a block of flats. Note the failure of the bush-hammering to remove the joint lines of the steel shutter panels, numerous mortar patches and construction joints at irregular heights. Shrinkage cracks have occurred below the two windows on the right

preferably as a spray, is needed to wash away the cement paste and clean up the surface of the aggregate as it becomes exposed.

Patching scrubbed work is difficult if the patches are to be indistinguishable from the rest of the surface. The best method is to patch immediately after scrubbing, using the largest aggregate which can be managed. After a number of hours depending on the weather, the patches should be scrubbed to the same texture as the general surface.

Two methods of avoiding the ragged corners which this type of work produces are to use rounded mouldings in the corners of the formwork or to have rebated corners and stop the scrubbing at the edge of the rebates.

For buildings higher than one story there are practical difficulties which increase the cost of scrubbing and make it almost impossible to produce a clean, uniform surface. This type of finish is, in fact, more practicable for precast concrete units than for monolithic concrete. A good example of precast unit construction with an exposed aggregate finish is shown in Fig. 15.11.

Fig. 15.10. Failure of bush-hammering to mask the effect of shutter slipping, even where there is no accompanying loss of fine material

Fig. 15.11. A good example of an exposed aggregate finish on precast concrete units forming the parapet of a bridge

Fig. 15.12. Scrubbed gravel concrete after weathering for some years

Fig. 15.13. Surface obtained by scrubbing after the use of a chemical retarder on the shuttering. Note the difference in texture between the bottom pour and the top pour. Slight lipping is seen at the joint

Removal of the cement skin is open to the objection that the surface is not then so impermeable and therefore less resistant to the weather. An added disadvantage is that the rough surface texture is apt to become grimy in a polluted atmosphere. Figure 15.12 shows the appearance of a scrubbed gravel concrete after a number of years' weathering in a polluted atmosphere. The gravel has become dull and lifeless, while a number of holes in the surface show that some disintegration has taken place.

Etching with hydrochloric acid produces a surface similar to that given by scrubbing, except that the aggregate usually appears brighter. Plenty of water must be used after etching in order to prevent a continuation of the erosion. Acid treatment is undesirable on reinforced work with little cover or on low-weight concrete.

Preparations containing substances such as sugar, which retard the setting of cement, may be painted on the inside of the formwork. When the formwork is stripped the surface skin may then be scrubbed away to expose the aggregate. The preparations at present in use are not entirely satisfactory either on account of staining or failing to produce a uniform surface (see Fig. 15.13). The latter trouble seems to be due to the ease with which the paint rubs off when fixing the reinforcement or when filling the forms. Neither are such preparations satisfactory for producing a rough surface as a key for renderings, since there is always a risk that traces of retarder or weakened cement at the surface will seriously reduce the bond. The staining may also penetrate the rendering.

Aggregate transfer method

Another method of producing an exposed aggregate finish which has given satisfactory results involves the sticking of the aggregate on to the formwork. The adhesive is one which softens under the action of water from the concrete, so that removal of the formwork, and a wash down, leave the aggregate exposed.

Grit-blasted finishes

Grit-blasting is not very effective as it tends to erode the softer portions of the concrete surface and so produce a non-uniform appearance. The process is also relatively expensive.

Finishes produced by grinding

The surface skin may be removed by the use of grinding stones or carborundum discs. A grinding stone is necessary for removing projections; discs may then be used over the surface as a whole. Any making good must be done before grinding, and every effort should be made to see that the patches match the main concrete.

Several concrete structures in London have been finished with carborundum discs, and on all of them the patches are becoming increasingly obvious, even though they were not discernible when the concrete was new.

Carborundum discs were used to grind down the precast concrete slab exterior of a hotel in London which had become discoloured after about three years' weathering. Although the surface was afterwards given a coat of cellulose, it again lost its brightness and grime collected,

Fig. 15.14. An example of profiled precast horizontal cladding units

Fig. 15.15. Coarse pebble exposed aggregate slabs

particularly at the joints. An example of weathering of a ground surface is given in Fig. 15.16.

Precast slabs

Precast concrete slabs can be finished by any of the methods outlined above and have the advantage over cast in-situ concrete in that they are made under more closely controlled conditions such that fewer defects are likely to occur. In addition, with precast units the joints are generally a feature rather than a disfigurement. Examples of precast slabs are shown in Figs. 15.14 and 15.15.

Grouting and cement washes

Cement washed or grouted surfaces cannot be considered as good practice. They are usually of a much richer mix than the concrete underneath, and after exposure to the weather the differential movements due to moisture and temperature give rise to fine cracks, and eventually the skin peels off or is washed away. Cement grouts are satisfactory for filling small blowholes, but they should not be used for filling larger patches or to mask honeycombing.

Painting concrete surfaces

Paints based on linseed oil cannot satisfactorily be applied to new concrete, since the linseed oil is saponified by alkalis from the cement. However, linseed oil is sometimes advocated for improving the weather resistance of old concrete, and experience in the U.S.A. indicates that a linseed oil application can effectively reduce deterioration due to the use of de-icing salts.

Paints having a plastic binder and giving a rough sanded finish have been more successful. They should be applied to a dry surface in warm weather, and may then be expected to last five to fifteen years. Much depends on surface preparation and the possibility of moisture getting behind the paint film.

Repairs

Repairs to concrete surfaces may be avoided to a large extent by arranging the shuttering so that it may be removed easily without the use of crowbars or other tools likely to damage the green concrete. Honeycombed patches should not occur where the supervision is good.

Small blowholes are almost unavoidable, but they may be filled without much fear of them becoming a source of trouble. Larger patches are usually troublesome and since they become more apparent as time goes on, every effort must be made to make a good job of them. Patches which are carelessly done begin to crack from the main concrete after a few years' weathering and finally become detached.

Repairs should be made as soon as possible after the forms have been removed and before the concrete becomes too hard. Stone pockets, segregation patches and damaged areas should be chipped out and the edges undercut slightly to form a key. The area to be patched should be cut to a size which will marry with the general appearance without undue disfiguration, since the patch must be expected to have a slightly different colour and texture from the remainder of the concrete.

Fig. 15.16. Repairs to concrete surfaces
Patches which have become increasingly obvious with exposure to the weather. The surface was originally ground with carborundum discs and the patches hardly discernible. Shrinkage cracks are also apparent.

All loose material should be washed out before patching. No excess water should be left in the cavity, but the concrete should be damp while retaining some of its natural suction.

A good bond between the patch and the parent concrete may be obtained by sprinkling dry cement on the wet surface just before applying the patching material. Other methods are to throw mortar forcibly on to the wetted concrete, or to brush in a coat of thick cement grout of about 1:1 cement:sand. Before this has dried, the remainder of the patch should be filled with mortar or concrete, depending on the size of the repair.

The filling material for very small patches should consist of mortar of the same materials and proportions as those used in the mixing of the concrete. The mortar should be mixed as dry as possible, well compacted into the cavity, and finished in a manner which most nearly conforms to the adjacent surfaces. The colour match may often be improved by adding a proportion of white cement to the mix.

If larger repairs are to be made to hardened concrete it is usually necessary to provide some formwork bearing tightly at the edges of the cavity. For such repairs it is necessary to chip out concrete to a depth of at least 4 in., and preferably 6 in. in old concrete. An upward slope from the back of the hole may be necessary at the top to facilitate placing of the concrete. Mortar should be scrubbed into all surfaces with a wire brush before placing the concrete. Any of the original reinforcement that may have been damaged by corrosion or from other causes should be adequately spliced with new steel so as to maintain the original strength. In some cases it may be advisable to include additional reinforcement in the patch.

Patches must be kept moist for several days or all the care taken in the preparation will be defeated. Deliquescent salts such as calcium chloride may be used to hasten hardening and help to keep the patch in a moist condition.

Pneumatically applied cement mortars (shotcrete) are frequently used in the repair of concrete structures. When properly proportioned, mixed, placed, and cured, the coating is dense, hard, and very strong. Normally this work is done by specialist firm.

Epoxy resins are now used successfully for concrete repairs. Generally, preparation should be similar to that for other methods of repair, except that the concrete should be as dry as possible at the time the repair is made. The epoxy-bonded mortar or concrete is compacted into the hole and finished as with cement patches. Temperatures are important since epoxy resins cure and harden more quickly as the temperature increases. Hardening is too slow for the practical use of epoxy resins below about 5°C. Epoxy repairs are particularly useful on concrete roads. Their effective life is uncertain, since plastics age and shrink to some degree with time.

Repairing and Filling Cracks

In such structures as dams, reservoirs and culverts, the formation of cracks due to shrinkage of the concrete or to settlement of the structure is liable to have serious consequences. Sealing of these cracks is difficult, perhaps the best method being to pump liquid cement under pressure into specially prepared boreholes.

Liquid cement will not, however, penetrate cracks less than about $\frac{1}{50}$ in. wide. For smaller cracks, chemical solutions may prove satisfactory as a means of reducing seepage of water. The plastics chemist is providing an increasing variety of materials for injection processes. Careful selection is needed if such a material is to provide long-term impermeability.

Other methods have been suggested from time to time including the use of lead-wool caulking, red lead and other materials, but have not proved satisfactory. The cutting of a groove which is patched with cement mortar is more likely to result in two cracks at the edges of the groove than to provide a permanent seal.

It is of little use repairing cracks due to subsidence without treatment of the foundations.

REFERENCES

15.1 Blake, L. S., Kinnear, R. G. and Murphy, W. E., Recent research into factors affecting the appearance of in situ concrete, *Symposium on Surface Treatment of In Situ Concrete*, Sept. 1964. Cement and Concrete Association.

15.2 *A Guide to Specifying Concrete*, Part 1—*Type 1 Specification*, A Report of the Concrete Specification Committee, Institution of Civil Engineers.

CHAPTER 16

Formwork

THE economical design and construction of formwork is of great importance since its cost may be one-third, or even more, of that of the whole concrete structure.

In addition, the appearance of the finished concrete surface and the speed with which the construction is done are largely dependent on the use of properly constructed forms arranged in a manner to give the most efficient usage.

It is not possible here to cover the many problems which commonly confront the designer and builder of the formwork. The design of the various members of the formwork should, nevertheless, receive careful attention, and generally it should not be left to the foreman carpenter to build the forms haphazardly from timbers of his own choice. The design should be made with a knowledge of the strength of materials, the loads to be carried and the general pattern required on the surface from an aesthetic point of view.

Design and planning

The hydrostatic pressure of wet concrete, which causes most of the bulges and collapses of formwork, depends mainly on the rate of filling, the rate of setting and hardening of the cement, temperature, the richness of the mix, and the water/cement ratio. In other words, the pressure is dependent on the comparative rates of filling and initial stiffening and setting of the concrete. The rate of pouring is therefore important, especially in walls and columns, since the faster the forms are filled, the higher the hydrostatic pressure becomes at the bottom.

Other factors, including the method of compaction, the size and shape of the formwork, also influence the pressures. Closely spaced reinforcement may be expected to have a similar effect.

Recent research by the Cement and Concrete Association on behalf of the Civil Engineering Research Association, has provided information which enables the formwork designer to predict loads on the forms with greater accuracy than was previously possible. The Report[16.1] on this work includes the pressure design chart reproduced in Fig. 16.1 built up from the test results. Its use is described in the Report as follows:

The radial lines represent rates of placing *R* from 1 ft/hour to 50 ft/hour while the full spiral lines represent different temperatures of concrete from 40°F to 90°F. The broken spiral lines *d* are arching limits for minimum section dimensions from 6 in. to 18 in., whereafter arching effects are no longer applicable to most concretes.

The diagram is used by selecting the appropriate rate of placing and temperature of concrete and then measuring the pressure radially, from point O in accordance with scale *P*, to the point at which the lines representing the above criteria intersect. The basic pressure thus derived is then modified by multiplying by a factor *F* (described in the next section)

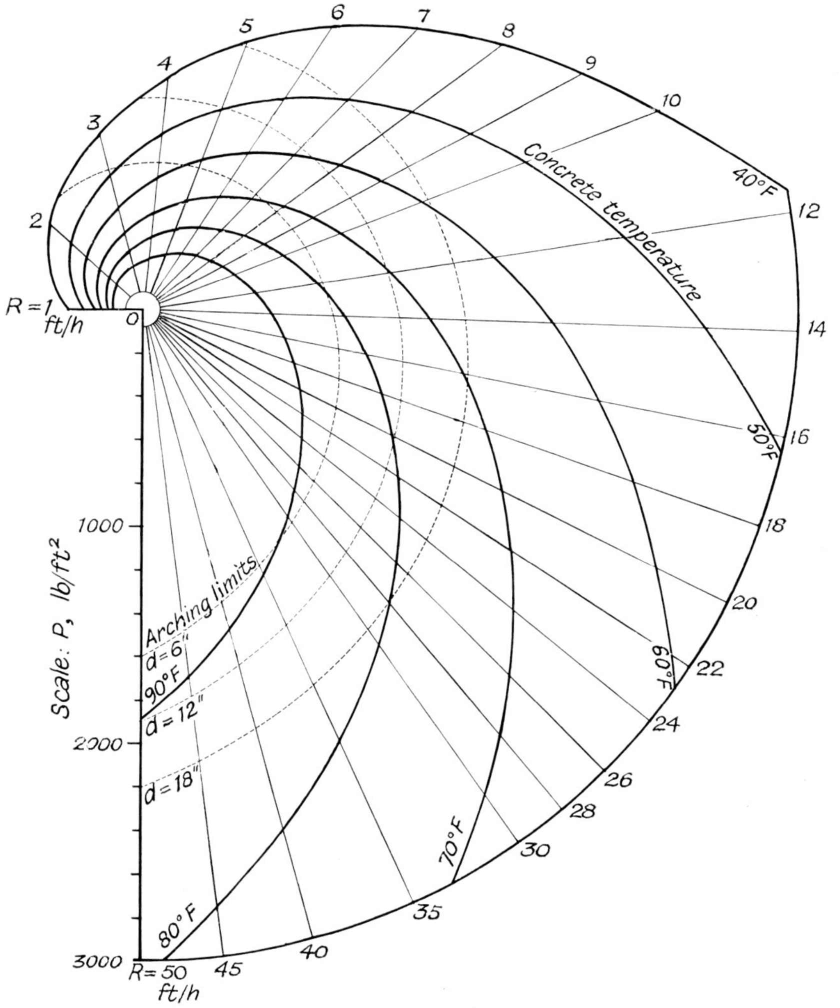

Fig. 16.1. Pressure design chart

which takes account of concretes of different workability and different degrees of continuity of vibration. The arching limit for the appropriate section is then checked on the same radial 'rate of placing' line and the lesser value used to determine the design pressure.

The pressure design chart is bounded by two arbitrary limits, a maximum pressure of 3,000 lb/ft² and a minimum temperature of 40°F. Circumstances outside these limits are very rarely encountered in practice but may be catered for by use of the pressure formulae in a previous section of the report.

Correction Factor, F

The basic design pressure P_{max} is finally corrected by a factor F (see Fig. 16.2) to compensate for concretes of different workabilities and for different continuities of vibration. (The design chart in Fig. 16.1 is based

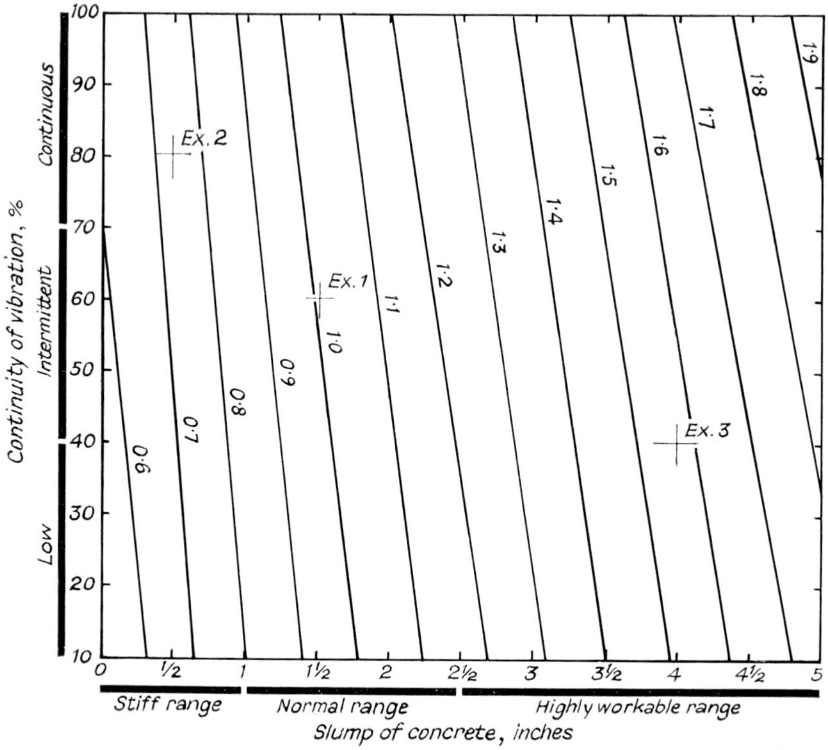

Fig. 16.2. Values of factor F

on a slump of $1\frac{1}{2}$ inches and a continuity of vibration of 55 %). An addition of 200 lb/ft² is made after the application of Factor F, to take account of pressure surcharges due to impact.

The shape of the design curve is determined only by P_{max} and the total height of the structure. The upper portion of the curve is triangular in shape and P_{max} was found to occur for all practical purposes at a depth of P_{max}/Δ ft below the top of the lift. The shape of the curve below P_{max} depends upon the degree of movement of the forms relative to the partially stiffened concrete. Where the formwork is highly rigid little reduction in pressure below P_{max} occurs; where the formwork flexes under load, a redistribution of pressure occurs and the pressure influence line may begin to tail off below P_{max}. The mechanism of the pressure reduction is highly complex, and in the absence of detailed information regarding the degree of rigidity of the forms it is recommended that a rectangular shape be adopted for the lower part of the pressure curve. It should be noted, of course, that P_{max} cannot be greater than the total head of concrete H multiplied by its density, i.e.

$$P_{max} \ngtr \Delta H$$

A worked example as given in the Report follows:

Calculate the design pressure on the formwork for a concrete column $2\frac{1}{2} \times 4 \times 16$ ft high, to be cast at a vertical rate of placing of 12 ft/hour in the summer at an estimated concrete temperature of 60°F. A single 3 in. poker vibrator will be used with about 60% continuity to compact concrete having a slump of $1\frac{1}{2}$ in. The concrete will be discharged from a skip without using trunking. The concrete will be site mixed and there will be no appreciable delay between mixing and placing.

$\Delta = 150$ lb/ft^2
$H = 16$ ft (therefore maximum possible hydrostatic pressure
$= 150 \times 16 = 2400$ lb/ft^2)
$R = 12$ ft/hour (therefore placing time is $\frac{16}{12}$ hours $= 80$ min)
$d = 30$ in.
concrete temperature = 60°F
slump = $1\frac{1}{2}$ in.
continuity of vibration = 60%

From Fig. 16.1, intersection of R of 12 ft/hour and concrete temperature of 60°F gives $P = 1{,}630$ lb/ft^2.

From Fig. 16.2, Correction factor $F = 1.01$.

$$\text{Design pressure} = (P \times F) + \text{impact allowance}$$
$$= (1{,}630 \times 1.01) + 200 = 1{,}850 \text{ lb/ft}^2$$

The shape of the design curve is, therefore, as shown in Fig. 16.3.

There are several requirements in the design of the formwork which must be met in the interests of efficiency and economy. The formwork must be designed to carry all the necessary loads with a reasonable margin of safety. This will cover all the details from the forms to the joists, girts, studs, shores, yokes, walings, props, etc.

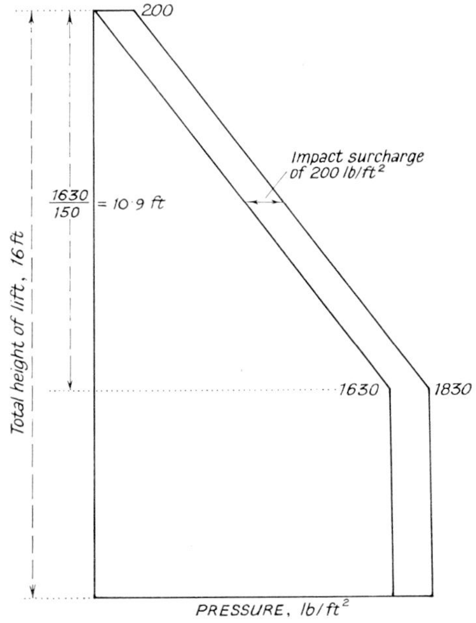

Fig. 16.3. Design curve for worked example

Every effort should be made to avoid the continual remaking and changing of forms, in order to prevent undue waste of timber. To this end, the formwork should, as far as possible, be made simple and to a symmetrical design. Frequent changes in sizes of columns and beams should be avoided whenever practicable, and in many cases it will be found that a standardization of size, which may result in a slight excess of concrete, will save many times the cost of the carpenter's work and timber.

It is advisable to standardize the sizes so that local standard widths of timber may be employed. Economy and efficiency in stripping is very important, since, if the formwork is so built that it has to be torn apart before being used for the next lift, not only will there be a loss of materials and labour, but probably damage to the concrete, owing to difficulty in moving.

For further information on the details of design, reference should be made to the several publications on the subject.[16.2, 16.3, 16.4]

General requirements

General requirements for forms are:

(1) The sheeting should be durable and rigid, and should be braced sufficiently to prevent bulging or twisting.

Sheeting to concrete walls and columns is frequently seen to be buckled, twisted and with untrue edges. The sheeting should be thick enough to withstand the pressures of wet concrete, and the edges should be true. Warping can be avoided by using seasoned timber and by careful storage when not in use.

(2) The forms and falsework should have sufficient strength to support the loads which they are called upon to carry, without undue deflection. The load may, in certain cases, include not only that of the concrete, but also that of men, equipment and materials used in the construction. Sufficient allowance should be made for any settlement or deflection, so that the hardened concrete conforms to the specified line and level. The engineer may also require a camber to allow for the elastic deflection of structural members, and that due to creep of the concrete. A Committee of the Reinforced Concrete Association (now the Concrete Society)[16.5] recommends that, in the absence of any specified camber, the soffit of all beams more than 16 ft in span, other than pre-stressed beams, shall be laid to a camber the amount of which shall equal in inches the square of the span in feet divided by 50 times the depth of the beam in inches.

(3) Forms should be correct to line and adequately braced. Reasonable tolerances are necessary in order to avoid the considerable expense involved in working to close limits, and it is advisable to specify them whenever possible. The tolerances given in Table 16.1 for the finished concrete, are intended only as a guide to general practice, and may be varied according to the particular conditions of the individual structure or component. For further information, reference should be made to the *Guide to Specifying Concrete*[15.2] from which the figures in Table 16.1 were extracted. A.C.I. recommendations[16.6] are very much more detailed, but are generally of the same order of magnitude.

To avoid buckling, the section of struts, posts, ledgers and shores,

TABLE 16.1 *Examples of Permissible Tolerances for Various Classes of Finish*

Type of structure	Type of irregularity	Tolerances (in.) Type of finish formed F1	F2	F3
Buried concrete in (foundations, culverts, retaining walls, etc.)	Departure from alignment and grade as shown on the drawings	$+2$ $-\frac{1}{2}$	—	—
	Variations in cross-sectional dimensions (1)	$+\frac{1}{2}$ $-\frac{1}{4}$	—	—
	Abrupt (3)	$\pm\frac{1}{2}$	—	—
	Deviation from templet in long dimensions (2)	$\pm\frac{1}{2}$	—	—
Exposed concrete in (columns, piers, retaining walls, abutments, etc.)	Departure from alignment and grade as shown on the drawings	—	$\pm\frac{1}{2}$	$\pm\frac{1}{2}$
	Variations in cross-sectional dimensions (1)	—	$+\frac{1}{2}$ $-\frac{1}{4}$	$+\frac{1}{2}$ $-\frac{1}{4}$
	Abrupt (3)	—	$\pm\frac{1}{4}$	$\pm\frac{1}{8}$
	Deviations from templet in long dimensions (2)	—	$\pm\frac{1}{2}$	$\pm\frac{1}{4}$

Notes

1. These tolerances refer to the cross-sectional dimensions of structural members such as walls, columns, beams, etc. where, for structural reasons, it is desirable to keep the tolerances within closer limits than those for alignment and grade. Tolerances smaller than those shown in the Table may be necessary for precast concrete columns, beams, and wall panels manufactured for assembly on site.
2. Gradual irregularities caused by misalignment of formwork from the dimensions shown on the drawings and measured from a templet of specified length.
3. Offsets and fins caused by displaced or misplaced formwork sheathing, lining or supports by loose knots in forms or by otherwise defective form material.

should be as nearly square as possible, and the sizes should be large enough to carry the loads and forces which are transferred without bending or buckling. Reduced sizes of struts are permissible if they are securely braced together at intervals along their length by horizontal or diagonal lacings.

(4) The junction between the formwork and any previously laid concrete should be tight in order to avoid any leakage of cement paste or mortar and subsequent ridging at this point. Lipping due to slight movement in steel forms is shown in Fig. 16.4.

(5) The joints between the various members should be tight enough to prevent any material leakage of the cement mortar, as such leakage causes defects in the appearance of the finished work.

(6) Forms and formwork must be of reasonable size and weight for easy handling with the equipment available. The forms should be easily and speedily erected and should be designed so that striking can be effected easily and in the required sequence.

(7) The pattern of the formwork, arrangement of vertical stop ends and the general quality of the form surface should conform with the requirements of the finished appearance of the structure. Straight and true angles, arrises or edges are particularly important.

Fig. 16.4. Concrete wall at a power station. Slight lipping has occurred at several of the joints where the shutter panels were not tight against the concrete previously placed

(8) The bearing faces of all timbers in struts, bearers and cleats should be cut square to the line of thrust with an accurately made joint. Instances have occurred when a strut has failed without warning due to a gradual slipping at the face of the bearing. Nailing of cleats should only be resorted to when absolutely necessary.

Folding wedges are to be preferred over the single type, but unless they are cut true they can be a source of constant trouble. The rule that wedges should only be used at the lowermost end of any strut or bearer, is a sound one, since the loads are more concentrated at these points, and consequently there is a reduced risk of slipping.

Common faults in formwork are illustrated in Fig. 16.5; the correct method of erection for the same wall is shown in Fig. 16.6.

Other items also result in movement, such as the omission of washers under the bolt heads (with the result that under a concrete pressure the bolt head becomes embedded in the wet timber), the use of wedges with too steep an angle, or wedges which are driven between badly cut timbers, and the loosening of members by vibration when this method of compaction is adopted.

Slipping, movement, or bending of formwork, whether in steel or timber, usually results in a serious loss of fines, leaving honeycombed patches. Further, the bulge which occurs is unsightly and very difficult to hide without cutting back the concrete to some depth, re-forming, and providing a new face.

Forms for prestressed concrete construction should be designed in such a way that the restraint due to elastic shortening during the stressing operation is minimized. Also any restraint to the shrinkage of the concrete should be kept to a minimum.

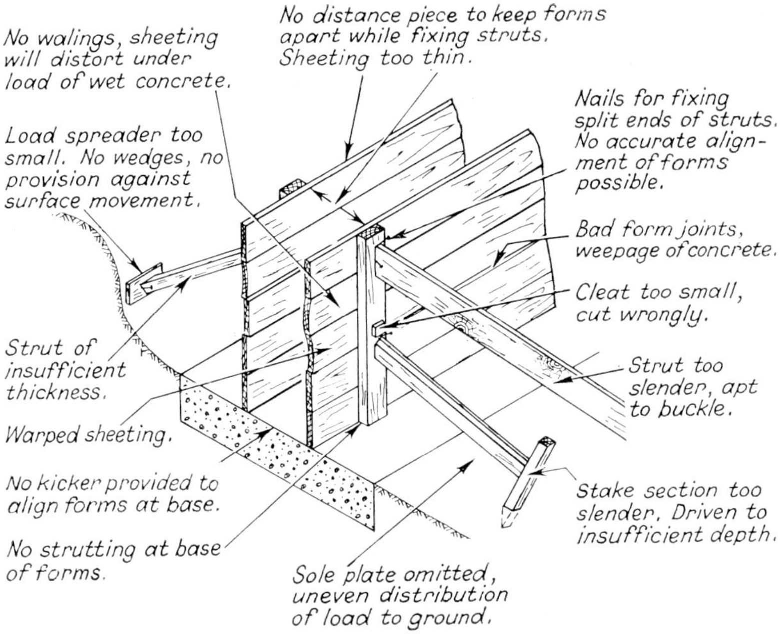

Fig. 16.5. Faults in formwork erection

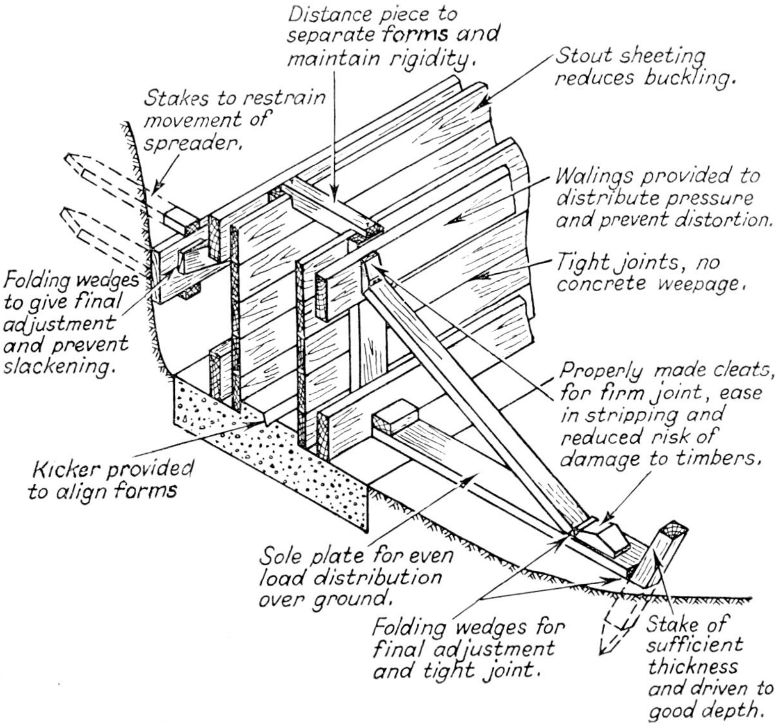

Fig. 16.6. Good formwork design

Timber for formwork

Timber which is only partially seasoned, i.e. with a moisture content not lower than about 20 per cent, is the most satisfactory for use in formwork. Kiln-dried timber has a tendency to swell when soaked with water from the concrete, and if the boards are tight jointed this swelling causes bulging and distortion. On the other hand, green timber dries out and shrinks, causing fins and ridges on the concrete, although this tendency can be checked to some extent by keeping the boards wet until the time of placing.

When using seasoned timber the joints should not be made too close, since provision should always be made for some slight swelling without any undesirable results. If the surface finish is important old and new timber sheeting should not be mixed indiscriminately. Old timber which has previously been used in concrete forms is usually more absorbent and has a rougher surface than new timber, and therefore gives a different surface texture.

Redwood (fir) is generally regarded as the best all-round material for formwork, having a higher resin content than whitewood (spruce), which tends to make it more durable, and gives more uses. American southern pine and Canadian redwood are also satisfactory, but are little used in Great Britain. Hemlock is sometimes used but the redwoods are preferred.

For such purposes as supports and long span-bearers which will be heavily loaded, particularly in the larger sizes of timber, Douglas fir is particularly suitable and is commonly used. American pitch pine, or Southern long-leaf pine are also very good, especially for marine work.

Hard woods are difficult and costly to work, and are therefore not used in formwork to any great extent. They are, however, very valuable for caps and wedges and for details where a large number of repetitions are to be made.

Timber may be rough sawn or wrought (planed). The main carcasing timbers such as props, joists and posts are usually rough sawn. Wrought timber gives more accurate assembly and a smoother surface finish if that is desirable. As indicated in the chapter on Surface Finishes, the carefully planned use of rough sawn boards gives an interesting finish which weathers well. If rough sawn boards are used, mortar tight joints are best obtained by planing the edges.

Wrought timber may be prepared in various ways, such as all four sides, one side and one edge; one side and two edges, etc., but it is usually better from the point of view of uniformity, size and adaptability to use timber wrought on all four sides. This applies especially to sheeting and studs. Freedom from knots and coarse grain is necessary in the sheeting if a smooth finish is required.

General sizes of timber

It is essential to obtain some degree of standardization in the sizes of the various scantlings of timber to be used, in order to avoid undue waste through repeated cutting to odd sizes. The following sizes of timber give a general indication of present-day practice:

Sheeting. By using boards 5 in., 6 in., 7 in., 8 in., and 9 in. wide, various heights of centering can usually be obtained without excessive sawing. Narrower boards, 3 in. or 4 in. wide, are commonly used in the Scandinavian countries, giving an improved appearance, Fig. 15.4. In general, sheeting is 1 in. to 2 in. thick, depending on the loads to be carried. The thinner sheeting (1 in.) is generally used for floor soffits and for the sides of beams and column boxes in the lighter types of construction. For heavy construction the thickness will be increased to $1\frac{1}{2}$ in. These thicknesses are for rough-sawn boards and should be reduced by $\frac{1}{8}$ in. when determining the wrought sizes. Two-inch timbers are often used for the soffits of beams in order to obtain economy in strutting after the side forms have been removed, and generally for very heavy construction, especially where much repetition is possible. The boarding may be tongued and grooved, squared, or bevel-edged. Tongued-and-grooved boarding avoids leakage of cement and gives a smoother surface, but is troublesome in that after having been used several times the boards no longer fit closely and in many cases the tongues are broken off.

Joists are commonly 6 in. or 7 in. × 2 in., but may vary from 4 in. × 2 in., to 9 in. × 4 in., according to the class of work.

Posts may be 3 in. × 3 in., up to 6 in. × 6 in.

Studs, walings, headtrees, transoms, and subsidiary bearers generally may vary from 4 in. × 2 in. to about 6 in. × 6 in. Sections about 3 in. × 2 in. are useful for general bracing and light forms.

Lengths of timber should always be ordered as nearly as possible to the length required, except in the case of sheeting which can be obtained in random lengths and cut in accordance with requirements. Every effort should be made to so arrange the work that cutting of timber is kept to a minimum since the cost of timber is considerable in reinforced concrete construction and a surprising amount of waste can occur.

Nails, wire and bolts

Common wire cut steel nails are used in sizes ranging from 2 in. to 4 in., a common rule being to use a nail at least twice as long as the thickness of the timber nailed on. The holding power of nails is indicated in B.S. Code of Practice C.P. 112 (1967).[16.7]

Nails driven at the ends of the timbers tend to split the wood with subsequent loosening of the formwork. Holes for large nails in such positions should be bored with a suitably sized drill instead of driving, and the heads should be left proud to facilitate easy removal on striking.

Tying wire, which is sometimes used for tying wall forms, may be black annealed wire, Imperial Standard Gauges between 9 and 12. The use of this wire for external surfaces is not considered good practice since rusting appears where the wire has been cut off level with the surface of the concrete.

Bolts are generally used in heavy wall construction in sizes from $\frac{1}{2}$ in. to $\frac{3}{4}$ in. They are usually drawn after use and in such cases are well greased or fitted with cardboard sleeves. Some special types of bolts are jointed an inch or two from the form surface, the central portion being left in the concrete. The holes left in the surface are then patched up.

Bolts used without large washers, are apt to bite into the timber and cause gradual movement, which will affect the finished dimensions of the concrete wall.

Other equipment

Many devices have been developed in order to facilitate the erection and removal of forms. These include such devices as the many and various sizes and shapes of clamps, column yokes, adjustable shores, centerings for floor slabs, and jacking arrangements. Most of these are satisfactory and economical, provided their use is limited to work for which they are suited and that care is taken to see they are not lost or damaged, since their initial cost is usually high.

Stop ends

Vertical stop ends present problems when they have to be constructed in and around congested reinforcement. Often pieces of timber cut to shape are wired to the reinforcement or wedged in place. Another method is to use expanded metal wired between the reinforcement and to leave it in place. In order to avoid corrosion it should not project beyond the reinforcement, and timber should be used near the surface.

Vibrated concrete

Vibratory methods of placing concrete produce some vibration in the formwork, especially when external vibrators are attached to the forms themselves. This vibration increases the tendency for joints to open and close, the nails to draw out, wires to stretch, wedges to loosen, and bolts to slacken. Extra care must, therefore, be given to these points. The following precautions are generally advisable:

(1) Tongued-and-grooved sheeting should be substituted for squared-edge sheeting since it gives tighter joints and a stiffer form.

(2) Particular care should be taken to see that all joints between different sections of form and between the formwork and concrete previously laid are tight, in order to prevent the escape of mortar.

(3) Wires should be avoided as far as possible, bolts being substituted, the nuts well tightened and watched for any slackening.

(4) Wedges are preferably nailed to prevent them from becoming loose. If used in conjunction with steel forms, they should be continuously watched.

(5) When external vibrators are used the form bracing should be increased by about 25 per cent, preferably by closer spacing.

Examples of formwork design

A few examples of formwork design are given in Figs. 16.7 to 16.18, as an illustration of the general arrangement adopted for sections which commonly occur in practice.

Basement wall construction.

A typical arrangement of the timbering and formwork for a basement wall during construction is given in Fig. 16.7. In this instance concreting has been completed to the underside of the second bracing which will

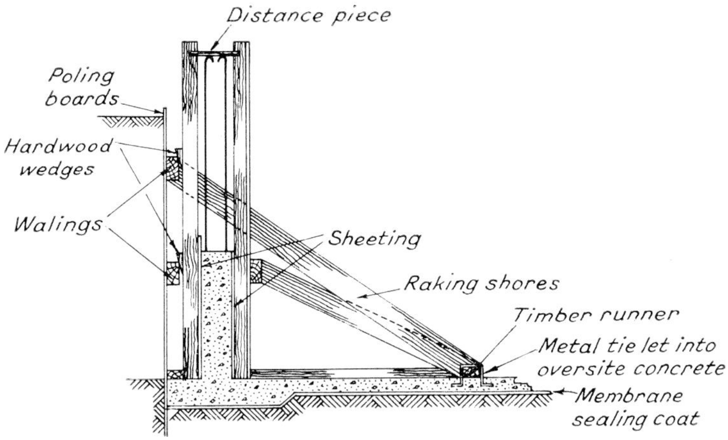

Fig. 16.7. Type of formwork used for a retaining wall

next be removed and will take a new bearing on the right-hand side of the concrete shutter.

It is important that the raking shores should not be at a steep angle, otherwise any horizontal force also produces a vertical force which may cause the formwork to lift.

An example of formwork to the walls of a services duct is given in Fig. 16.8.

Fig. 16.8. Construction of formwork to the walls of a service duct

Beam and slab forms.

The formwork to beams is usually constructed so that the sides may be struck before the soffit, since the latter must be left in place until the beam can safely support its own weight. The slab formwork is usually supported by that to the beam.

Examples of forms for beams and slabs are given in Fig. 16.9. Figs. 16.10 and 16.11 show the construction of the forms on the site.

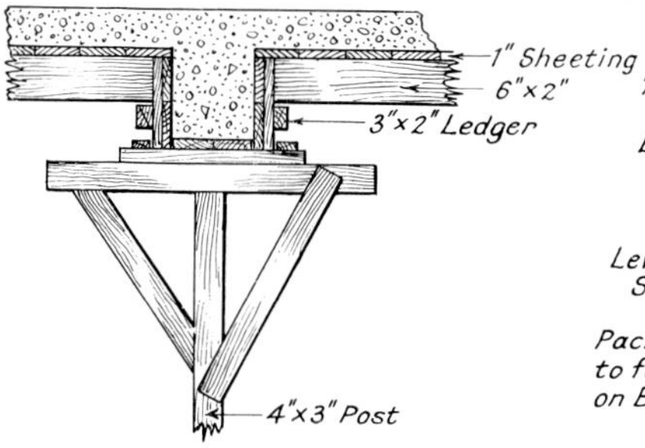

(*a*) Beam box with slab formwork supported by bearers fixed to sides of beam.

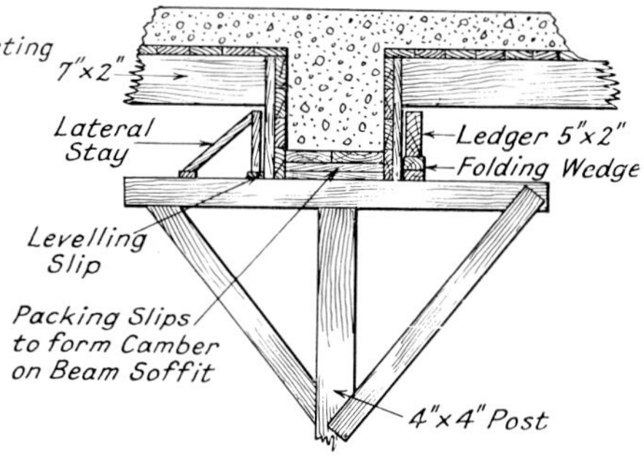

(*b*) Beam box with alternative methods of carrying slab formwork on bearers spanning between props.

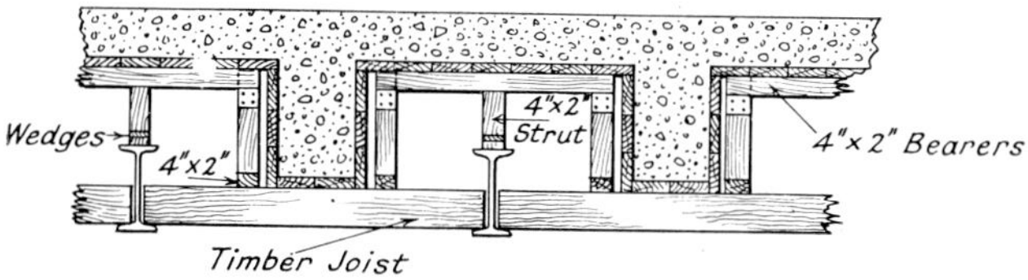

(*c*) Formwork supporting a heavy beam and slab construction.

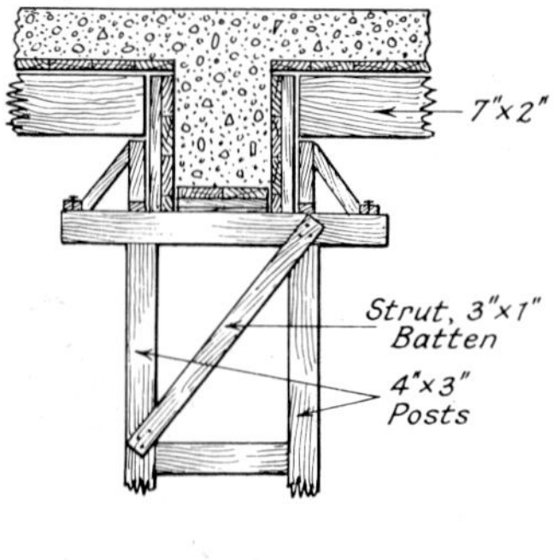

(*d*) Use of twin posts for beam formwork.

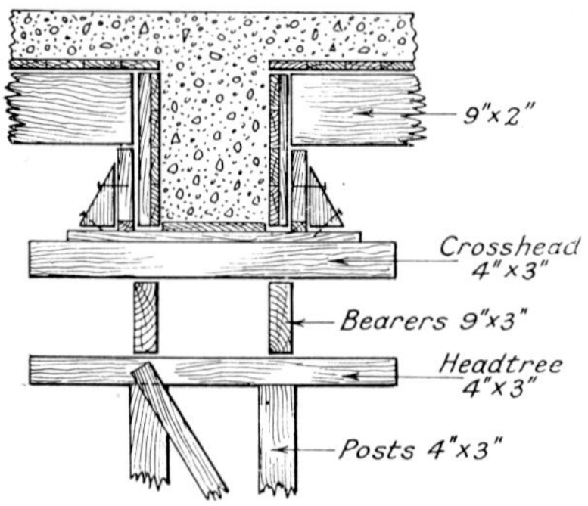

(*e*) Formwork for heavy beam.

Fig. 16.9. Examples of formwork design for beams

Fig. 16.10. The construction of wooden forms for reinforced concrete beams and slabs

Column forms

A common arrangement for a column form is given in Fig. 16.12, and is seen in practice in Fig. 16.13. The two sides of the column are held together by bolts and the two opposite sides by hardwood wedges between the bolts and the form. The sheeting runs vertically and is made up into panels. Provision is made at the head to link up with the formwork for the horizontal beams. In this illustration the formwork allows placing only from the top, inferring lifts up to, say, 10 ft in height. This is practicable provided the reinforcement allows the use of some form of trunking to break the free fall of the concrete. Otherwise it is desirable to arrange one side of the formwork in removable panels, say, 3 ft high. The concrete can then be placed in 3 ft lifts, tamped, rodded or vibrated

Fig. 16.11. Combination of wooden and steel shuttering for floor and beam construction

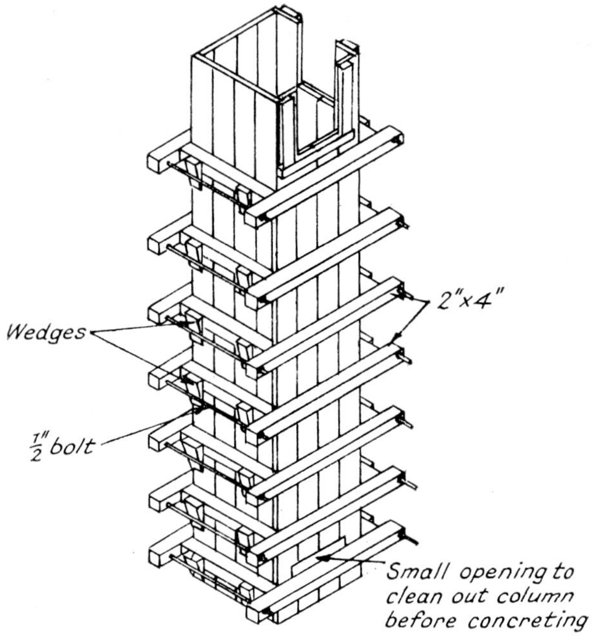

Fig. 16.12. Column form

Fig. 16.13. Formwork for a column. The posts carrying the soffit of the secondary beam formwork can also be seen

at each lift. Even this method presents its own difficulties in that it is not easy to introduce concrete from the side if the reinforcement is congested.

Wall forms

Figure 16.14 shows a simple wall form held together with bolts fitted with sleeve pieces. Figure 16.15 is an instance of the use of wire ties to hold wall forms together.

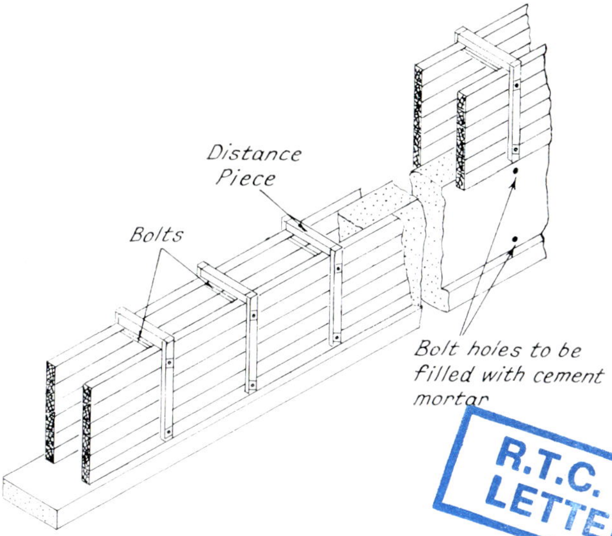

Fig. 16.14. Simple bolted formwork for walls

Wire ties should never be used for uncovered external walls since the ends of the wires rust and produce stains, or if cut back the patching necessary always shows and is unsightly. Trouble may also arise due to the rusting of the ends of wire ties if gypsum plaster is used directly on to the concrete wall; staining and spalling may occur. Wire ties are apt to stretch slightly and so allow leakage of cement mortar at the junction with the previous lift, resulting in lipping and honeycombing.

Other forms

Moulds for a bridge parapet and cornice are shown in Fig. 16.16.

An example of the formwork for a wave wall is given in Fig. 16.17. The formwork used in the construction of a pier is shown in Fig. 16.18.

Special form linings

In order to obtain smooth surfaces free from the ridges and fins left by ordinary boards, special form linings such as paper, hessian, plywood, plastic sheeting, fibre boards and sheet metal have been used.

Paper is the cheapest type of form lining and is useful, when the finish is unimportant, to cover slightly open joints in formwork through

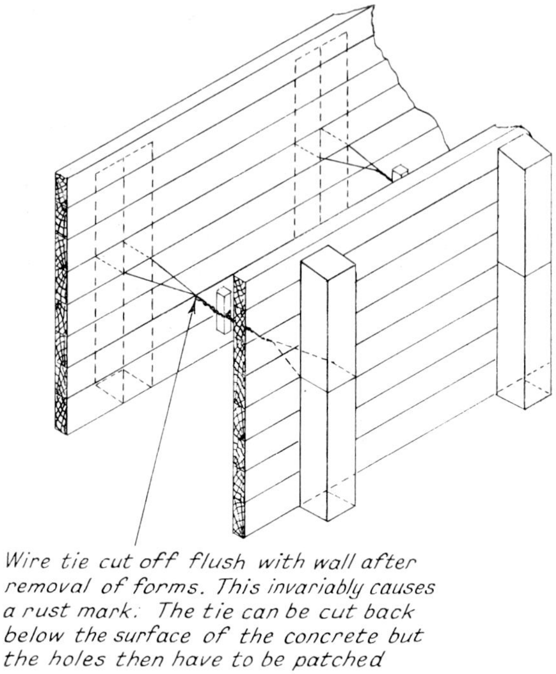

Fig. 16.15. Use of wire ties in a wall form

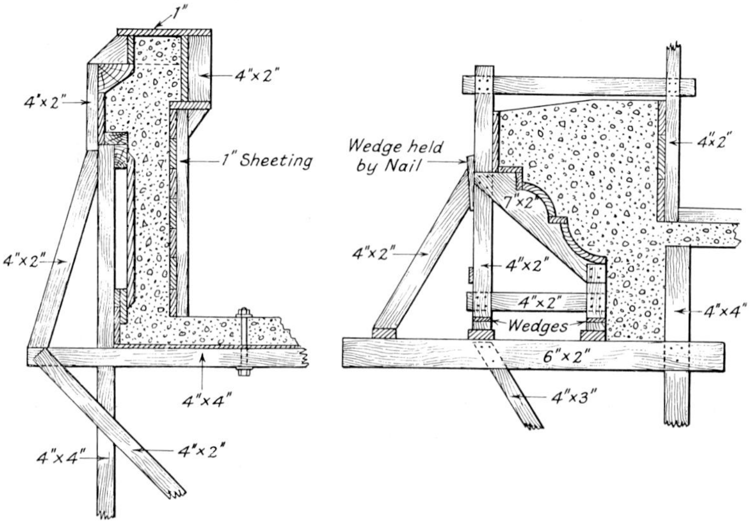

(*a*) Part formwork for parapet of bridge (*b*) Cornice form

Fig. 16.16

Fig. 16.17. Wave wall formwork in steel

Fig. 16.18. Various types of formwork in the construction of a pier. Part of the work was covered at each tide

which cement mortar would flow. It is, however, difficult to avoid cockling and to prevent concrete from getting under the paper at edges and laps.

Hessian stretched over the formwork in panels gives a slightly rough texture and hides most of the board marks. Some difficulty may be experienced in getting it stretched tightly and in stripping it cleanly from the concrete.

Plywood is a very satisfactory form lining especially if it is made from Douglas Fir and is resin bonded. The ply should be not less than $\frac{3}{16}$ in. thick for a small number of re-uses and small lifts. For a large number of re-uses and higher lifts greater thicknesses are desirable, $\frac{5}{8}$ in. (approx.) being commonly used. The thicker ply has the advantage of needing support at less frequent intervals.

Plywood has a tendency to split and fray at the edges after a number of uses, and should, therefore, be fixed in panels conforming with the general pattern of the structure and with suitable arrangements round the edges of the panel to mask the joints, as, for instance, by means of rebates. Plywood sheets should have a small gap between them to allow for swelling, and should be painted with a flat paint before use.

Plywood has considerable strength in itself and close boarding is not required behind it. Some indication of the spacing of the supports may be obtained from the following figures when using $\frac{3}{16}$ in. ply. For floor slabs 3 in. thick the supports should be spaced 5 in. apart, while if the slab is 10 in. thick, the spacing should be reduced to 3 in. A suitable spacing in wall forms where the lift is not more than 3 ft high is 2 in. With $\frac{5}{8}$ in. ply these spacings can be multiplied by about ten quite safely.

Plywood has an advantage over some other types of form lining in that it can be bent to curved surfaces. It is sometimes sprayed with a plastic based paint to give easy striking and a smooth finish to the concrete. Plastic sheeting may also be used to cover the plywood, to which it is usually stuck.

Sheet metal is sometimes used to cover forms which are to be used repeatedly. It gives a good surface providing it is oiled properly before use, and may be used for a large number of repetitions, but is costly to work in the first instance. Galvanizing is an advantage to avoid rusting. If the sheet metal is only supported at intervals, it is apt to belly after some time.

Fibre boards, made from wood fibres and resinous glues highly compressed, are sometimes used instead of plywood. They have the disadvantage that they tend to swell at the edges with repeated use, and the surface becomes rough. The number of re-uses obtainable is far less than with a good resin-bonded plywood. It is advisable to use only sufficient 1 in. panel pins along the top and sides of each panel to hold the sheet in position. Some manufacturers recommend fixing the sheets with panel pins in the middle only. The same remarks apply to chipboards, an additional disadvantage being weakness in bending.

Absorptive form linings consisting of insulating board $\frac{1}{2}$ in thick have been used to eliminate blowholes on the surface of the concrete. Absorptive form linings can only be used once and are too expensive for general use.

Precast concrete slab linings

Precast concrete slabs are sometimes used as a facing to monolithic concrete and are generally about 2 in thick. They are usually supported by narrow boards, one near the top and the other near the bottom of the slab, nailed to studs situated at the joints in the slabs. The purpose of the boards is solely to keep the slabs in their proper positions, and the slabs themselves should be reinforced to support the weight of wet concrete across the span between the boards. In some instances precast facing slabs are built into a masonry wall as shown in Fig. 15.14, and are then only given support at intervals along the length, and the lift of concrete placed behind them is kept fairly small.

Moving forms

Moving or slip forms are commonly used for such structures as storage bins and silos, and in certain cases for walls. Either they are made to give lifts of about 2 ft to 4 ft, or they are constructed to move continuously during concreting, thus avoiding loss of time during stripping and resetting. Continuous pouring day and night is an advantage structurally when watertightness is important, as with oil and water tanks.

Moving forms in general are built in sections and are raised on a centre post, on several posts placed around the wall, or by using two sets of forms and raising one on top of the other. Lifting is effected by a block and tackle or by jacking. In the case of continuously sliding forms, the forms are lifted by screw-jacks operating on rods or pipes which are left in the concrete.

Precautions to be taken with continuously sliding forms are that they should be well oiled and accurately centred before commencing to place the concrete. The jack rods or pipes must bear on the foundation slab and must be kept vertical. The concrete used should be fairly dry, with small-sized aggregate and well-graded sand, a slight excess of sand being preferable. The forms should be filled about three-quarters full before starting to raise them, and care must be taken to see that the concrete is not dragged up with the form and is sufficiently stiff at the bottom not to flow out. The raising of the forms should be kept about 12 in. to 15 in. ahead of the concreting, the level of the concrete being uniform all round the form.

An example of slip form construction is given in Fig. 16.19.

Metal forms

Various types of metal form have been developed for use in constructions which allow a large number of repetitions. Steel forms have been commonly used for tunnels, sewers, culverts and circular columns, and often for floor and column construction. Steel moulds are used extensively for the manufacture of many types of precast concrete products. Figure 16.20 shows Blaw-Knox steel wall forms being used for the wall of a factory building. It will be noticed that the steel panels themselves are supported by steel bearers and posts of standard angle sections.

A further example of the use of steel forms in the construction of a large culvert is shown in Fig. 16.21. The internal and external forms

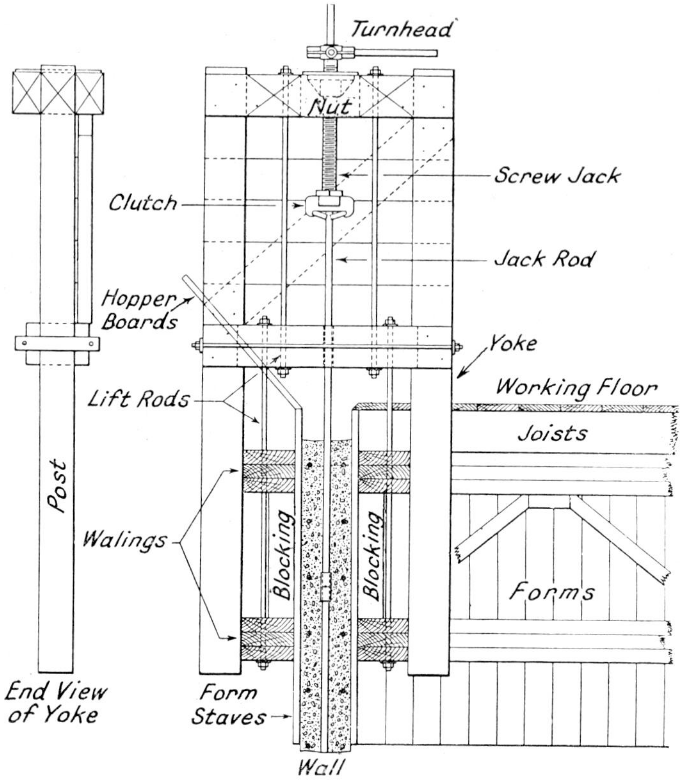

Fig. 16.19. Section of slip forms showing lift rods, yoke, jack and jack rod

Fig. 16.20. Steel wall forms

Fig. 16.21. Travelling formwork for a large concrete culvert

in this instance are carried on steel railway tracks along which they are moved as concreting proceeds.

Some care is necessary in using steel shuttering, in that the lift for which it has been designed may only be applicable if the rate of placing is slow. Failures of standard panel units through lack of appreciation of this point are not uncommon.

It is advisable when using steel forms to protect the back from hot sun.

Vacuum forms

The vacuum process is specially useful for placing concrete on a steeply sloping surface, as for instance the side of a reservoir, or a canal lining. The vacuum is applied to special top shutters, which consist of a stiff frame with a $\frac{5}{8}$ in. thick waterproof plywood face. Holes $1\frac{1}{2}$ in. diameter are bored through the plywood at 3 ft centres, and a $1\frac{1}{2}$ in. diameter pipe is attached to the back to coincide with the holes. Extra weight is often given to the shutter by filling the frame with concrete.

On to the front of the plywood is attached, (i) a sheet of $\frac{1}{4}$ in. gauge mesh; (ii) a screen of fly gauze, and (iii) a nylon cover stretched across. The vacuum is applied through the $1\frac{1}{2}$ in. diameter pipes, extracting air and water through the nylon and mesh. A suction equivalent to 20 in. of mercury is applied by the exhauster pump. Vibration of the concrete during placing helps to seal any small air leaks at the shutter, and hastens the process. The shuttering may be removed within $\frac{1}{2}$ hour of placing the concrete. See also p. 194 and Fig. 12.9.

Steel supports to formwork

Special types of floor centres have been developed for the support of sheeting, such as the Hico beam shown in Fig. 16.22. This beam is

Fig. 16.22. Hico floor beams in place ready to receive floor sheeting

made up from short sections, with adjustable links in the bottom tension member, by which adjustment can be made to the line and level. Such beams eliminate the necessity for intermediate strutting for most floor constructions.

Tubular steel scaffolding owing to its simplicity and adaptability, is extensively used instead of timber for the support of formwork. It has, however, some disadvantages compared with timber in that it has a higher initial cost and it is relatively weak when used as a beam. There is also more danger from loose couplings.

Tubular steel scaffolding is of greater advantage for structures of more than one storey, an example of the use of tubular supports in the construction of a viaduct widening being given in Fig. 16.23.

Fig. 16.23. Tubular scaffolding used as a formwork support

Removal of forms

The length of time that forms should remain in place depends upon the strength of the concrete and the loads to which it will be subjected on removal of the formwork, and will vary with different parts of the structure. In general the formwork can be removed earlier with richer mixes and with rapid-hardening cement.

Forms can be removed more rapidly in warm weather than in cold weather and unless special precautions are taken, it is necessary when the temperature remains round about freezing-point to leave the forms in place for some time. Formwork should not be removed without specific instructions from the engineer in charge, who should make sure that the concrete is strong enough to carry its own dead load and any other loads upon it. In C.P. 114(1965) it is stated that in no circumstances should forms be struck until the concrete reaches a cube strength of at least twice the stress to which the concrete may be subjected at the time of striking. Some authorities recommend three times. While no specific times can be given for the removal of formwork the following may be useful as a guide for structures in which normal Portland cement is used. When the temperature is low but does not fall below 2°C (36°F), the times should approach the higher end of the range given; when the weather is warmer i.e. around at 15°C (59°C) the times at the lower end of the scale may be used. The times may be reduced by about one third if rapid-hardening Portland cement, or ordinary Portland cement to which 2 per cent of calcium chloride has been added, is used. Reference may also be made to Figs. 4.9 and 4.10 for the rate of strength gain when using different types of cement.

(*a*) Forms to walls, columns and beam sides, 1 to 4 days according to conditions. Twelve hours may be sufficient if the temperature exceeds 15°C (60°F).

(*b*) Soffits of slabs and beams should be left supported as long as possible. Minimum times in days are:

	Props left	Removal of props
Slabs	3–10	7–14
Beams	7–14	14–28

The time of removal should be lengthened as the span is increased. The use of props for the support of beams is illustrated in Fig. 16.24.

(*c*) Arches and similar structures—7 days minimum for spans or diameters up to 10 ft and thereafter one day per foot increase up to 30-day maximum.

The figures given assume that the concrete will have the strength required by the specification and that cubes are tested before the formwork is struck. In case of doubt it is advisable to err on the side of leaving the forms in place too long, especially as regards the soffits of beams and slabs.

A.C.I. recommendations[16.6] are not very different from the times given above.

Release agents

Numerous materials have been painted on forms to prevent sticking of the concrete and to facilitate removal, such as whitewash, soft soap,

Fig. 16.24. Props supporting reinforced concrete beams after removal of the formwork

grease and oils. Oils are generally regarded as the most satisfactory, and various types have been used extensively on wooden, metal, concrete, and other form materials. The requirements of good release agents are that they shall prevent sticking, they shall reduce to a minimum absorption of water by the formwork, and they shall not harm the concrete either by staining or by softening of the surface.

The Cement and Concrete Association have undertaken comprehensive investigations into factors affecting the appearance of in-situ concrete. These were reported at a Symposium on Surface Treatment of In-Situ Concrete, held in London in September 1964. Their main conclusions on release agents are given below.

Straight oils may be of vegetable, mineral or fish base and should contain no additives. While they are particularly useful for impregnating timber surfaces and so give to the concrete uniform colour with little staining, they do tend to leave such blemishes as blowholes and pocking.

Oil-phased emulsions, which consist of water globules encased in a continuous phase of oil, are considered to be the most suitable for general use. They should be used as sparingly as possible, and then yield concrete of uniform colour, good durability, resistant to efflorescence, and with few mechanical defects such as blowholes. Staining is minimal. Some types are, however, unsuitable for use on steel formwork, since they have poor anti-corrosion qualities.

Water-phased emulsions consist of oil globules encased in a continuous phase of water. They are not generally satisfactory.

Oils can be purchased containing emulsifying additives, but no water, with the intention that this shall be added on site. Mixing with water on site, however, may not be sufficiently uniform to avoid variations in colour and finish on the concrete surface. Without water they may be used on steel formwork where their anti-corrosive properties may be advantageous.

Similar oils may also be used sucessfully on concrete moulds. Plaster-of-Paris mouldings should be given one or two coats of shellac before oiling in order to reduce absorption.

The surfaces to which the release agent is applied should be clean and smooth. The oil may be applied by means of an oily rag, a brush, or a spray, but whichever method is adopted, the forms must be covered fully but thinly. A satisfactory rate is 300 to 400 ft^2 per gallon. The important requirement that no oil should be allowed to get on to construction joints or reinforcing bars limits the use of a spray once the formwork has been fixed in place.

Surplus oil should not be left on formwork, especially if the concrete is to be plastered or rendered. Not only is there a risk of formation of a soft surface skin, but there is the probability that the surface will be water-repellent, both of which reduce the adhesion of plaster or rendering mortar, if these are to be applied.

REFERENCES

16.1 *The Pressure of Concrete on Formwork*, C.E.R.A. Research Report No. 1, April 1965. Civil Engineering Research Association, London.

16.2 WYNN, A. E. and MANNING, G. P., *Design and Construction of Formwork for Concrete Structures*, 1965. Concrete Publications Ltd., London.

16.3 SNOW, SIR FREDERICK, *Formwork for Modern Structures*, 1965. Chapman & Hall Ltd., London.

16.4 HURD, M. K., *Formwork for Concrete*. American Concrete Institute Special Publication No. 4, 1964. Distributed in United Kingdom by Cement and Concrete Association.

16.5 Specification Clauses for Structural Concrete, Technical Committee of Reinforced Concrete Association, *Structural Concrete*, **3**, No. 3, May/June 1966.

16.6 A.C.I. 347–63, *Recommended Practice for Concrete Formwork*, A.C.I. Standards, 1965. American Concrete Institute.

16.7 C.P. 112(1967), *The Structural Use of Timber in Buildings*. British Standards Institution.

CHAPTER 17

Reinforcement

THE design of the various parts of a concrete structure must be related to all the requirements necessary for satisfactory service. In some structures the only consideration may be that the members should be capable of carrying the loads imposed. In others there may be a necessity to limit the deflection of certain parts. In sea structures adequate cover must be given to the reinforcement to prevent corrosion and subsequent spalling. Similar considerations apply to reinforced concrete in factories where corrosive gases are present. Cracking should be controlled to reduce the size of the cracks and thus avoid an unsightly appearance, as well as to reduce the risk of corrosion of the steel. Again, cracking must be avoided in water-retaining structures and oil-tanks.

Reinforced concrete design is a subject in itself and reference should be made to standard textbooks.[17.1,17.2] The basic principles involved in designing simple concrete members to satisfy the requirements for load-bearing capacity are outlined below, and some notes are given on the control of cracking.

Various requirements concerning design considerations, including permissible stresses in concrete and steel reinforcement, size of aggregate in heavily reinforced sections, distances between bars, cover, bond and anchorage, and also concerning workmanship, inspection and testing, are given in the B.S. Code of Practice for the Structural Use of Reinforced Concrete in Buildings.[7.5] As the Code deals adequately with these aspects of reinforced concrete, brief reference only is made in this Chapter, except for such matters as cover, which are dealt with more fully as they are matters of site supervision and design.

Elementary principles of reinforced concrete design

The necessity for the use of concrete and steel together arises from the nature of the stresses which may result from loading. These are compressive, tensile and shear stresses, and they may occur in different members singly or all together.

Concrete has an adequate compressive strength, but its tensile strength is low. Tests have shown that the tensile strength is, in fact, only about one-tenth of the compressive strength, and for the purposes of reinforced concrete design it is ignored. On the other hand, steel has a high tensile strength, as well as a high compressive strength when adequate lateral support is provided. Suitable combinations of these two materials, therefore, provide resistance to both compressive and tensile stresses. Shear will be mentioned later. A further necessity for the efficient use of combinations of concrete and steel is that there shall be a good bond between them, and this will be considered after an explanation of the development of stresses in simple members has been given.

Beams

A load acting on a beam gives rise to tensile, compressive and shear stresses, so that a consideration of the design of a beam provides a convenient introduction to the elementary principles of reinforced concrete design. When a beam simply supported at both ends is loaded, the upper layers are reduced in length while the bottom layers are extended.

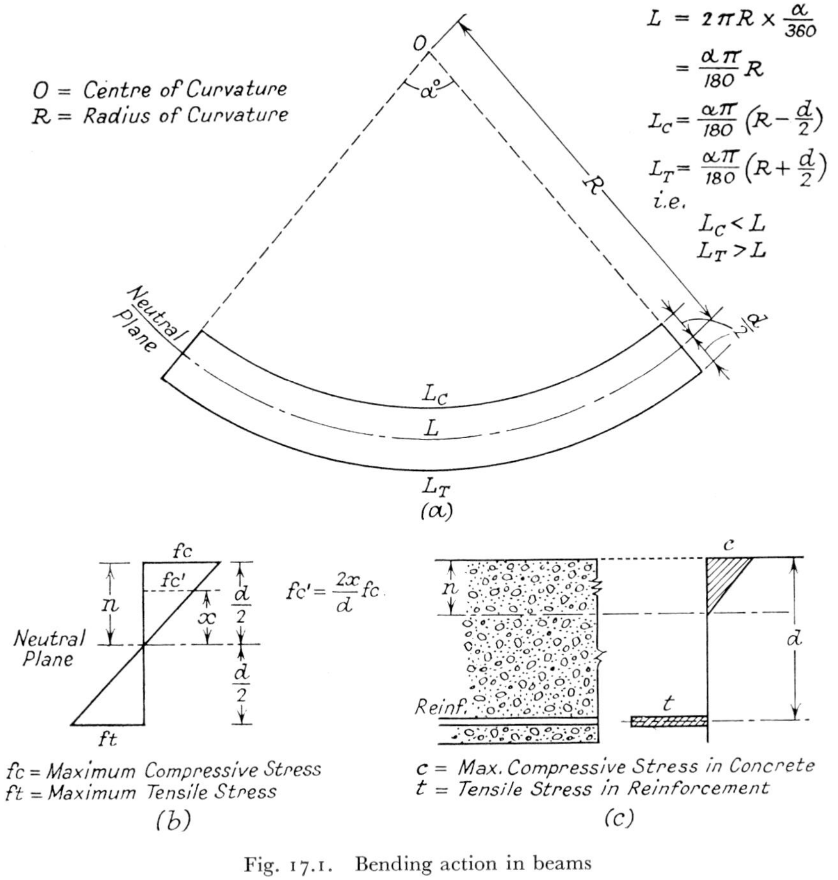

Fig. 17.1. Bending action in beams

Between the top and bottom of the beam there is a plane whose length remains unaffected, as indicated diagrammatically in Fig. 17.1 (*a*), and this is called the neutral axis. It should also be noted that the alteration in length increases with the distance from the neutral axis.

Those layers of a beam which are shortened or compressed must be subjected to a crushing action, while those which are extended must be subject to tensile forces. That portion of the beam above the neutral axis is therefore in a state of compression, and that portion below is in a state of tension, the magnitude of the stresses being directly proportional to the distance from the neutral axis in accordance with the stress diagram, Fig. 17.1 (*b*).

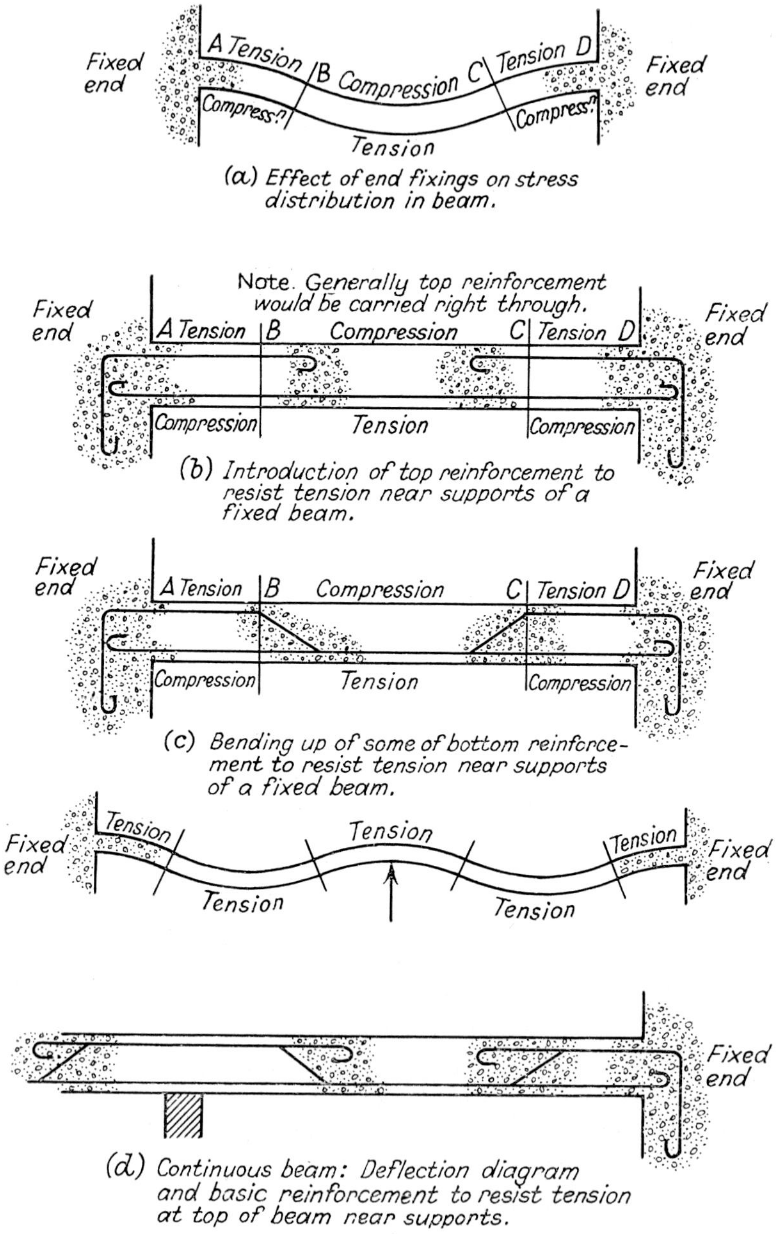

Fig. 17.2. Reinforcement in fixed and continuous beams

For any beam which is symmetrical about the neutral axis, which is then at the center of the depth, the maximum compressive stress is equal to the maximum tensile stress. This applies to any simply supported plain concrete beam of rectangular cross-section. The uppermost fibres will be subjected to the maximum compressive stress while the lowermost fibres will be subjected to the maximum tensile stress, and both stresses will be equal in magnitude.

Since concrete is much weaker in tension than in compression, it is obvious that failure under a comparatively small load will occur in the outer tension fibres, while the compression fibres would be capable of sustaining a much greater load. The beam would thus be uneconomic in design and of little practical use.

Consider now the effect of placing steel bars in the lower half of the beam near the outer tension fibres. The strength of the beam is no longer limited by the tensile strength of the concrete, and provided the bars are of the right sectional area the beam will be as strong in tension as in compression. The new stress diagram is shown in Fig. 17.1 (*c*). In order to obtain the greatest advantage from the reinforcement, it should be placed as near the bottom of the beam as possible. Some cover is, however, necessary to prevent corrosion and to provide the necessary fire resistance.

The area of steel required to give the necessary tensile resistance in a singly reinforced rectangular beam is generally less than 1 per cent of the area of the concrete. Beam sections are generally designed so that the allowable working stresses in the concrete and steel are reached simultaneously. It often happens, however, that owing to limitation of space the size of beam is restricted so that additional steel must be placed in the compression zone to avoid overstressing the concrete.

If a beam is fixed at the ends instead of being simply supported, fixing moments are introduced and the beam takes the form shown in Fig. 17.2 (*a*). The top surface of the beam is seen to be convex from *A* to *B*, and *C* to *D*, indicating that it is extended over these portions and therefore in a state of tension. From *B* to *C* the top surface is concave, indicating compression. It therefore becomes necessary to introduce top reinforcement near the supports to take care of the tensile forces there. This may be achieved as shown in Fig. 17.2 (*b*) or by bending up some of the bottom bars as shown in Fig. 17.2 (*c*). Continuous beams deform in the manner indicated in Fig. 17.2 (*d*) and should therefore be provided with top reinforcement near and over the supports.

A cantilever is stressed similarly to the end portion of a fixed beam, as shown in Fig. 17.3 (*a*). The position of the main reinforcement is shown diagrammatically in Figs. 17.3 (*b*) and (*c*). Mistakes are more often made in the positioning of reinforcement in cantilevers than in beams and floors, and special care should be taken to see that it is near the top and not half way down or near the bottom. Reference should always be made to the designer's detailed drawings and specification.

Shear stresses are usually greatest near the supports of a beam, and act horizontally and vertically as illustrated in Fig. 17.4 (*a*) and (*b*). These shear stresses produce tensile and compressive stresses at an angle of 45° to the planes of shear stress, Fig. 17.4 (*c*), and it is therefore necessary to

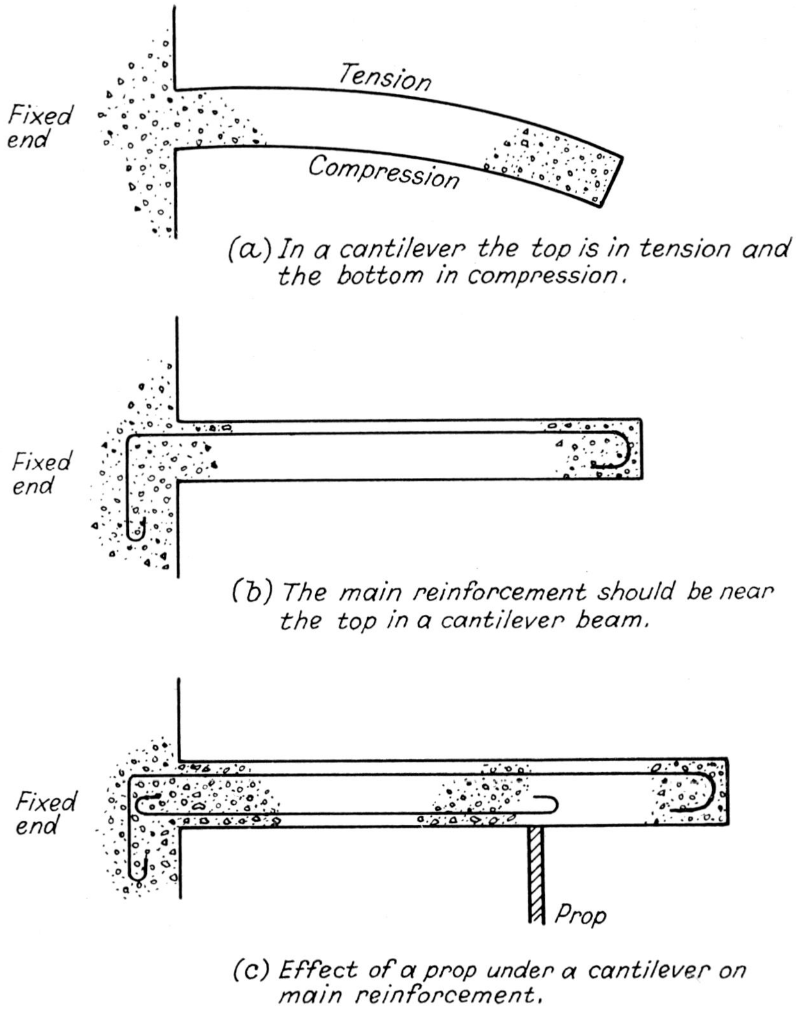

Fig. 17.3. Reinforcement in cantilever beams

ensure that the beam offers sufficient resistance, especially near the supports. This may be accomplished in two ways: (i) by means of stirrups, usually placed vertically as shown in Fig. 17.4 (*d*), and (ii) by bending up a proportion of the bottom reinforcing bars in the manner shown in Fig. 17.4 (*e*). The concrete itself may be taken to resist the shear forces provided the permissible shear stress given in the Code of Practice[7.5] is not exceeded.

The object of stirrups is, then, to resist shear stresses when they are beyond those which the concrete is capable of resisting. Stirrups will normally be spaced at closer intervals near the supports of the beam since the shear forces are at a maximum at these points. The shear force is zero at the centre of a symmetrically loaded beam. The stirrups must pass

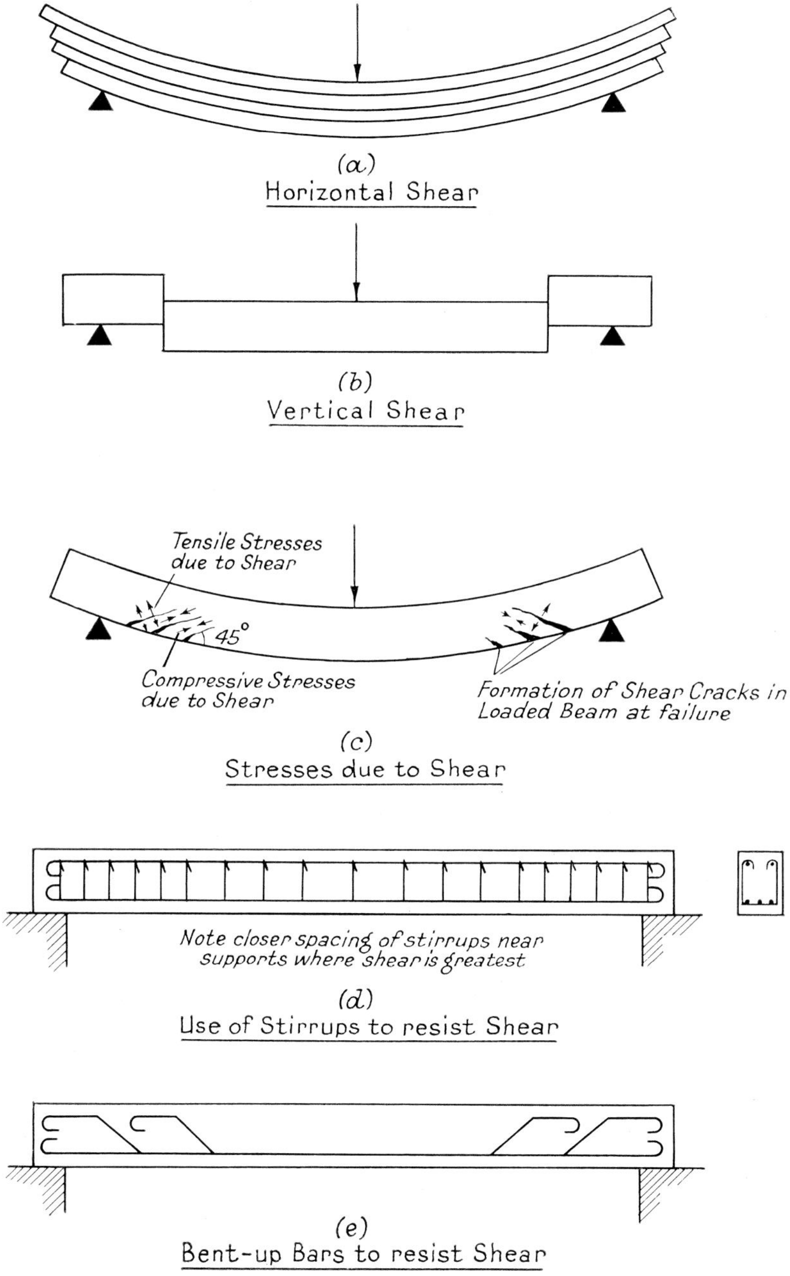

Fig. 17.4. Shearing action in beams

Note that top layers in (*a*) are actually compressed. In this diagram in which no resistance is assumed between layers (as in a pack of cards) each layer should be of equal length.

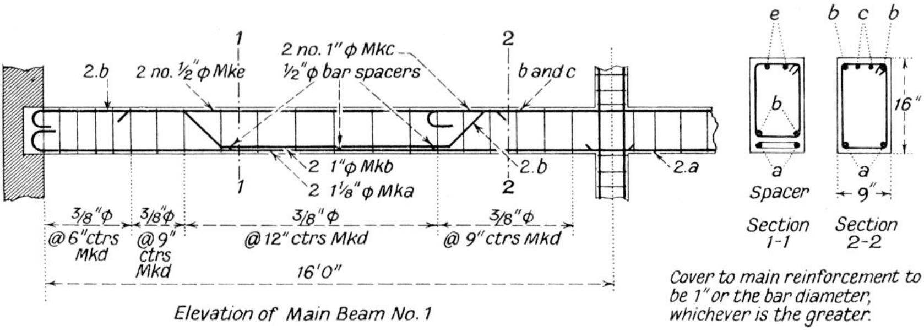

Schedule No. ....

Location	Mark	Type* and size	No. off	No. in each	Total no.	Length ft	Length in	Shape All dimensions are in accordance with B.S. 1478 unless otherwise stated
Main beam No. 1	a	MR† 1⅛	1	2	2	18	2	17'3"
	b	1	1	2	2	23	0	17'3"; 4'3"; 1'4"; 12¼"; 6'9"; 4'3"; 1'4"
	c	1	1	2	2	11	0	9'6"
	d	⅜	1	20	20	4	2	13¾"; 6¾"
	e	½	1	2	2	10	0	Straight

* Preferred abbreviations for mild steel, high yield steel, round and square bars are M, H, R and S.
† MR denotes mild steel, round bars.

Fig. 17.5. Typical reinforcement diagram and standard form of schedule (B.S. 1478: 1964) for a reinforced concrete beam

round the main tensile reinforcing bars, and must be effectively anchored at both ends in order to develop the full working stress.

The points enumerated above are illustrated in Fig. 17.5, which gives a typical arrangement of the reinforcement in a beam.

Columns

Plain concrete columns are capable of carrying considerable axial loads, but in order to reduce the cross-sectional area, especially of long columns and those subjected to eccentric loading, it is normal practice to use reinforcement. When a column is loaded it is compressed, and if the concrete and steel are bonded together, both must compress equally. It therefore follows that the compressive stresses in each must be proportional to their respective moduli of elasticity. The maximum stress which can be developed in the steel is therefore *m* times the safe compressive stress in the concrete, where $m = \dfrac{E_s}{E_c}$, E_s and E_c being the moduli of elasticity of the steel and concrete respectively.

The Code of Practice[(7·5)] limits the cross-sectional area of the longitudinal reinforcement to between 0·8 per cent and 8·0 per cent of the gross cross-sectional area of the column. The limits of bar sizes are given later. When the proportion of longitudinal reinforcement approaches 8 per cent, serious difficulties arise in concreting, especially at sections where splicing occurs and a lower percentage is to be recommended. The Code recommends that, where bars from the column below have to be lapped with those in the column above, the percentage of steel should usually not exceed 4 per cent. A smaller size of aggregate and a somewhat richer mix may be advisable at such positions.

The compression reinforcement requires lateral support to prevent it from buckling outward under load. This is provided by the use of transverse reinforcement in the form of lateral ties, so disposed that every longitudinal bar is held against outward buckling. The ends should be anchored.

In order to resist buckling of a reinforced concrete column it is necessary to place the bars as near the outer surface as possible, consistent with adequate cover.

Slabs

Slabs are sometimes supported on only two sides, and at others on all four sides. They may be simply supported or fixed at two or more of the edges. In all cases it is usual to provide reinforcement running in both directions. If the slab spans in one direction only, the bars at right-angles to the direction of spanning are known as distribution steel. If the slab spans in two directions main tensile reinforcement will run in both directions. Top reinforcement is, of course, necessary if the slab is fixed or continuous over a support, as in the case of beams.

General

It is not possible to deal with other reinforced concrete sections, but the brief description of the elementary principles of reinforced concrete design should be sufficient not only to indicate the necessity of fixing reinforcing bars accurately in the positions determined by the designer, but to enable the concrete engineer to make a rough assessment of the seriousness of any errors. It is most important that the requirements given on the working drawings by the designer should be closely studied.

Bond and anchorage

An effective bond between the concrete and the reinforcement is required since the efficient use of combinations of steel and concrete is dependent on the transfer of stresses from the concrete to the steel. The bond strength, or the measure of the effectiveness of the grip between the concrete and the steel, is best defined as the stress at which a very small slip occurs. Bond is due initially to adhesion and frictional resistance, but as soon as slipping begins the adhesion fails and the subsequent bond is due to friction and mechanical resistance.

The bond strength is improved by increasing the crushing strength of the concrete but is not directly related to it. The development of bond strength is most rapid under moist curing conditions.

Deformed bars provide increased bond strength although slightly more slip is likely to occur before the full bond strength is developed. In C.P. 114 a deformed bar is defined for the purpose of assessing bond strength as one for which it has been proved that the bond strength exceeds that of a plain round bar by 25 per cent or more. For such a bar bond stresses may be increased by 25 per cent.

Bars in tension must extend for a sufficient distance from any stressed section to develop a satisfactory resistance to failure of the bond between the bar and the concrete. Values for the permissible bond stress and methods of calculating the bond stress are given in the Code of Practice.

Hooks at the end of bars should have an inner diameter of at least four times the diameter of the bar; except that when the hook fits over a main reinforcing bar the diameter of the hook may be equal to the diameter of such bar. Hooks at the ends of high-yield bars should have an inner diameter of at least six times the diameter of the bar. The length of the straight part beyond the end of the curve to the end of the hook should be at least four times the diameter of the bar.

Control of cracking

The tensile strength of the concrete is generally ignored in normal reinforced concrete design and it is assumed that the concrete is cracked at the working load. A certain amount of cracking is, therefore, normal in the parts of reinforced concrete members in tension, but if the cracks are distributed along the length of the member they can be made small enough to be of no account. On the other hand, if the cracking is not controlled, one or more large cracks may occur, causing unsightliness and a route for the penetration of water to the reinforcement, with consequent corrosion. Cracking can also occur due to tensile stresses set up in restrained members by shrinkage, Fig. 17.6. This shrinkage may be the normal drying shrinkage of the concrete and there may also be shrinkage due to thermal movements. Shrinkage cracking in such parts of a structure as wall panels is unavoidable if there is an external restraint.

The formation of cracks is accompanied by a local failure of bond between the steel and concrete, so that a high total bond resistance is likely to impede the development in width of cracks more than a low total bond resistance. From this it may be concluded that the development in crack width is retarded by the use of bars of small diameter (i.e. with a large relative surface area), and a high percentage of steel. A high bond stress is also required, this being related to the physical properties of the concrete.[17.3]

In general, the probability of cracks forming as a result of bending of concrete members is greater as the stress in the tension steel is increased. Reinforcement may, however, be used to control the width of cracks, the greater the quantity of reinforcement that is provided, the smaller being the width of individual cracks, while the total width of cracking remains constant.

The use of higher stresses with high tensile steel has led to a closer

Fig. 17.6. Shrinkage cracks in a reinforced concrete plinth
Note the darkened patch above the grill and the cracking of the trowelled-on sill.

study of crack formation. High tensile steel is usually produced as deformed bars, giving increased bond resistance, which also has the merit of distributing cracks. These are in consequence smaller, although greater in number. It is, however, necessary to limit tensile stresses in the steel to prevent unduly large cracks forming, and a number of crack formulae have been developed for this purpose.

One such formula is given in the C.S.A. Standard A23.3[17.3] as follows:

"If in working stress design the computed tensile stress in reinforcing steel exceeds 24,000 lb/in^2, or in ultimate strength design the specified yield strength or yield point exceeds 40,000 lb/in^2, crack widths shall be investigated and:

(*a*) The width of flexural cracks calculated by the following formula shall not exceed

(i) Members exposed to the weather 0.008 in.
(ii) Members not exposed to the weather 0.012 in.

$$w_{\max} = f_s \sqrt[4]{\frac{A_{ct}}{N}} \times 10^{-7}$$

f_s = calculated stress in reinforcement,
N = no. of bars constituting total tensile reinforcement,
A_{ct} = area of concrete surrounding the tension reinforcing bars and having the same centroid as the total tensile reinforcement."

The crack widths given above are similar to those given in the Principles of the *Recommendations for an International Code of Practice for Reinforced Concrete*[2.17] which states that the following limits may be generally accepted for the maximum width of cracks:

(*a*) For normal structures in a normal atmosphere 0.3 mm (0.0117 in.).

(*b*) For internal structural parts in a humid or aggressive atmosphere, and external parts exposed to the weather, 0.2 mm (0.0078 in.).

(*c*) For internal or external structural parts exposed to a particularly aggressive medium or where watertightness is needed, 0.1 mm (0.0039 in.).

The International Code concludes, however, that in view of all the uncertainties it is not advisable, for design purposes, to carry out complete theoretical calculations for cracking. The Regulations will have to provide simple formulae.

An investigation of crack control in concrete beams made on behalf of the Civil Engineering Research Association[17.4] concluded that the maximum crack width on the surface of the effective tension zone was predicted most accurately by the formula:

$$w_{\max} = KC\frac{f_s}{E_s}\left(\frac{d - d_n}{d_1 - d_n}\right)$$

where C is the distance of the point of measurement of the crack from the surface of the nearest main reinforcement bar,
d is the distance from the compression face of the section to the point of measurement of the crack,
d_1 is the distance from the compression face of the section to the centroid of the main tension reinforcement,
d_n is the distance of the neutral axis from the compression face of the beam,
f_s is the mean stress in the reinforcement,
E_s is the modulus of elasticity of the reinforcement,
K is a constant of value 3·3 for deformed bars and 4·0 for plain rounded bars.

It is usual to provide both vertical and horizontal reinforcement near the outer and inner faces of wall panels. Extra shrinkage reinforcement should be provided at sudden changes of section as, for example, corners of door and window openings.

Types of reinforcement

Mild steel bars

Plain mild steel bars manufactured in accordance with B.S. 785 are normally employed for concrete reinforcement and are required to have an ultimate tensile strength in the range of 28 to 33 ton/in^2. The bars can be readily bent without fracture provided that the radii of the bend or hooks are not less than the standard appropriate to the particular bar diameter. If heating is employed to aid bending of the larger diameter bars, cooling by water quenching should be avoided. Permissible stresses are given in B.S. Code of Practice C.P. 114.[7.5]

High tensile deformed hot-rolled bars

At the present time there is no British Standard for this type of bar, but it is expected that it will be covered by an addition to B.S. 785 in the near future. Hot-rolled bars produced by all steel manufacturers are claimed to have a minimum yield stress of 60,000 lb/in^2, but other properties such as elongation and bend radius requirements vary from one manufacturer to another depending on the constituents of the steel, the rolling process and the type of deformation produced by the patterned rolls. With this type of bar longitudinal ribs run straight along their length.

Heating of bars to aid bending, or welding at splices and junctions should not normally be permitted. If it is essential to heat or weld the bars this must be done only under close technical supervision, as is the case with cold-worked bars to B.S. 1144; reversed bending or rebending must be prohibited in all circumstances.

Regular tests for tensile strength should be made because there is a risk of the inclusion of mild steel in deliveries of these bars.

High tensile cold-worked bars

The manufacturing process for this type of steel involves the twisting or stretching of square or deformed section mild steel rods manufactured in accordance with B.S. 785. The bars may generally be recognized by ribs or other features forming a pattern along their length. The "cold working" given to the mild steel bars raises their tensile strength and yield strength. The increase in tensile strength is gained at the expense of some ductility, so that fracture of the bars may take place during bending unless proper precautions are taken.

Only correct sizes of mandrel appropriate to the bar diameter should be used in the bending machine, and it is most important to avoid re-bending or reversed bending. For example, cranked bars projecting from a partly completed structure should not be temporarily bent one way and then bent back again to their final position, to give clearance for a construction operation. This operation would involve bending the bars at least twice and possibly three times, and fracture would then be likely.

Heating should not, in any circumstances, be employed during the bending of high-tensile cold-worked bars. If this is done the tensile strength of the steel will revert to the B.S. 785 quality. Similarly heat produced in welding bars at junctions or splices can weaken the steel; therefore welding should only be carried out under skilled technical supervision.

When high-tensile cold-worked bars are purchased in accordance with B.S. 1144, the purchaser should check that the original steel satisfied B.S. 785 or an approved equivalent standard.

Mesh reinforcement

Mesh reinforcement to Part 1 of B.S. 1221 is fabricated from hard-drawn wire complying with Part 2 of B.S. 785, or square twisted bars complying with B.S. 1144. The coiled hard-drawn wire is straightened mechanically, and the straightened lengths are cut and electrically welded to produce the various sizes of mesh. In other processes, square section rods are twisted and cut to length for the fabric. The straightening and welding processes can alter the mechanical properties of the wire; therefore the British Standard requires the supplier to furnish data on the steel manufacturing process and the results of mechanical tests on the wire before fabrication as well as mechanical tests on pieces cut from the welded fabric. The latter test pieces should include at least one transverse weld within their length.

The cold working given to the steel in the wire drawing and twisting (if any) processes raises the tensile strength at the expense of some ductility. However, because mesh reinforcement is normally fabricated from small diameter bars there is not usually the same risk of fracture on reverse bending as occurs with the larger cold-worked high-tensile bars. There is however a risk of fracture at carelessly made bends where damage has occurred to the steel forming a notch or a sharp kink.

Mesh reinforcement is used mainly as reinforcement for slabs and roads.

Working drawings for reinforced concrete members

Working drawings should provide information concerning the reinforcement to be placed in the various members and should be sufficiently detailed to enable the reinforcement to be correctly scheduled in accordance with B.S. 1478:1964. The method for preparing such details cannot be discussed fully in this book but the following points may be used as a guide.

All the reinforcement in a beam should be shown on the elevation together with an indication of the number and size of bars in each layer, and any other necessary dimensions. See Fig. 17.5 as an example. It is wise also to make a large scale drawing of the reinforcement at laps, intersections, etc., to ensure that there is room for the bars at these points, and that the intended positioning can be maintained.

The size of stirrups should be specified and the spacings shown.

Cross-sections of the beam should show the overall size, the various bars, and the cover of concrete to be provided.

Columns should be shown in elevation in a similar manner to beams giving overall sizes and cross-sections at various points. The drawings should show the size and spacing of the main bars and the shape and distances between the transverse ties.

In the case of slabs and wall panels it is usual to draw in only a few of the bars of each type in full, together with their sizes and spacing. An indication should be given as to whether the bars are top bars or bottom bars and whether they are main bars or distribution bars. Sections should also be given showing the thickness of the slab, the concrete cover and the laps, bends and cranks in the bars. In all cases the positions and lengths of laps and other such details should be given.

Requirements for the cover to the reinforcement are given on p. 271.

Bending schedules

Bending schedules are prepared for all the bar reinforcement in order to show the size, number, total length and bending dimensions. Each bar should be identified on the drawing, the bending schedule, and the labels attached to the bars, by the same mark, which should be a symbol or number and as simple as possible. The bar bender should be given all the dimensions necessary to enable him to bend the bar as accurately as possible to the shape required. Important items are the sizes of cranks and the dimensions of hooks. It should be noted that the dimensions of the parts to be cranked are measured to the outside of the bar, whereas the dimensions of stirrups and binders are given from inside to inside and not from centre to centre. An example of a bending schedule is given in Fig. 17.5.

Inspection of reinforcement on arrival

Reinforcing bars should be inspected on delivery for damage in transit or heavy rusting. Some definition is required of the term heavy rusting, since ordinary rust which adheres tightly to the bars has been found to have no detrimental effect on the bond strength. Objectionable rust and scale may be defined as that which can be removed by firm rubbing with a piece of sacking or a hard brush.

The diameters of the bars should be checked to ensure that they are correct, and that corrosion has not caused a material reduction in the cross-sectional area.

On a site near a chemical works, a reduction of bar diameter of $\frac{1}{32}$ in. was measured after a period of storage of six months in the open.

Checking diameter of deformed bars

Nominal diameters of deformed bars refer to the equivalent diameter of a circle having the same cross-sectional area as the bar. When measuring bars to check their diameter it should be remembered that the raised ribs can cause the greatest measurable diameter to be 10 per cent more than the nominal and the manufacturers' recommendation should therefore be adhered to. If there is any doubt on the actual nominal diameter, a sample length should be weighed to check that the weight per foot is appropriate to the specified nominal diameter.

Tests on reinforcement

Occasional laboratory tests should be made on samples of reinforcement to ensure that it complies with the relevant standards. These tests should be made more frequently in the case of bars or fabric manufactured from high-tensile steels, since the risks of failure, should the steel be of an inferior grade or quality, are higher than for mild steel.

Certificates should be obtained from manufacturers. This may not be easy to arrange because of the number of hands through which the steel may pass from the original manufacturer to the final purchaser. For example, billets of steel may be purchased by a rolling mill for the production of hot-rolled high-tensile steel. The bars so produced may then pass into the hands of a steel stockist before it is purchased by the contractor. In the case of hard-drawn wire mesh fabric, the steel may pass from the rolling mills in rod form to a wire drawer, after which the coils are straightened and resistance welded to produce the fabric reinforcement. Further complications arise when fabricators carry out their rolling and re-rolling and welding processes on imported Continental steel.

Because of the above circumstances, the identity of the originally manufactured steel billet can easily be lost, and in the event of failure of test specimens there is none of the basic data on which the results of mechanical tests can be judged. It is for this reason that British Standards require steel suppliers to furnish information about the original steel manufacturer, the steelmaking process employed, and the analysis of the original cast. Steel fabricators do not always seem to be aware of the effects of their processes on the mechanical properties of the steel. For example, the tensile strength of a hard-drawn wire may be lowered quite appreciably in the process of straightening the wire from coils.

Bending reinforcement

Bends and hooks in reinforcing bars should be made in accordance with the requirements of B.S. 1478. The radius of the bend is dependent on the steel stress at the point of the bend, on the permissible concrete stress, and the position of the bend. The danger of splitting of the concrete due to compression induced by the bend is dependent on its position. Splitting is unlikely if the bend is protected by other reinforcement, but care should be taken in narrow beams and similar members.

Reinforcement is usually bent in hand-operated bending machines of the type shown in Fig. 17.7. On works contracts and in steel yards where large quantities of reinforcement are dealt with, mechanical bar-benders may be used. Inefficient equipment is liable to result in inaccurate bending and consequent difficulty in fixing. Skilled bar-benders should be employed on this work in order to avoid inaccuracies. Stretching of bars, normally occurs to some extent during bending, and after completing the first bar it is usually necessary to make minor alterations to the positions of the bends of subsequent bars in order to obtain closer agreement with the intended dimensions. This applies particularly if there are several bends in the same bar. The extent of the adjustment necessary depends on the type of machine and the method adopted to position the bars in relation to the mandrel of the machine.

Fig. 17.7. Bending steel reinforcement

Fig. 17.8. Fixing steel reinforcement to a beam

Bars should be bent cold whenever possible in a slow and regular movement. Rapid movement is likely to lead to fractures which may not be detected at the time the bars are bent. Bending hot, at a cherry red heat (not exceeding 1,550°F), may be allowed, except for bars which depend for their strength on cold working. Bars bent hot should not be cooled by quenching. In order to avoid fractures during cold weather, special precautions such as reducing the speed of bending or increasing the radius of bending may be necessary.

Attention is drawn in C.P. 114:1957, Amendment 1965, to the fact that, when reinforcement bars are bent aside at construction joints and afterwards bent back into their original position, care should be taken to ensure that at no time is the radius of the bend less than four times the bar diameter for mild steel or six bar diameters for high-yield steel.

Some form of marking should be adopted to enable the steel fixers to identify bars without difficulty. The bars should be stacked so that the various lengths and sizes can be found easily, and in such a way that damage is avoided.

Fixing reinforcement

It is very important that reinforcement should be fixed in the correct position and properly supported to ensure that displacement will not occur when the concrete is placed. Skilled steel fixers are essential if the work is to be carried out efficiently, especially with the complicated reinforcement which is required in many cases.

Reinforcement is, in some cases, built up on supporting frames and

Fig. 17.9. Reinforcement supported on "chairs"
These are small bars welded to form the supporting units and tied to the main reinforcement with wire.

then hoisted into position in the moulds, the erection being completed after it has been placed in position. This often facilitates the work and leads to greater accuracy, although some tolerance must be allowed as suggested under "Cover," page 271. It is not always realized just how difficult it is to align a mass of reinforcement when efforts to straighten a bar at one point are quite likely to result in further disturbance elsewhere.

Fig. 17.10. Use of concrete spacer rings for location of reinforcement in the shuttering

Supports of various types are obtainable which prevent displacement during concreting. Commonly used materials are concrete and plastic spacing blocks or rings, asbestos cement rings, metallic supports and other suitable devices, see Figs. 17.9 and 17.10. It is important that some such means is adopted to prevent movement of the steel which might reduce the cover and lead to spalling after exposure to the weather. Examples of displacement of reinforcement revealed during the reconstruction of gas industry plants(17.5) are shown in Fig. 17.11.

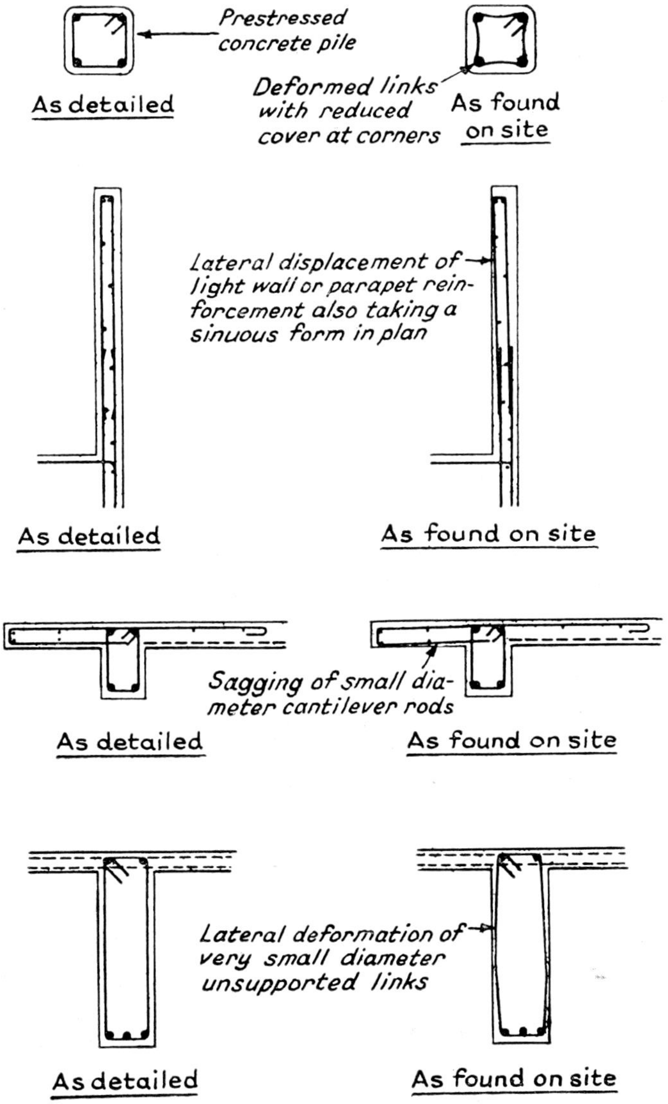

Fig. 17.11. Typical examples of defects exposed in concrete work during reconstruction

Reinforcing rods are usually held together by soft iron wire ties. The tying wire is ordinarily between 16 and 18 Imperial Standard Wire Gauge. Patent spring wire fixings are available for use instead of tying wires. They have the advantage that they can be clipped on quickly.

Certain limits to the sizes of reinforcement are given in Code of Practice C.P. 114, to which reference should be made for such information when required.

Distance between bars

In order to facilitate placing of the concrete the minimum lateral distance between reinforcing bars should be the diameter of the bar, or $\frac{1}{4}$ in. more than the maximum size of the coarse aggregate, whichever is the greater. If the bars are of different sizes the distance should be at least the diameter of the larger bar. A greater distance should be provided when convenient. However, a closer spacing of two-thirds of the nominal maximum size of the coarse aggregate is acceptable for groups of bars, provided a sufficient space, normally 3 in., is left between groups to allow the insertion of a vibrator.

The vertical distance between horizontal main reinforcing bars should be at least $\frac{1}{2}$ in., or the maximum size of aggregate if this is greater than $\frac{1}{2}$ in., except at splices or where transverse bars are in contact.

Other requirements concerning the distance between reinforcing bars are given in C.P. 114.

Corrosion of reinforcement

Reinforcing bars are normally protected from corrosion by the impermeability of the surrounding concrete, and its alkalinity. The latter

Fig. 17.12. Failure of a beam and corbel supporting a seaside promenade. Failure was caused by penetration of sea salts and lack of adequate cover to the reinforcement, resulting in rusting, expansion and spalling of the concrete

property has probably saved many reinforced concrete members, in which the concrete has not been sufficiently impermeable to prevent the entry of moisture, or where fine cracks have occurred.

The object of specifying a minimum cover is to provide a practical guide based on general experience. However, such cover may not give sufficient protection in the presence of salts. Sea salts present in the aggregate, or the use of brackish water or sea water for mixing or curing, increase the tendency for corrosion to occur. Overdoses of calcium chloride have the same effect. Serious failures have been recorded, even in arid climates, where the only moisture present is likely to have been that due to condensation.

Maritime structures exposed to alternate wetting and drying, should be constructed with strong, dense concrete, and adequate cover given to the reinforcement, if failures such as that illustrated in Fig. 17.12 are to be avoided.

Cover to bars

Adequate cover to reinforcing bars is necessary if trouble from corrosion and consequent spalling of the concrete is to be avoided. It is therefore particularly necessary to see that the proper cover is provided, and to ensure by inspection of the reinforcement before the concrete is placed that no displacement is likely to occur.

C.P. 114[7.5] requires that the thickness of concrete cover to the reinforcement (exclusive of plaster or other decorative finish), should be:

(*a*) for each end of a reinforcing bar in a column, not less than 1 in. nor less than twice the diameter of such bar;

(*b*) for a longitudinal reinforcing bar in a column, not less than $1\frac{1}{2}$ in., nor less than the diameter of such bar. In the case of columns of minimum dimension of $7\frac{1}{2}$ in., or under, whose bars do not exceed $\frac{1}{2}$ in. diameter, 1 in. cover may be used;

(*c*) for a longitudinal reinforcing bar in a beam, not less than 1 in. nor less than the diameter of such bar;

(*d*) for tensile, compressive, shear or other reinforcement in a slab, not less than $\frac{1}{2}$ in., nor less than the diameter of such reinforcement;

(*e*) for any other reinforcement, not less than $\frac{1}{2}$ in., nor less than the diameter of such reinforcement.

For all external work, for work against earth faces, and also for internal work where there are particularly corrosive conditions, the cover of the concrete should not be less than $1\frac{1}{2}$ in. for all steel, including stirrups, links, etc. An exception is made to this requirement where the face of the concrete is adequately protected by a suitable cladding or protective coating. A cladding is recommended where severe corrosive conditions occur. Protective coatings, as, for instance, paints of all types including bituminous preparations, and sprayed plastic skin types, have not as yet proved to be permanent. If the cover is reduced, regular maintenance of protective coatings of these types becomes necessary. Below ground, asphalt tanking, and similar relatively thick coatings give adequate protection, provided leaks do not develop where services pass through; if this happens water may enter any porous areas of the concrete.

The need for adequate cover, wherever concrete is exposed to the weather, cannot be too strongly emphasised. In providing such cover, it must be remembered that reinforcement cannot be bent and fixed with great accuracy, except at an unjustifiable cost, particularly when the larger sizes of bar are used. It is suggested that a working tolerance of 25 per cent of the sizes of the bars is allowed, the cover given in the Code of Practice being specified as the minimum at any point. For concrete likely to be produced under difficult conditions, whether due to climate, materials, or shape of member, it may be wise to increase the cover by 50 per cent or more.

Fig. 17.13. Insufficient cover to reinforcement on concrete parapet

Some local bye-laws differ in their requirements from those of the Code of Practice; for instance, the Greater London Council requires a minimum cover of 3 in. for all reinforced concrete directly in contact with earth.

Failure to provide adequate cover has resulted in the corrosion of the reinforcement, and the consequent spalling of the concrete on many structures. Examples of such failures are shown in Figs. 17.12, 17.13 and 24.2. Reference may also be made to Fig. 17.11, which gives examples of misplacement of reinforcement recorded in concrete work for the gas industry.

The required amount of cover to the reinforcement may not always be governed by the need to guard against the risk of corrosion, and many authorities have their own requirements with regard to resistance to fire. This may result in a greater thickness of cover than that given in the Code of Practice. If abrasion or other damage is expected, sufficient cover should be provided to give a reasonably long life. This may be difficult to assess, and it may be necessary to take other measures either to protect the concrete at vulnerable points, or to allow repair when necessary.

Welding

Welding is now used fairly extensively in the precast concrete industry, but is not seen to the same extent on site work, except on the Continent. The use of welded fabric for roads, floor slabs, and similar purposes is widespread.

Welding is of two kinds:

(1) Tack or positional welding between rods crossing more or less at right angles so as to fix them in position, and for which no special precautions in regard to stress are necessary.

(2) Butt welds between rods in line, stresses being transferred across the section.

Gas, electric resistance, and metal arc welding are the alternative methods available, but the last-named is the best and the most economical for mild steel, being used almost exclusively in practice nowadays. Cold-worked high-yield bars can be resistance welded without loss of strength. Metal arc butt welding of cold-worked high-yield steels is also possible without loss of strength but it is necessary to allow the weld to cool between runs in order to control the heat input, and the area of metal must be increased proportionally when mild steel electrodes are used.[17.6]

With suitable combinations of pressure and current it is possible to flash butt weld all types of steel without loss of strength but the process is essentially one for the factory.[17.6] Skilled welders are essential.

Butt welding of main bars is of little advantage as regards any direct saving in cost, but may lead to a saving due to the elimination of laps, additional bars, hooks, etc. and to reductions in the cross-sections of some members where the area is determined by the spacing of the reinforcement rather than strength requirements. Congestion caused by hooks and lapped bars may be relieved. A further advantage lies in the rigidity of welded reinforcing frameworks consisting of intersecting bars, stirrups, etc.

In order to take advantage of the use of welding the design must be adjusted to exploit every possible reduction in the size of members, and the elimination of steel included solely to provide rigidity to the reinforcement framework.

Bar ends to be welded should be clean and cut to an approved shape, except in the case of small bars up to $\frac{1}{4}$ in. in diameter, which can be welded with square-cut butt ends. Ends cut to a single or double **V** are considered most suitable, the latter being preferred for bars over $1\frac{1}{2}$ in. in diameter. Butt welds should be reinforced generally as shown in Fig. 17.14, the weld metal being built up to a thickness of 10 per cent greater than the thickness of the bars being welded. Excess metal is then dressed off.

The bars must be kept in alignment during welding by suitable supports. When space permits, jigs of the type shown in Fig. 17.15 may be used.

For bars larger than 1 in. in diameter, and where a single **V** is used, a bent plate below the joint as shown in Fig. 17.14 may usefully be employed. The plate should be about $\frac{1}{32}$ in. clear of the joint at the gap.

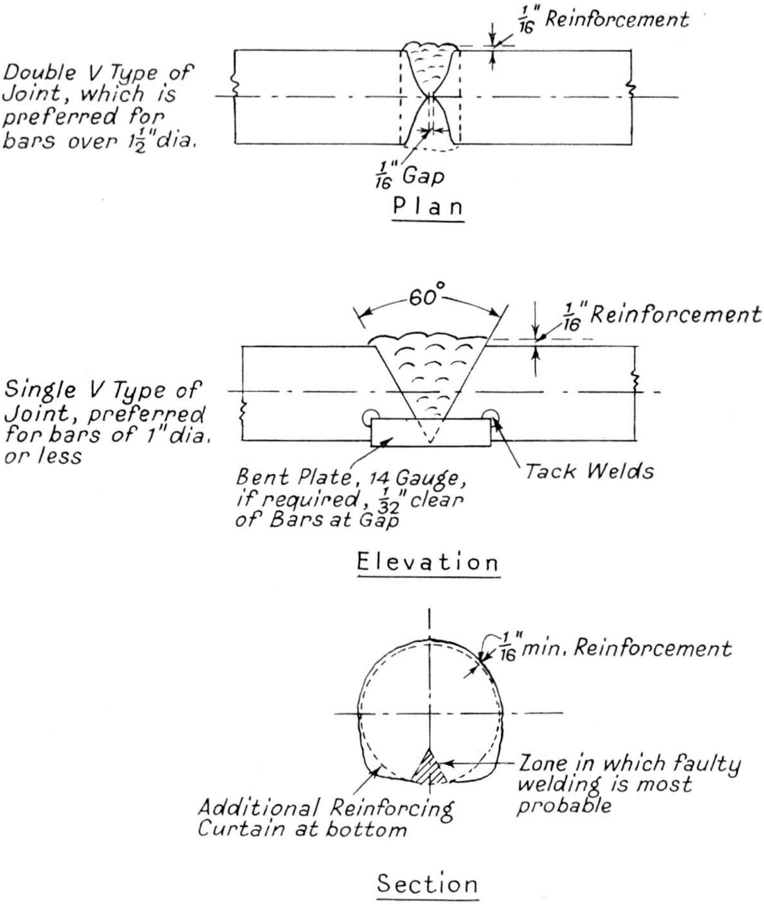

Fig. 17.14. Illustrations of welding methods for butt welds

Cuerel[17.7] gives the following precautions which, if observed, will ensure sound welds:

(1) An experienced inspector or clerk of works should be employed.

(2) The skill of the welder should be tested by a series of preliminary sample welds.

(3) Suitable electrodes and equipment should be used.

(4) Careful inspection should be made in order to ensure satisfactory welding procedure, especially in regard to penetration.

(5) The use of a sufficiently high welding current is essential, since with ordinary mild steel round bars the ill effects of excess current are negligible compared with the possible dangers from using too little current. Lack of penetration is almost invariably the cause of failure. Penetration depends to a considerable extent on the skill of the welder; but if ample current is used, there is less danger of lack of penetration.

(6) The inspector should check the amount and disposition of the reinforcement to the weld.

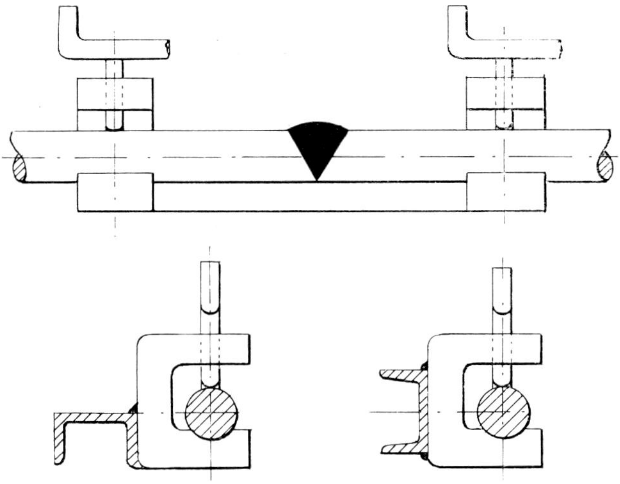

Fig. 17.15. Jig for holding reinforcing bars during welding

(7) Perhaps more for the psychological effect than the actual direct benefit, occasional welds should be cut out and tested to destruction.

The introduction of gamma-ray equipment has given a valuable non-destructive method of examining welds.

Final inspection

The concrete engineer should see that all reinforcement is fixed in the correct position, and properly supported in the forms by means of built-in concrete spacing blocks or rings, asbestos–cement rings, steel chairs, or other suitable devices.

He should also check the position of splice lengths, and the lengths and diameters of the hooked ends.

The rods should be rigidly held by the tying wires. Welded joints should be carefully inspected to see that they are properly made.

Embedded metalwork

It is not always appreciated that embedded metal work apart from steel may require special protection to prevent corrosion. Moisture is necessary for corrosion to proceed, so that parts embedded in impervious concrete with a reasonable cover are not likely to be attacked. Some parts may, however, be embedded in the surface, and these should be protected. Copper is not normally affected but may be corroded if calcium chloride is present. Lead, zinc, aluminium, and cadmium plated parts are susceptible to corrosion in fresh concrete and in moist concrete at all ages. They should therefore be protected by shellac, varnish or other inert paint.

Corrosion of metals in concrete is accelerated by the presence of stray electric currents, but again moisture must be present if any appreciable action is to occur.

REFERENCES

17.1 Reynolds, Chas. E., *Reinforced Concrete Designer's Handbook.* Concrete Publications Ltd, London.

17.2 Manning, G. P., *Reinforced Concrete Design.* Longmans, Green & Co., London.

17.3 C.S.A. Standard A23.3–1966, *Code for the Design of Plain or Reinforced Concrete Structures.* Canadian Standards Association.

17.4 *Crack Control in Concrete Beams*, C.E.R.A. Research Report No. 6, Dec. 1966. Civil Engineering Research Association.

17.5 Gardner, S. V., *The Use of Concrete as an Engineering Material in the Gas Industry*, Oct. 1954. Reinforced Concrete Association.

17.6 Lancaster, R. I., Steel reinforcement, *Structural Concrete*, **3**, No. 4, July/August 1966, p. 171.

17.7 Cuerel, J., *Welding in reinforced concrete.* Reinforced Concrete Association Technical Paper No. 2, 1937.

CHAPTER 18

Prestressed Concrete

In a normal simply supported reinforced concrete beam the reinforcement is designed to take the tensile forces in the bottom of the beam which, together with the compressive forces in the concrete in the top of the beam resist the bending caused by the load. When the reinforcement is in tension the concrete which is bonded to it in the bottom of the beam is also in tension. As the tensile strength of concrete is low, fine cracks form in the bottom of the beam before its safe working load is reached and when the load is removed, there is only partial recovery, that is to say, cracks once formed do not disappear and, apart from being unsightly, they reduce the protection given to the reinforcement by the concrete cover. If the working load is exceeded, but not so much that the steel stress reaches the yield point, the lack of recovery and consequent widening of the cracks, will be very much more noticeable although the structure may still be able to carry the working load with safety.

When a beam is prestressed (i.e. stressed before the working loads are applied) generally the whole section is put into compression. In one of the earliest known prestressed beams this compression was obtained by jacking both ends of the beam horizontally off firm anchorages at each end. Nowadays, although the principle is the same, this compressive force is obtained by tensioned wires within the beam which are either bonded to the concrete throughout the length of the beam, or are held between firm anchorages embedded in each end of the beam. This then is the starting point comparable with the no-stress condition of the unloaded normal reinforced concrete beam. When the load is applied to the prestressed beam the force which would cause tension and cracking in the bottom of a reinforced concrete beam only reduces the compression which the prestress has already provided. At the same time the compressive force at the top of the beam caused by the bending moment, is added to that caused by the prestress. The effect of this principle is that full advantage can be taken of the potential compressive strength of good quality concrete, and its low tensile strength is not of any consequence. The resistance to shear is greatly increased, and shear reinforcement, so common in normal reinforced concrete in the form of stirrups, is seldom required. Lighter members can be used than with normal reinforced concrete, and this is a particular advantage with bridges, long span roofs and similar structures, in which the dead weight of the members themselves is a major part of the total load which the structure is designed to carry.

A prestressed member is generally designed to carry the working loads without tension in the concrete occurring. However, if the member is over-loaded it has the capacity of almost complete recovery, provided that the yield point of the steel has not been reached, and tension cracks caused by the overload virtually disappear.

Circular structures such as liquid retaining tanks usually require prestressing in two directions, circumferentially to resist hoop tension and vertically to resist the moments imposed by the various loads including prestress.

Reinforcement

The steel reinforcement for prestressed concrete must be capable of resisting high tensile forces without excessive "creep" occurring, i.e. without stretching when such forces are maintained. Any appreciable yielding of the steel would relieve the prestress and the beam would then behave as a normal reinforced concrete beam. As such, it would be very much weaker, and quite unable to carry the loads for which it was originally designed. Ordinary mild steel does not meet these requirements and "high-tensile" steel is therefore necessary in prestressed concrete. High-tensile steel does not show a definite yield, as in the case of mild steel, and when held under constant stress it will show creep, with time, similar to that of concrete. However, the loss of stress due to creep of high-tensile steel is relatively small.

Prestressing steel may be hard-drawn wire, strand, twisted wires, or alloy steel rods similar in size to normal mild steel reinforcement. In order that the high-tensile properties are fully maintained, prestressing reinforcement should not be heated for bending or die drawing purposes.

Ordinary mild steel reinforcement, not stressed above 20,000 lb/in^2 is used to resist the bursting forces induced by loads at the anchorages; such reinforcement is usually spirally wound round the line of action of the tendons.

Prestressing steel and anchorages should be free from loose mill scale rust, oil, grease or any other harmful matter. The steel should not be kinked or pitted. Cleaning of the steel when necessary should be done by brushing, immersion in suitable solvent solutions, or any other means which will not adversely affect the tendons.

All prestressing steel should be stored in a ventilated weather-proof shed, protected from ground damp. Steel should always be carefully inspected before use to ensure that it is not pitted.

The cost of a prestressing cable complete and in place in a concrete member may be taken as roughly 3 to 4 times the cost of the wire as delivered to the site.

Quality of concrete

In order to obtain the maximum advantage from prestressing, it is necessary to use high strength concrete, and specification requirements of compressive strength of 6,000 lb/in^2 or more, at 28 days are not unusual. The production of such high quality concrete requires the adoption of strict control measures in order to reduce variations in strength to a minimum.

Cement used for prestressed concrete should normally comply with the requirements of B.S. 12 Portland Cement (Ordinary and Rapid Hardening). Other cements may be used in special circumstances but high alumina cement is usually inadvisable.[18.1]

From the data on mix design, Chapter 7, it will be seen that rich stiff mixes are necessary in order to obtain satisfactory strengths. Vibratory methods of placing are usual.

Three properties of concrete must be borne in mind in applying prestressing methods, namely, the drying shrinkage which may continue for some time, and the creep under load, which have been described in Chapter 2, and the elastic deformation of the concrete at load transference.

The use of a high-strength concrete accompanied by a low water/cement ratio reduces the movements due to these factors and increases the efficiency of the prestressing system. These movements result in a shortening of the concrete which tends to relieve the prestressing effect of the wires or cables with time and it is necessary to increase the tension beyond that considered necessary for design purposes, in order to provide for this gradual loss.

The loss in prestress is greatest at early ages, but the rate of loss gradually diminishes with time. From tests which have been made over periods of two years and more, it appears that the limit of loss of prestressing due to the above factors will be of the order of 16 per cent.

Because of the dangers associated with the corrosion of steel, particularly of thin wires, calcium chloride should never be used in concrete or grout likely to be in contact with the prestressing reinforcement.

Pre-tensioning

In pretensioning, the reinforcement, which usually consists of high tensile steel wire or 7-wire stress-relieved strand, is stressed and then the concrete is cast round it. The tension is maintained until the concrete has attained sufficient strength to resist the compression imposed on it, and to transfer by bond the prestress into the concrete. The wires are then cut from their anchorages.

Pretensioning is seldom applicable to structural members cast on the site, but lends itself to the factory production of precast units, ranging from small floor beams and railways sleepers, to hollow piles which may be over 100 ft in length. For factory production, the most usual method of pretensioning is known as the long-line system, in which wires are stretched between anchorages at opposite ends of a long stretching bed on which forms are constructed. Stop ends fitted round the wires form small gaps at distances apart which give the desired length for the individual beams. The stressing bed is of concrete or steel and is provided with heating ducts in order to increase the rate of hardening of the concrete and so allow the wires to be cut between each unit at as early an age as possible. A factory operating on the long-line system is seen in Fig. 18.1.

When producing individual units by the pretensioning method, the resistance to the tension in the wires is taken by the moulds themselves, which must be designed and built to withstand the prestressing loads without distortion. Special measures are necessary to ensure that the formwork does not restrain shrinkage of the concrete, otherwise cracking can occur before the formwork is removed.

The use of individual moulds has two advantages, namely, greater flexibility in size of unit produced, and the loss of only a single unit, if the anchorage should slip, as compared with a whole line of units in the long line process. Slipping of one or more prestressing wires and unequal

Fig. 18.1. Prestressed beams being manufactured by the long line process. The stressing gear is shown in the foreground

stressing can occur, and are difficult to detect in prestressed units. Tests should be made on a proportion of the units manufactured.

Post-tensioning

In contrast to the pretensioning method, the reinforcement to be used for post-tensioning is initially prevented from bonding with the concrete by some form of sheath, or by other means, or is threaded through holes cast in the concrete, as seen in Fig. 18.2. The wires or cables or steel bars are stressed by jacking directly off the ends of the structural member after the concrete has hardened sufficiently. Various patent methods such as those illustrated in Fig. 18.3 are used for anchoring the wires within each end of the member before releasing the jacks, but the essential in each is a device which grips them by a wedging action, after tensioning has been completed by a hydraulic jack. Diagrams illustrating the Freyssinet type of cable anchorage and jack, are given in Fig. 18.3. The Lee McCall system differs from these methods, in that $\frac{7}{8}$–$1\frac{1}{4}$ in. diameter alloy steel rods are used instead of cables, and the ends of the rods are threaded to take a nut, see Fig. 18.4. It should be noted that the prestressing is dependent very largely on the efficiency of such anchorages, although it

Fig. 18.2. Post-tensioned roof beams. The end of the beam on the left has holes ready to receive the cables; in the centre beam the cables are threaded through ready for stressing; on the right the cables have been stressed and the wires trimmed off

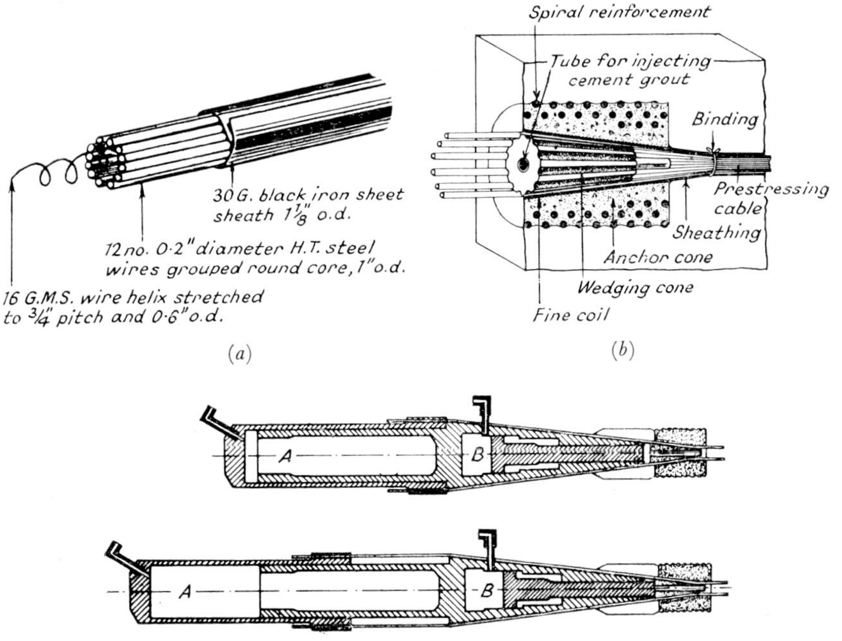

Fig. 18.3. Freyssinet method of anchoring group of prestressing wires, and type of jacking used for stressing cables

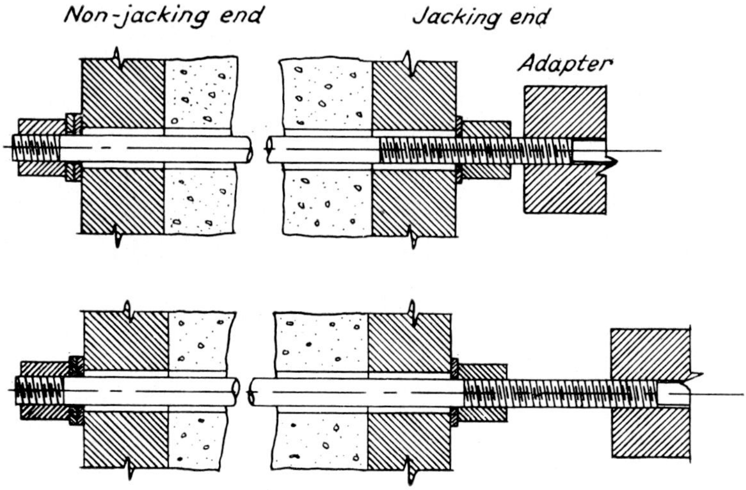

Fig. 18.4. The arrangement of the Lee McCall prestressing units before and after stressing

is usual to grout up the cable passages, immediately after the full prestress has been applied. Failures have occurred through slipping at anchorages and the breakdown of the bond between cables and concrete produced by grouting.

When post-tensioning cables, the force applied should be increased to the design load and then reduced nearly to zero. The cables are then marked and the force again increased until the desired extension is attained; this may require a force greater than the design load. The reason for stretching cables to give a predetermined extension rather than to achieve a certain stress at the jacking end, is that there is some loss of stress due to friction along the cable.

In the Magnel-Blaton system the wires forming the cable are stressed two at a time in order to achieve uniformity of stressing.

An advantage of post-tensioning is that as the reaction from stressing the wires is taken on the concrete, there is no loss of stress due to elastic deformation, as with pretensioning. Furthermore, as the concrete has hardened, some of the shrinkage in the concrete has already taken place, although there remain creep losses in the steel and concrete.

Whereas the capital investment for prestressing is generally high, it is much lower for post-tensioning equipment.

Grouting and sealing

Post-tensioned cable passages are usually grouted up immediately after stressing and anchoring. The grout, consisting of neat cement and water, should be of thin consistency at first in order to clear out air. It should then be thickened as much as the pressure requirement to push it through the cable passage will allow. Grout is forced in at one end until thick grout issues from the other; the ends are then sealed off to ensure that none of the grout flows out. The total quantity is small and even a small loss of grout may mean a long ungrouted passage.

The water in the grout cannot evaporate, and under winter conditions, bursting has occurred caused by freezing of water in cable passages. Rusting is unlikely because the passageway is virtually sealed to air which is also required for rusting to proceed.

It is also necessary to patch up the recess in the end of the structure where the anchorages are fixed. This can easily be a source of weakness if moisture can gain access to the ends of the cables, and particular attention should be given to the prevention of this possibility. An ordinary mortar patch is unsuitable and is certain to be pushed off by corrosion of the reinforcement ends.

Other methods of prestressing are described in the literature on the subject. Reference should also be made to C.P. 115:1959, The Structural Use of Prestressed Concrete in Buildings, and C.P. 2007:1960, Design and Construction of Reinforced and Prestressed Concrete Structures for the Storage of Water and other Aqueous Liquids.

Safety

All the equipment for prestressing operations should be in a good and safe condition. The accuracy of the equipment used for measuring the prestressing force should be checked before its first use and subsequently at intervals of a month or so. If any doubts arise a check should be made immediately. Calibration errors should not exceed 5 per cent.[18.1]

Wires do break occasionally and safety precautions against injury should be rigorously maintained.

REFERENCES

18.1 COMITÉ EUROPÉEN DU BÉTON, *Practical Recommendations for the Design and Construction of Prestressed Concrete Structures*, Provisional Edition, FIP-CEB Joint Committee, June 1966.

CHAPTER 19

Road Construction

WELL designed and well constructed concrete roads have a long life and low maintenance costs. The riding quality has at times been compared unfavourably with the best asphalt or tarmacadam roads, mainly because of unevenness at the joints. However, with the introduction of cut joints, and improved methods of "planing" the surface, the riding quality has been greatly improved.

In this chapter it is only possible to survey the more important aspects of concrete road construction. More detailed information may be obtained from publications dealing more specifically with the subject.[19.1,19.2]

Road base

The main concrete slab is generally laid on a base, sometimes also a sub-base, consisting of such material as compacted hoggin, sand, slag or crushed rock. In recent years increasing use has been made of dry lean concrete (Fig. 19.1) and stabilized soil. Base materials are generally compacted by rolling, but vibrating plate compactors are sometimes used for loose granular materials which have no cement binder.

On all base materials it is advisable to lay waterproof paper or plastic sheeting as an underlay for the concrete. A waterproof underlay prevents the escape of cement paste into the permeable material of the base; it reduces the friction between the concrete slab and base, and thereby reduces the tendency for cracks to form in the concrete slab during movements caused by temperature or moisture changes in the concrete; it prevents any loose material on the surface of the base from becoming mixed with the lower layer of freshly placed concrete; and it eliminates the need to wet the base immediately before placing the concrete.[19.3] It also prevents sulphates and other deleterious salts from moving up into the concrete with their detrimental effects. Sometimes the road base is

Fig. 19.1. General view of dry lean concrete being laid in 27 ft 6 in. width using box spreading machines and compacting with an 8 ton vibrating tandem roller and with a small vibratory roller for edge rolling.

given a bituminous surface dressing as soon as it is laid, in order to protect it from the ingress of water or, where the base has a cement binder, to act as a curing membrane. Although this dressing has limited effect in reducing friction between the slab and the base, a subsequent underlay is not generally considered necessary.

Concrete for road construction

Concrete used in pavement construction needs to be of high strength and low water content in order to carry the imposed loads without cracking, to resist abrasion, to have a low drying shrinkage, and to be frost-resistant even when it has been treated with salt for de-icing purposes. Air entrainment is usually recommended to improve the resistance to frost and salt.

The compressive strength (by 6 in. cubes) of concrete used in road pavements and bridges, where exposed to frost and salt action, should not be less than 4000 lb/in^2 at 28 days; preferably, it should be 4500 lb/in^2 or higher. See p. 124 for recommended minimum cement contents, and p. 108 for the effect of the water/cement ratio on durability. The entrained air should be 5–6 per cent for aggregates of maximum size, 1½ in. to ¾ in.

The general principles involved in good concrete production described in other chapters apply equally well in road construction.

Slab thickness and reinforcement

The main concrete slab may be from 5 in. to 12 in. in thickness, depending upon the intensity of the traffic, the quality of the subgrade and the type or thickness of the base. The thickness also depends to some extent upon whether or not the slab is reinforced. In most countries, including the United Kingdom where the traffic density is high, reinforcement is considered to be desirable in order to prevent the opening of cracks, and to allow wider spacing of joints, but in some European countries all concrete roads have been constructed unreinforced. As traffic increases it is noticeable that these countries are finding it necessary to introduce or increase the reinforcement. The present trend in reinforced roads is towards a single layer of reinforcement placed near the top of the slab, generally 2½ in. from the top in slabs 8 in. to 12 in. in total depth.

The road thicknesses required and the quantity of reinforcement recommended are given in Road Note No. 29.[19.1] Reference may also be made to A.C.I. Standard 325–58,[19.4] which gives recommended practice in America for the design of concrete pavements.

Joints

High stresses or uncontrolled cracking can result from the warping and changes in length which accompany changes in the moisture contents and temperatures of concrete slabs. Joints are provided to reduce these stresses and to control the cracking. The different types of joint are shown in Fig. 19.2. The requirements of a good joint include the following:

(1) There should be good alignment of the surface on each side of the joint.

(2) The joint should be waterproof and should exclude small stones and

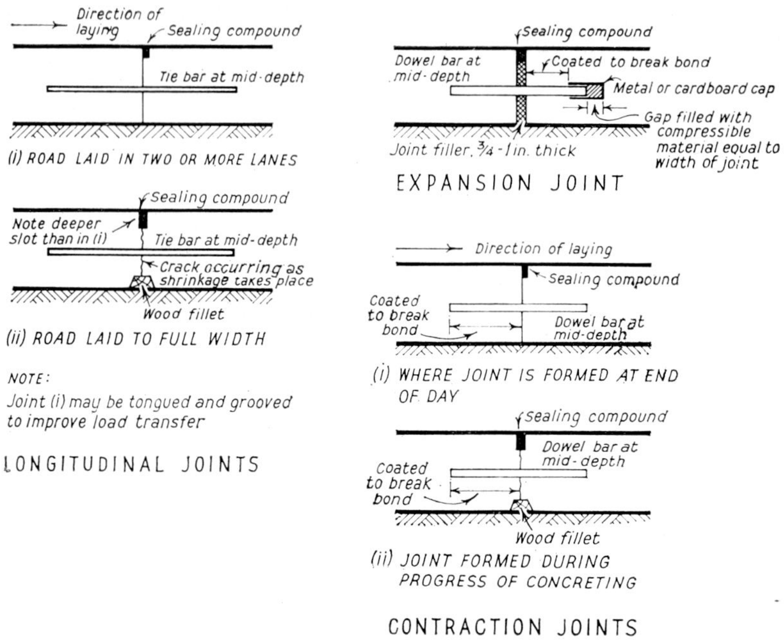

Fig. 19.2. Examples of joints used in road construction. The reinforcement should stop short of joints in all cases

dirt. The edges of the concrete should preferably be slightly rounded to reduce fretting.

(3) Jointing material should allow contraction and expansion without itself being extruded or permanently compressed.

(4) The joint should be capable of distributing the traffic load between adjacent slabs, but should not unduly restrict any curling of the slabs.

(5) The formation of the joint should not be difficult to carry out in practice, and it should not easily be displaced by the passage of vibratory equipment.

(6) The formation of the joint should interfere as little as possible with the continuity of the work.

Longitudinal joints

Concrete roads are often constructed in widths of 10 ft to 12 ft which have the advantage that they correspond with the widths of traffic lanes and are free from longitudinal cracking, but with present day plant, slabs may be constructed in single widths of up to about 30 ft. In any slabs more than 15 ft wide, whether or not they are reinforced, it is advisable to provide a central longitudinal joint, or to increase the amount of transverse reinforcement over the centre third in order to control longitudinal cracking. Longitudinal joints are formed in a similar manner to contraction joints (see paragraphs to follow). Tie bars consisting of $\frac{1}{2}$ in. diameter rods, each 3 ft 6 in. long are spaced at 2 ft to 2 ft 6 in. centres, should be provided at mid-depth of the slab across longitudinal joints. For further details reference may be made to Road Note No. 29.[19.1]

Transverse joints

The spacing of transverse joints is influenced by the thickness of the slab and the weight of reinforcement, the temperature during the setting and initial hardening of the concrete, and the restraint provided by the sub-base or subgrade. In unreinforced concrete the spacing should be close enough to minimize the risk of uncontrolled cracking. In reinforced slabs the joints assist in controlling the opening of any cracks which form, and the expansion joints permit thermal expansion to occur without undue compressive stress in the concrete.

Recommendations for the spacing of transverse joints are given in Road Note No. 29, 1965.(19.1) These are related to the traffic intensity, which also affects slab thickness.

A brief summary of the recommendations is given under the headings Expansion Joints and Contraction Joints. It is noted that, as concrete made with limestone aggregate expands less than other types, less provision for expansion may be needed.

Expansion joints

Expansion joints in reinforced concrete roads may be spaced 120 to 240 ft (maximum) apart in summer, but the spacing should be reduced to a maximum of 80 to 160 ft in winter. The lower figure in each case is for very light traffic (0–45 commercial vehicles* per day) while the higher figure is for roads subject to heavy traffic (more than 3,000 commercial vehicles per day). The spacing should be reduced in unreinforced concrete pavements to about three-quarters of the spacing for reinforced concrete pavements which carry light traffic, and to half for pavements carrying over 450 commercial vehicles per day.

An expansion joint provides a complete separation between the concrete slabs on either side; the gap formed should ideally be of sufficient width to allow compression of the infilling material without crushing it sufficiently to destroy its elasticity, and to avoid permanent exudation of the sealing material at the surface. This ideal is not achieved with materials so far developed, and indeed continual movement involving an elastic compression of 50 per cent or more, continuing over several years, coupled with other requirements in the case of the sealing compound, such as the exclusion of stones, and strong adhesion to the concrete, may prove difficult to achieve. A compromise must therefore be made, and regular maintenance of the joints is necessary.

Expansion joints are usually formed by the inclusion in the slab of a filling material extending from the bottom to within 1 in. to $1\frac{1}{8}$ in. of the surface (more in thick airfield slabs). The joint filler should be 1 in. thick if the maximum spacings given under the heading Transverse Joints are adopted (see Road Note No. 29) but may be reduced in proportion to the expansion joint spacing if a closer spacing is adopted. The thickness should not be less than 1 in. Materials commonly used are impregnated fibre board, soft knot-free wood, and cork base boards.

The gap above the filler is sometimes formed after the passage of the compacting machinery, by driving a tee iron into the surface. While this

* Defined as a goods or public service vehicle of upladen weight exceeding 30 cwt

method is acceptable for estate roads, it is not suitable for main trunk roads as it causes a slight irregularity at the joint which affects the riding quality of the surface. A better method, suitable for large roadworks, is to use a machine fitted with a long vertical vibrating plate which is lowered into the concrete before it has hardened, to form a slot or groove of any depth required. Another method is to saw a slot in the concrete after it has hardened. In Fig. 19.3 a longitudinal joint is being formed in this way. This method has the merit of avoiding any ridge being formed near the joint. Unless the sawing is completed while the concrete is relatively soft, the cost is high; if the concrete is too soft incipient cracks

Fig. 19.3. A joint sawing machine in operation on a concrete road slab

are formed and the corners break away under traffic. Hard aggregates such as flint are also costly to cut. Careful control is necessary.

The gap above the filler should be made $\frac{1}{4}$ in. wider than the filler itself, a sealing compound to be poured in after hardening of the concrete. Bitumen has been used in the past but is not elastic and exudes in hot weather. Newer materials have been developed which provide greater elasticity, provided they are not overheated when being prepared for pouring. Other materials have been developed to give greater resistance to the hot blast from jet aircraft. Strong adhesion is difficult to achieve in wet winter weather conditions, and it may be necessary to fill the joint temporarily, until drying out occurs. In Switzerland neoprene rubber strips are forced into the joint with a simple machine. This method appears to have possibilities.

Failures of roads at expansion joints may occur because the joint is not vertical, because the gap for the sealing compound is not made vertically above the filling material, or because of insufficient care in cleaning out concrete above the jointing material before filling with the sealing compound. In the latter case, as soon as expansion occurs, the concrete adjacent to the joint spalls off, as shown in Fig. 19.4. Diagrams showing faulty constructions are also given in Fig. 19.5.

Fig. 19.4. Failure of the joint in a concrete road, caused by not cleaning out properly before pouring the sealing compound. See Fig. 19.5 (ii)

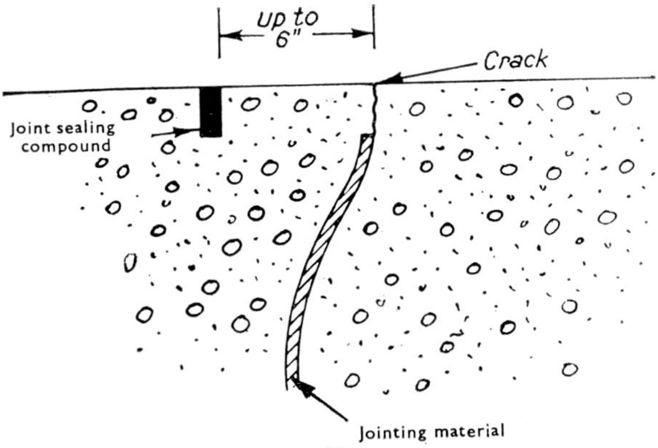

(i)

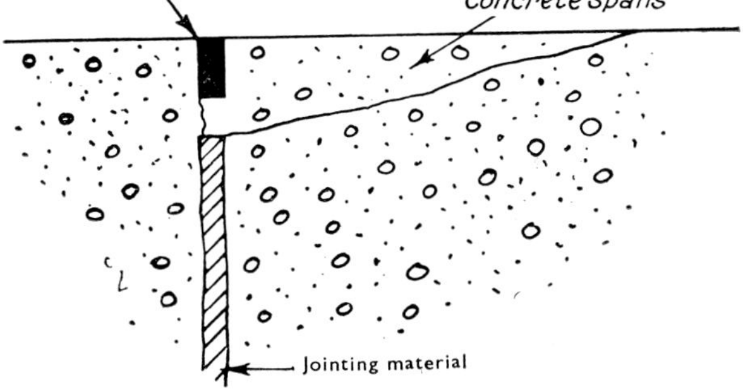

(ii)

Fig. 19.5. Two common causes of failure at expansion joints in roads

Except for joints in unreinforced slabs where the traffic is less than 150 commercial vehicles per day and for all joints in 4 in. thick reinforced slabs, dowel bars should be provided at 12 in. centres along expansion joints. They should be fixed straight and parallel with one another and with the centre line and surface of the road. Suitable sizes and lengths are given in Road Note No. 29.[19.1] Failures brought about by mis-alignment of dowel bars occur more often than they need if sufficient care is exercised. They should be rigidly held in place so that they are not disturbed during concreting. In major roadworks it is now common to use wire cradles to support the joint filler and the dowel bars. The sliding half of the dowel bar should be coated with a thin film of bitumen, or covered with a close fitting plastic sheet to prevent bond with the concrete; on the end it should be provided with a steel or cardboard cap, the end 1 in. of which should be filled with soft material to allow closure of the joint.

Experiments have shown[19.5] that the base fixing for expansion joint assemblies should not be more than 2 ft apart. The dowel bars should fit tightly in the joint filler, the stiffness of which has a considerable influence on the overall stiffness of the assembly. The importance of quality of workmanship in fabricating the assembly is emphasized.

On small scale works, alternate bay construction is sometimes adopted, in which case the construction of expansion joints is simple, since the transverse forms provide a vertical surface against which the joint filler can be fixed.

Contraction or dummy joints

Contraction joints, or dummy joints relieve warping stresses, and localize crack formation due to contraction, whether due to drying shrinkage or temperature changes. Except where the intensity of traffic over reinforced concrete roads is small (150 commercial vehicles per day or less) and in 4 in. thick reinforced concrete slabs, it is advisable to provide dowel bars similar to those in the expansion joints, but without the cap at one end. Reinforcement and tie bars should not be carried through contraction joints, unless the latter are closely spaced.

Contraction joints in reinforced concrete pavements should be spaced not more than 40 ft apart for traffic densities up to 450 commercial vehicles per day, and 60 ft apart when the traffic density is greater than this. Contraction joints in unreinforced concrete pavements should have an average spacing of 15 ft. Contraction joints should be made $\frac{3}{4}$ in. to $\frac{7}{8}$ in. deep for spacings up to 60 ft and 1 in. to $1\frac{1}{8}$ in. deep for a spacing of 80 ft. The width should not be less than $\frac{1}{4}$ in. for a 15 ft spacing, varying up to 1 in. minimum for an 80 ft spacing.

The A.C.I. recommendation[19.4] is that the groove should be between $\frac{1}{6}$ and $\frac{1}{4}$ the thickness of the pavement. If a parting slip is installed at the bottom of the pavement the depth of groove may be reduced to $1\frac{1}{2}$ in. and the depth of groove plus the height of the parting slip should not exceed $\frac{1}{3}$ the pavement thickness. Contraction joints may be formed by the methods already described for expansion joints. The gap is filled with a sealing compound similar to that used for expansion joints.

Road forms

Steel forms are generally used for the construction of roads. They are usually about 10 ft long and so designed that they can be held to the sub-base or subgrade of the road by pins driven in at 3 ft to 4 ft centres. For minor roads, for which the concrete is spread by hand and compacted with a small manually operated tamping beam, the forms can be of relatively light construction (Fig. 19.6). For major roads and airfield

Fig. 19.6. Manually operated wooden tamping beam
The beam to which two electric vibrators are attached, used in the construction of minor roads

pavements, heavy vibrating and finishing machines are used and these require forms of relatively heavy construction. Generally, these forms have a wide flat base and, as in the case of those in Figs. 19.8 and 19.9, have provision for attaching a rail on which the spreading and compacting machines can travel. There are notable advantages in using rails instead of allowing the machines to run on the edge of the forms. These include the prevention of distortion of the forms, more even spreading of the weight of the machinery over the bed supporting the forms, and the provision of a surface for the machines to run on which is free from the concrete droppings generally to be found on the top edge of the form.

It is important that road forms are set accurately to line, as this directly affects the riding quality of the road surface. Modern motorway specifications require that forms with rails shall be laid to within $\pm\frac{1}{8}$ in. of the level shown on the drawings, and that the rails themselves shall be set so as to ensure an accuracy of finished surface to within $\pm\frac{1}{8}$ in. under a 10 ft straight edge. A less strict tolerance than this is permissible on minor roads.

It is also important that forms be uniformly and firmly supported to prevent them deflecting under the loads involved in concreting the slab. They should be laid on a continuous bed of cement mortar or concrete; when heavy machines are used, a substantial concrete foundation may be needed for the forms.

An interesting development for large scale works in the United States and recently introduced into Europe, is that of the *slip-form paver*. This is a spreading and compacting machine which requires no fixed forms because it trails a suitable length of forms behind itself. It nevertheless requires a uniform running surface on each side of the construction width.

Spreading and compaction

For small road works it is usual to mount vibrating equipment on a wooden or metal beam, which serves to screed off, compact, and finally level off the surface. The uncompacted concrete should be levelled evenly and to sufficient height above the forms to allow for the reduction in volume from the loose state as deposited. On small jobs, the concrete is usually spread by hand and is preferably levelled with the aid of a template moved along the forms. The reinforcement is laid in place after a first layer of concrete has been spread. Vibratory equipment for compacting concrete on small scale works usually consists of a wooden or metal beam on which is mounted either electric, pneumatic or petrol-driven vibrators. Figure 19.6 shows electric vibrators attached to a wooden tamping beam being used in the construction of a minor road. The compaction may also be achieved by the use of a compacting beam such as that shown in Fig. 19.7, which utilises a petrol driven vibrating unit. Two such vibrating units are sometimes used on a long beam. With these types of compactor it is necessary to leave the uncompacted

Fig. 19.7. Manually operated petrol driven vibrating tamper compacting a mix of stiff consistency

concrete an inch or so above the side forms and then to lower the beam vertically until the concrete has been compacted. The beam is then raised and lowered vertically on the adjacent strip, and so on, working in strips the width of the compacting beam. With both these types of equipment the final levelling is obtained by tilting the beam slightly and pulling it slowly along the forms. Two extra men are usually required for this operation. Any lifting and replacing of the beam during final

Fig. 19.8. The main units in a concreting train on the construction of a modern road. On the left is the box spreader, and on the right the compacting and finishing machine followed by a wet surface profileometer recording the evenness of the finished surface

levelling, concrete on top of the forms, screed pins insufficiently hammered in, or uneven joints between forms, cause ridges, and careful attention should therefore be given to avoid their happening.

On large jobs a high standard of riding quality is generally required. If the concrete is dumped in heaps between the forms, the position of the heaps is reflected in the level of the finished surface whatever method of spreading and compacting is used. Modern specifications therefore require that the concrete be spread to the required surcharge level, by a method which prevents the concrete being compacted while being spread. In one type of spreader the concrete is fed into a trough running transversely between the forms; a worm at the bottom of the trough evenly distributes the concrete across the road. The most common type of spreader now used is known as the *box spreader* seen on the left of Fig. 19.8. It has its own motive power to move along the forms. The box, a long trough parallel with the forms, is filled at one side of the road, generally from side tipping lorries; it then moves transversely and, as necessary, longitudinally, to distribute the concrete evenly to approximately the required surcharge.

Reinforcement may be laid after a first layer of concrete has been spread, or by a travelling jig which holds the reinforcement in position.

The spreader in the latter case deals with the whole depth of concrete in one layer.

Modern compacting machines such as that shown on the right of Fig. 19.8 and to the left of the train in Fig. 19.9, generally incorporate strike-off paddles to trim the uncompacted concrete to just the surcharge required, a heavy vibrating beam to compact the concrete, and a finishing beam known as the "shuttle smoother" because of its shuttle movement over the surface. The riding quality may be immediately assessed by means of a wet surface profilometer, such as that developed by the Road

Fig. 19.9. Box spreader to the right followed by the compacting machine on the left. The strike-off paddles, which leave the uncompacted concrete at the surcharge required, are seen in the front of the compacting machine

Research Laboratory, and seen in the foreground of Fig. 19.8. This is a most useful tool, in that it indicates unsatisfactory areas in time for corrective measures to be taken.

Further improvement in riding quality has been achieved by finally "planing" the surface with straight edges, operated by hand or mechanically.

It is necessary when vibrating concrete in the construction of roads, to use a mix consistency which will be plastic enough to enable either the manually operated or the mechanical compacting machine to compact fully, without leaving honeycombed patches along the forms and at the bottom of the slab. The large vibrating and finishing machines are particularly sensitive to the consistency of the mix, and it is necessary to adjust this to suit the machine. With the type of machine shown in Figs. 19.8 and 19.9 a workability of barely $\frac{1}{2}$ in. slump and a compacting factor of about 0·82 gives the best results; for hand manipulated screeds a higher workability is needed (see Table 7.9). Mixes that are too dry for the method of compaction leave the concrete honeycombed at the bottom and are extremely difficult to finish true to level and free from ridges and other blemishes. On the other hand, wetter mixes are apt to sag and overflow the forms, part of the camber thus being lost. Wet mixes are, of course, a disadvantage also on account of the lower strength of the concrete and the higher drying shrinkage, and the increased risk of frost damage.

Curing

In paving work, it is necessary to adopt a form of treatment which does not cause any damage to the newly finished surface, especially during the period before appreciable drying out has taken place. In summer weather,

Fig. 19.10. Wetting hessian covering to a concrete pavement by means of a water spray wagon operating from one side

to keep off the wind and sun's rays, curing should commence immediately after finishing the surface, by the erection of coverings on frames, or by means of a resinous curing membrane sprayed directly on to the surface. If a bituminous surface layer is to be applied subsequently, early curing can be effected by spraying the sealing coat on to the surface. Omission of this early curing is a contributory cause of the cracking shown in Fig. 2.5 which occurs in road slabs during warm dry weather, even within half-an-hour of finishing. Protection by coverings on frames should be followed as soon as the concrete has hardened sufficiently, by other forms of treatment such as spraying water on a cover of hessian, coconut matting, or sand laid directly on the concrete. Spray wagons of the type shown in Fig. 19.10 are very useful for road work. The wagon seen has a capacity of 1,000 gallons, and by means of a small centrifugal pump at the rear, can spray a distance of 25 feet.

Waterproof paper, plastic sheeting and tarpaulins are effective without spraying, provided the concrete is already wet and they are overlapped and secured at the edges to prevent wind from blowing underneath.

Ponding of horizontal surfaces provides a very effective means of curing, but it is not very practicable, since the small banks of clay usually built round the edges of each bay are not easily maintained in a condition such that the area remains filled with water.

During recent years, *sprayed resinous curing membranes* have been used over large concrete surface areas as the sole curing agency. They have the advantage that once applied they need no further attention, and while it has not been established conclusively that such use of curing membranes is completely effective, it is widely recognized that, in practice, they are probably more effective than methods of curing which require continuous spraying with water—a process not always strictly adhered to.

On roadworks they are normally applied through a motorized pump, and spraying lance, or on large road or airfield works automatically by a machine mounted on a bridge towed by the finishing plant or having its own motive power. It cannot be emphasized too strongly that application of a curing membrane must follow as soon as possible after the final finishing operation on the concrete, i.e. as soon as free water brought to the surface by the finishing process has been re-absorbed by the concrete or has evaporated. Under windy conditions special care in spraying is necessary to ensure complete coverage of the whole area to be cured. Further comment is given in Chapter 13.

REFERENCES

19.1 *A Guide to the Structural Design of Flexible and Rigid Pavements for New Roads*, Road Note No. 29, Ministry of Transport, Road Research Laboratory and Highway Engineering Division, 1965. H.M.S.O.

19.2 *Concrete Roads. Design and Construction*, Road Research Laboratory, 1955. H.M.S.O.

19.3 *Guide to Concrete Road Construction; Questions and Answers*, 2nd ed., Nov. 1958. Cement and Concrete Association.

19.4 A.C.I. 325–58, *Recommended Practice for Design of Concrete Pavements*, A.C.I. Standards, 1965. American Concrete Institute.

19.5 WEAVER, J., *An Examination of the Stability of Expansion Joint Assemblies under Loads applied by Concrete Road Making Machines*, Technical Report TRA/396, August 1966. Cement and Concrete Association.

CHAPTER 20

Piles, Fluid Retaining Structures, Dams, Tunnels and Floors

DIFFERENT forms of construction have their own special features and while the technique employed may vary considerably, the production of good concrete is dependent on the principles already described. Variations in method of construction and special requirements in certain circumstances make it necessary, however, to modify the proportions and workability of the mix, or to make other changes. Concrete piles, fluid retaining structures, dams and tunnel linings are examples of constructions in which special care is necessary to obtain satisfactory results.

Several types of flooring system are available for the production of hollow concrete floors and a brief description is given for reference.

Concrete piles

Concrete piles may be either precast or cast in-situ. Precast piles are usually cast in suitable moulds on a platform on the site. In some cases, especially in the shorter lengths, they may be cast at a central depot and transported to the site after hardening, but the risk of damage in handling is thereby increased considerably. In-situ piles are cast in position.

Precast concrete piles

The most important considerations in the design of precast concrete piles, whether using mild steel or prestressed high tensile steel reinforcement, are the magnitude of the stresses likely to arise during handling and driving. The quantity of reinforcement is usually governed by the stresses set up during handling, and it is very important that piles should be lifted at the points indicated by the designer. Care is also necessary when stacking that supports are provided at the same points. Usually these points are situated at about one-fifth of the length of the pile from each end. Holes are often formed in the pile by means of metal or cardboard tubes fixed in the mould to ensure suspension at the correct position. Care is especially necessary during the pitching of the piles.

Driving sets up very high stresses at the head of the pile and to a somewhat less extent at the toe. It has been found that failure due to crushing at the head of the pile is minimized by (1) efficient and properly placed packing, (2) the provision of adequate transverse reinforcement near the end of the pile, (3) the use of strong concrete at the head of the pile.

It is interesting to note in this connection that the amount of reinforcement has no important influence on the stresses set up. Impact stresses vary according to the type of soil through which the concrete pile is being driven, but in hard driving they may amount to as much as 3,000 lb/in^2. Since the impact strength of concrete is not much more than half of the crushing strength under a slowly increasing load, it is necessary to provide a cube strength of as much as 6,000 lb/in^2 near the head and toe.

Nominal proportions in common use are 1:1½:3 of cement:fine aggregate:coarse aggregate for the main part of the pile and 1:1:2 for one or two mixes for a short length near the head. Ordinary Portland cement is generally used. High-alumina cement has sometimes been used but involves the risks described in Chapter 4.

B.S. Code of Practice No. 4, Foundations[20·1] gives recommended proportions for the main body of the pile and minimum crushing strengths as quoted in Table 20.1.

TABLE 20.1 *Minimum Crushing Strength of Concrete (6 in. cube test) for Piles*

	Hard or very hard driving and for all piles in marine work. Concrete Class II, 1:1½:3 mix, driving stress over 1500 lb/in^2		Normal or easy driving. Concrete Class III, 1:2:4 mix, driving stress under 1500 lb/in^2	
Age days	Preliminary test	Works test	Preliminary test	Works test
7	3350	2630	2800	2100
28	4750	3750	4000	3000

Adequate cover to all reinforcement should be provided. The Code requires a cover of 2 in. at the head of the pile, and not less than 1½ in. at the sides, increased to at least 2 in. where piles are exposed to sea water or other corrosive influences.

When placing is carried out by hand it is necessary to use a slump of 2 in. to 3 in. in order that the concrete can be rammed properly round the reinforcement. However, in order to obtain high strengths it is better to reduce the water/cement ratio and the workability, and to obtain good compaction by the careful use of vibrating equipment.

Concrete piles are usually made in timber moulds set out and bedded on a level base preferably of concrete. The moulds should be tight enough to prevent the loss of cement and water and should have fillets fitted in the corners to form chamfers on the edges of the concrete.

Placing and compaction should be carried out in accordance with normal good practice taking the precautions which have already been described in Chapters 11 and 12. If form vibrators are used it is important to see that they are fixed at intervals of not more than about 4 ft in order to obtain good compaction throughout the length of the pile.

The curing of concrete piles is particularly important both from the point of view of strength and impermeability. Curing may be effected by wet sacking, straw, or other covering, or better still by flooding the forms. For this purpose the forms are sometimes made about an inch deeper than is necessary for concreting in order to make provision for flooding. As with other types of construction, special care should be taken if high alumina cement concrete is used. Moist curing of concrete piles should be continued as long as possible, suitable periods being 28 days for normal Portland cement concrete, 14 days for rapid-hardening cement, and 4 days for high-alumina cement. In summer or in warm climates these periods may be shortened considerably, even by half.

Prestressed concrete piles are now being used to an increasing extent, especially in the longer lengths. They give savings in reinforcement and weight. They are handled and driven in a similar manner to precast piles with ordinary reinforcement.

Pile driving

Precast concrete piles are driven into the ground by one of several types of pile-driving hammer. A common type comprises a falling weight or hammer which is lifted a predetermined distance and is then tripped so that it falls and hits the top of the pile. The weight is guided by a pile frame which is also used to retain the pile in the correct position and direction during driving.

Single-acting hammers have a heavy piston which is raised by steam or compressed air and then allowed to drop by gravity. An advantage of this type of hammer is that it is much faster than the ordinary drop hammer and causes little strain on the pile frame and other gear.

Double-acting hammers are similar to the single-acting type, except that in addition to being raised by steam or compressed air the piston is also pushed down under pressure.

Diesel pile hammers are designed to give an extra kick as the pile moves down under the impact of the piston, by the explosion of diesel fuel in the cylinder at the bottom of the stroke of the piston.

Fig. 20.1. Driving raking hollow precast concrete piles

The choice of hammer is determined by circumstances and driving conditions.

The pile head is protected from the full force of the hammer blow by means of a helmet. This is usually of steel with a hardwood timber or composite plastic dolly and some type of resilient packing between the helmet itself and the pile. The packing usually consists of such material as rope, sacking, hemp, felt, asbestos fibre, or bags of sawdust. The resilience of the packing has an important bearing on the impact stresses arising during pile driving and it is important to replace the packing as soon as it becomes hard. In general, high impact stresses can best be avoided by the use of a heavy hammer, a small drop, and a soft or resilient packing.

Recommended practice is given in the Code of Practice for Foundations, Part IV, Section 7.00,[20.1] and further information may be obtained from textbooks dealing with piling.[20.2]

Care should be taken when using a toggle bolt to hold a pile which is driving out of plumb, as cracking may arise if the bending stresses become excessive.

Pile heads should not, on completion of driving, be stripped by automatic hammers used downwards, as this practice is liable to cause fractures below the "cut-off" level.

In-situ piles

In-situ piles have been developed in order to overcome a number of disadvantages inherent in precast piles. A few short descriptions will illustrate the various methods adopted in construction. For further details of these and other in-situ piling systems reference should be made to relevant publications.

Bored piles

Piles may be formed by boring a hole in the ground with the type of equipment used in well sinking or with a mechanical auger, and filling the hole with concrete. Some of the newer mechanical augers produce shafts up to 7 ft in diameter, which are then filled with concrete. The hole is not cased unless the ground shows signs of caving in or there is a serious inflow of water from pervious strata. The larger shafts are sometimes belled out at the bottom by hand or by under-reaming cutters to form an enlarged toe.

The load-bearing capacity is often dependent largely on shaft friction, and in order to reduce softening of the soil round the hole the concrete should be placed as soon as possible after completion of the boring. A rich (say $1:1\frac{1}{2}:3$) and workable mix (slump 3–6 in.) is necessary in order to ensure that the hole is filled completely when compacted by rammer or vibration. If casing is used the concrete is required to slump down without arching as the casing is withdrawn, otherwise gaps, waisting or necking will result.

If the bottom of the hole is wet a batch of dry concrete should first be placed and well rammed. This may be sufficient to allow further concrete to be placed satisfactorily "in the dry." In water-bearing sands and gravels it is likely that water will rise in the hole, making it necessary to place the concrete under water, and tremie methods should be used (see p. 182 for details). Water flowing in fissures or channels in the rock surface or in bouldery material is particularly troublesome and may even wash the toes of the piles away, leaving the concrete arched over. In such circumstances it may be necessary to adopt precast piles, either driven or lowered into boreholes.

Franki piles

Franki Piles are constructed by driving a steel tube to the desired depth and then by ramming concrete down through this tube and at the same time withdrawing it slowly. The initial driving of the tube

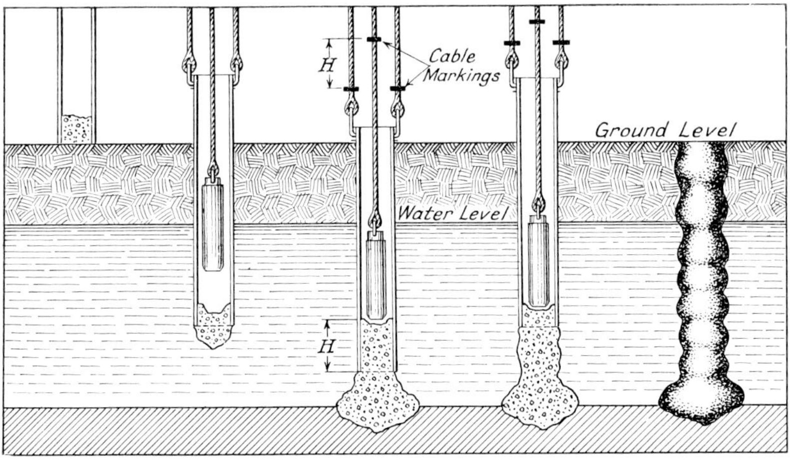

Fig. 20.2. Steps in formation of Franki pile

is achieved by filling the casing with a plug of concrete and then forcing this into the ground by a suitable drop hammer. The casing follows the block of concrete down in the manner shown in Fig. 20.2. When the penetration of the tube is sufficient, further movement is prevented, and the concrete is forced downwards and outwards by means of the drop hammer to form an enlarged toe, more concrete being added as necessary. The tube is then slowly withdrawn while successive batches of concrete are added and rammed into position. The use of stiff concrete coupled with heavy ramming produces a pile of high strength providing there is no infiltration of ground water. When necessary, a reinforcing cage is lowered into the tube as soon as the enlarged toe has been formed.

Prestcore piles

Prestcore piles are of a quite different type and are formed by assembling circular precast concrete units around a central steel core which is then lowered into a hole bored in the ground. When the whole pile has been assembled in this way, vertical reinforcing rods are inserted in pre-formed holes, the central core withdrawn and cement grout introduced under pressure. This is forced out between the joints of the precast units and

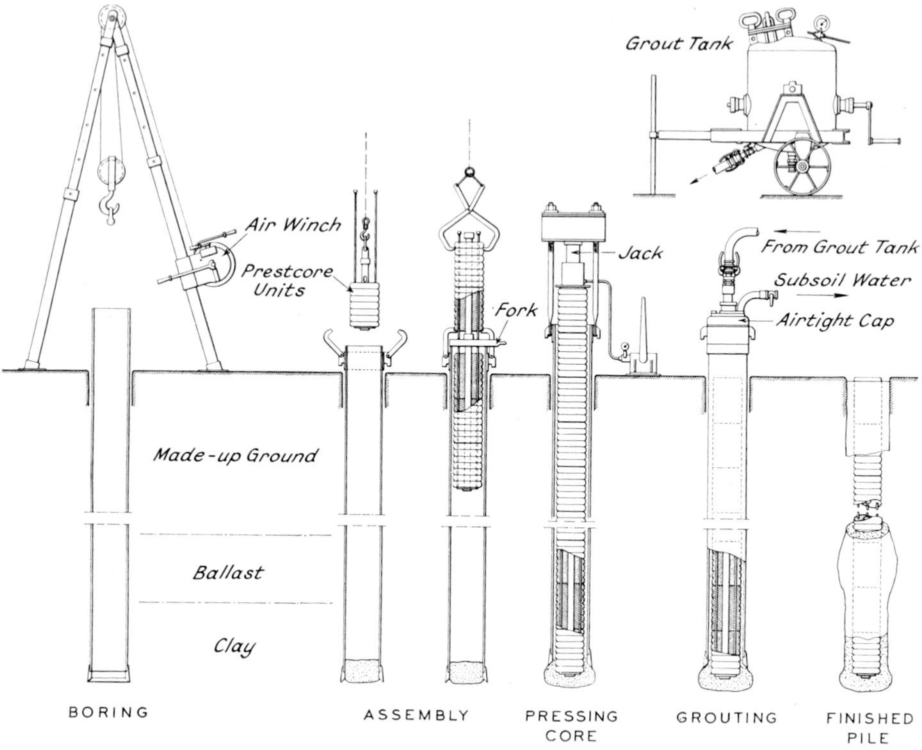

Fig. 20.3. Method of forming Prestcore piles

fills the space previously occupied by the tube lining the hole. In this way a solid pile is built up, as shown in Fig. 20.3.

The Prestcore system is designed mainly for circumstances in which piles have to be driven in limited headroom or without vibration. They also have advantages in squeezing ground and where chemical attack may be a danger.

Comparison of different types of concrete pile

For purposes of comparison in situ piles may be divided roughly into three classes:

(1) Those where the lining tube is left in.
(2) Piles hammered down, the tube being removed.
(3) Those which are formed by boring, the tube being removed.

There are certain advantages attached to each method, and these are summarized in Table 20.2. A comparison with precast piles is also given in the Table.

Concrete construction below ground water level

Structures such as sewage works settling tanks, underground pumping stations, culverts, basements, and the like, are very often built in reinforced concrete. When they are constructed below the ground water table,

TABLE 20.2 *Comparison of Different Types of Pile*

Type of pile	Advantages	Disadvantages
Precast concrete	The reinforcement can be placed accurately and carried right to the toe of the pile. The concrete can be placed under good conditions, a stiff mix being placed by vibration to provide high strength and density. They can be driven in water-bearing ground. They are suitable for use as columns above ground, and alternatively are easily extended.	They must be properly matured before driving, thus occasioning some delay. Space must be reserved, and a bed prepared for their manufacture, a point of some importance on congested sites. Pile driving presents difficulties in case of limited head room. The guiding and driving plant is heavy. The top of the pile has often to be cut away when the cast pile is too long, or when damage occurs due to heavy hammering. Possible damage during driving.
In situ piles: (1) General. (Applies to all types included below.)	They are cast in their final position. No cutting is necessary at the head. The piling plant is lighter than that necessary for precast piles. Certain types have special advantages in cases of limited head room, and where vibration might affect adjoining buildings.	The concrete is not cast under the best conditions. The reinforcement is in many cases difficult to place accurately. Possibility of gaps in pile. Difficulty in cleaning out bottom of hole before concreting.
(2*a*) Bored piles, lining tube left in.	The concrete cannot become contaminated with earth or water. There is no risk of trouble during removal of the tube. Moisture and soluble sulphates are excluded which might otherwise cause corrosion of the steel or disintegration of the concrete.	
(2*b*) Bored piles, lining tube withdrawn	The cost of the tube is saved. The concrete surface provides more frictional resistance.	Water and, to a less extent, earth, may cause weakness in the concrete. The tube sometimes presents difficulty in removal. If the concrete is not properly consolidated there is a danger of corrosion of the reinforcement. The reinforcement is difficult to position correctly.
(3) In situ piles hammered down, tube withdrawn.	The concrete must be of relatively stiff consistency, and this coupled with the heavy hammering gives a high strength. The dry concrete plug usually prevents the ingress of water and slurry. When in contact with clay it tends to absorb water and so improve the skin friction. The concrete is hammered out into any loose strata of soil, thus helping to consolidate the ground and increasing the resistance. The enlargement thus formed, increases the bearing capacity. The rough surface produced, gives increased skin friction.	Water and, to a less extent, earth, may cause weakness in the concrete if workmanship is poor. The tube sometimes presents difficulty in withdrawal. If withdrawn too quickly, waisting, necking or even gaps may occur. There is a danger of corrosion of any reinforcement if the concrete is not properly consolidated or, for some reason, the cover is insufficient. Heavy hammering may, under certain circumstances, cause vibrations detrimental to adjoining buildings and to piles already cast.

special precautions are necessary, mainly in connection with watertightness, percolation damage to setting concrete, and buoyancy. Methods of placing concrete under water are dealt with in Chapter 11.

(*a*) *Watertightness.* If the structure is designed to be tanked in asphalt, no special precautions are necessary in the concrete work other than to avoid damage to the asphalt layer, which can occur either during construction or by movement due to shrinkage of the concrete walls in contact with the asphalt.

The effects of shrinkage can be greatly minimised by the use of a low water-content in the mix, by careful attention to curing, and, in large structures, by leaving gaps to be concreted after a certain period has elapsed—usually two weeks or more.

Where no asphalt tanking is provided, more than usual care is necessary in the production and placing of the concrete. Methods of producing good concrete, and of avoiding weakness at construction joints have been fully described in the preceding Chapters. Reference should also be made to "Fluid Retaining Structures" below.

(*b*) *Percolation damage to setting concrete.* When constructing below the ground water table, the water level is lowered by pumping from open sumps, or by well-points. The walls are very often concreted in direct contact with the ground (i.e. outside formwork is not provided). In these cases, pumping from the sump or well-points must be maintained until the concrete has attained a sufficient degree of hardness and strength to resist the external water pressure caused by the rise in head of the ground water after pumping has stopped. A head of only a foot or so is sufficient to burst green concrete, and if this has occurred the only remedy is to break out all the damaged concrete and start again.

In rock excavation, ground water often issues from fissures in the form of springs, after the water level has been lowered by pumping from a sump, and although the pumping is continued after concreting, the flow of water from the fissures over the rock face is liable to cause damage to the setting concrete. If the springs are few in number, they can be dealt with by cutting a chase in the rock surface into which are placed agricultural drain pipes surrounded by chippings and led to the sump. Numerous springs over the whole excavated face can only be effectively stopped by pressure grouting with cement or cement: sand mixes, or in some cases, by plastering up crevices in the rock face with quick setting cement. If high-alumina cement is used for this purpose, it should not be allowed to come in contact with normal Portland cement concrete being used for structural purposes.

(*c*) *Buoyancy.* After completion of a watertight concrete structure below ground there is a risk of the whole structure floating to the surface after the water table has been allowed to rise to its normal level.

Calculations should be made to ascertain whether the dead weight of the structure exceeds the uplift due to buoyancy. If flotation is liable to occur, the interior may be kept flooded until superimposed loads are added to provide negative buoyancy. Cases of flotation of structures are by no means rare, and may be disastrous, since it is usually impossible to get them back to their original position.

Fluid-retaining structures

It is particularly important in fluid-retaining structures that cracking from any cause should be reduced to a minimum. The amount of shrinkage depends on the extent to which the structure is likely to dry between the time of construction and the time at which it is filled with liquid, and whether in the future there is a danger of it drying out; for instance, little shrinkage would occur in a tunnel which was not allowed to dry out from the time of construction until it was filled with water, but if it remained unfilled for a period of some months the shrinkage could be quite appreciable.

Shrinkage can normally be reduced by using a lean mix and by reducing the water/cement ratio to a minimum consistent with the requirements for adequate compaction. In water-retaining structures, however, strength and permeability considerations make it necessary to use a fairly rich mix, and it is therefore important that the water/cement ratio should be reduced to a minimum. Vibratory methods of compaction are an advantage, and all construction joints should be given particular attention. It is, in fact, generally considered better to pour continuously whenever possible in order to avoid the formation of lift planes altogether. A water bar of the type shown in Fig. 10.11 should be inserted at all lift planes, and at all joints, whether horizontal or vertical. Special care should be taken to ensure that the water bar is continuous at points of intersection and corners. For further information reference should be made to B.S. Code of Practice for Water-Retaining Structures.(10.3)

Adequate cover to the reinforcement and a thorough inspection to ensure that none of the reinforcing bars is likely to become displaced during concreting, is essential.

Dams

There are several features in the design and production of concrete for use in dams which require special consideration. These arise from the mass of concrete involved, with resultant temperature differentials arising from the heat of hydration of the cement, and shrinkage which may lead to internal or surface cracking, or cracking of the whole structure.

In many cases, the location of dams makes it necessary to utilize local aggregates which are not always of high quality. However, the limits suggested in Chapter 3 for amounts of salts and other deleterious materials should be observed. In the larger masses of concrete it is usual to use aggregates up to a maximum size of 6 inches. The U.S. Bureau of Reclamation specifies that this grading shall be within the limits shown in Table 20.3. It is considered preferable to have between 25 and 30

TABLE 20.3 *U.S. Bureau of Reclamation Specfication for Grading of Aggregates for Dams*

Screen size or number (American sizes—see Table 3.2)	Individual per cent by weight retained on screen
6–3 in.	13 to 30
3–1½ in.	20 to 35
1½–¾ in.	20 to 35
¾–3/16 in.	20 to 35
No. 4	0 to 5
8	5 to 15†
16	10 to 25†
30	10 to 30
50	15 to 35
100	12 to 20
Pan	3 to 7

† If the individual per cent retained on the No. 16 screen is 20 per cent or less the maximum limit for the individual per cent retained on the No. 8 screen may be increased to 20 per cent.

per cent of the 3 in. to 6 in. cobbles, and this usually gives the most economical concrete.

Low-heat cement is usually chosen in order to slow down the rate of heat evolution, as described in Chapter 4. Limits are specified for the heat of hydration of these cements. If pozzolanic materials are available within a reasonable distance, they are commonly used to replace about one third of the cement. A further slowing down in heat evolution and hardening results, but the pozzolans usually provide better resistance to chemical attack, and often give an improvement in the workability of the mix. Air-entraining agents are often used to improve the workability of the mix, and to improve the resistance to freezing and thawing.

Experience in America has indicated that cement contents as low as 250 lb/yd^3 of concrete give satisfactory results in the interior of the dam. For exposed faces the cement content is usually increased by 100 to 150 lb/yd^3. French experience suggests that a higher cement content i.e. 400 lb/yd^3 and more, is required to provide sufficient frost resistance, and going further north to Scotland a minimum cement content of 500 lb/yd^3 with a low water/cement ratio is advisable for the exterior face. The spillway section requires special consideration because of the abrasion problems, and dense concrete with a low water/cement ratio is necessary. This will usually lead to the necessity for a higher cement content, probably 50–100 lb/yd^3 more than for the other exposed faces. Erosion, frost action and chemical attack are discussed in more detail in Chapter 24.

Mixers of 4 yd^3 capacity are required to deal with the 6 inch cobbles, and sturdy transport and placing equipment must also be provided. It is usual to limit the height of lifts to 5 ft, although lifts up to $7\frac{1}{2}$ ft are allowed when measures are taken to cool the concrete after compaction. The rate of concreting is usually controlled in order to take advantage of the cooling which occurs from the exposed surface of the concrete.

In the cooler climates, metal forms have an advantage in conducting away heat more quickly than wooden forms, but may be a disadvantage in hot climates unless shading is provided.

Construction and contraction joints are provided at intervals in order to reduce the risk of random cracking. These joints may take a number of forms and various measures are adopted to prevent leakage through them. In dams it is not usual to provide a water bar at all horizontal joints, reliance being placed on the massiveness of the structure to prevent seepage, but a water bar is used at all vertical joints. Leakage at horizontal joints is not unknown, however. Another method of sealing joints sometimes used, more perhaps in cut-off walls through permeable ground than in the dams themselves, is the formation of a series of horizontal and vertical channels at the joints, which are filled with a soft bitumen. Provision is usually made for topping up at a number of points along the top of the wall. In the larger structures expansion joints should be provided at intervals. The subject is, however, too specialized for adequate treatment in this book, and is covered in various publications including those by the U.S. Bureau of Reclamation.[(20.3)] Contraction joints are often grouted with cement some months after construction has been completed, through pipes embedded in the concrete. If the grouting is performed when the concrete has cooled, and preferably in cool weather, such grouted joints are usually effective.

Measures to reduce the setting and hardening temperature of concrete in dams include both pre-cooling of the aggregates and the concrete mix and post-cooling of the compacted concrete by a piped cooling system. Pre-cooling is unnecessary if the air temperature at the time of placing is below 50°F. However, in many climates, pre-cooling is necessary and may be achieved by sprinkling the coarse aggregate with chilled water, by the use of refrigerated air blasts over the aggregate or in the mixer, by cooling the mixing water, by additions of chip ice to the mix, and by refrigerating the mixed concrete just before placing.(20.4)(20.5) Often a combination of these methods is used. It is usual to aim at a temperature of not more than 50°F in the concrete as it leaves the mixer. Further cooling at the Chief Joseph and Folsom dams ensured that the concrete was placed at a temperature below 50°F, and under those conditions post-cooling was considered unnecessary, although air temperatures were 80°F and more.

Post-cooling is usually achieved by pumping cool water through a grid of pipes embedded in the concrete. One-inch pipes are commonly used, being placed on the top of each lift of concrete after the concrete has hardened. The horizontal spacing of the tubes depends on temperature conditions, and is usually between 2 ft and 6 ft. Water is pumped through these pipes at about 4 gallons per minute, this having been found to be the most economical rate of flow for most purposes. This artificial cooling is started immediately after placing the concrete and is continued until a stable temperature condition is achieved. Sub-cooling or other control of the temperature of various parts of the structure may be accomplished and, if the cement grouting of the joints is undertaken under these sub-cool conditions, it is possible to achieve a more favourable stress distribution under the load as a result of the subsequent temperature rise.

As in other mass concrete structures, there is some risk of surface cracking when the concrete is placed in warm conditions, and cooling occurs rapidly during falling temperatures at certain times of the year. In climates where hot sunshine may be followed by clear, cool, or cold nights, measures should therefore be taken to prevent the surface from either becoming hot in the day time or cooling at night.

Large mass concrete structures

With proper control of the temperature it is possible to induce temperature stresses which will be beneficial to the structure rather than otherwise. In order to prevent uncontrolled cracking during the cooling of mass concrete structures it is necessary to provide contraction joints at intervals. The spacing of these joints is important since it is necessary to grout them up at a later stage when the major part of the contraction has taken place. An opening of $\frac{1}{50}$ in. or more is necessary for successful grouting with cement so that the joints should be spaced at distances which will provide openings of sufficient width, but which are not so large as to create the danger of intermediate uncontrolled cracking. See also p. 225 on the sealing of cracks.

Tunnels

Particular care is necessary when placing concrete tunnel linings to avoid segregation. If methods have to be adopted which are liable to cause separation, it is advisable to use a richer concrete with smaller aggregate.

This applies if a pneumatic placer is used, since a relatively high workability is necessary with this equipment. The concrete is shot between the shuttering and the rock; it is difficult to avoid diagonal flow, which is conducive to segregation. The placing pipe must be kept buried in the concrete.

The slight settlement, which occurs in wet concrete during the period between placing and setting, leaves a slight gap between the crown of the tunnel and the rock. This gap is filled with cement grout injected under a low pressure through holes drilled through the hardened concrete.

Shrinkage cracking of tunnels intended as water carriers is sometimes reduced by installing tight bulkheads at the ends of the tunnel, and maintaining a pond of water inside during the interval between completion and the placing of the structure in service. More often the tunnel remains wet from water accumulating from various constructional operations.

Floors on the ground

The minimum thickness for floors on the ground should be 4 in. This is generally sufficient for residential purposes. In public places this should be increased to 5 in. for foot traffic and light rubber-tyred vehicles, increasing to a minimum of 6 in. or more for industrial traffic.

Floors placed on the ground should be isolated by suitable movement joints from columns, beams, etc., supported on deeper foundations. Normally a vertical strip of bituminous material or other material, say, $\frac{1}{2}$ in. thick, and of the type used in road joints, will be satisfactory.

Control joints are necessary in large floor areas, and it is usual to form these between columns. If such bays are more than 20 ft in length (American practice allows a 25 ft limit[20.6]) they should be sub-divided by movement joints very similar to those used in road construction, see Chapter 19.

Vapour barriers are necessary in concrete floors on the ground. These are still commonly formed by applying a coating of bituminous paint or a layer of mastic. However, p.v.c. membranes about $\frac{20}{1000}$ in. thick, laid on the formation before laying the concrete, have proved satisfactory and are now widely used. The formation should be finished with fine material to avoid puncturing, the p.v.c. should be lapped and folded over to give a reasonable seal at joints, and some care should be exercised to avoid tearing during concreting operations. The p.v.c. should be turned up at walls to meet the d.p.c.

As in other forms of construction, concrete for floors should be dense and strong. A minimum compressive strength of 3000 lb/in^2 at 28 days should be satisfactory if a further wearing course is to be applied. For floors to resist wear, the minimum compressive strength should be increased to 4,500 lb/in^2, or more if heavy industrial trucking is involved. Good quality aggregate complying with the requirements of B.S.882, particularly as regards hardness and abrasion resistance, should be used.

The concrete should be spread evenly, avoiding segregation, and compacted by vibrating beam as in the case of roads. If no extra wearing course is to be applied, further levelling by means of straightedges (preferably metal) and by darbies of wood or light alloy, becomes necessary. It is important that excess moisture and laitance are not brought to the surface during these operations. Joints and edges are then given attention as necessary.

The final floating is delayed, depending on atmospheric conditions, until the surface is free from any water sheen and is firm enough to stand on. Disc type floats and trowelling machines provide light pressure to compact the concrete and give a dense smooth surface. If steel trowels are used by hand, considerable pressure is necessary to give a good wear-resistant surface. A second trowelling, an hour or so later, may be necessary and beneficial.

A report by an A.C.I. committee on proposed recommended practice[20.6] gives much detailed and useful guidance on concrete floor laying. A close similarity will be found with British practice as outlined above.

Concrete floors

Many flooring systems are available, both in-situ and precast, which are designed to reduce the dead weight of concrete. Cavities are formed in floors by various methods, for example hollow concrete blocks, hollow clay blocks, and retractable steel or wooden tubes. The use of retractable cores is restricted to precast units. Holes for services are sometimes formed by inflated rubber tubes. More often, however, services are laid in grooves formed in the upper surface, or simply laid in a cement : sand screed which is trowelled to provide a smooth level surface on which to lay the floor finish.

An effective saving in weight for heavily loaded and long span concrete structures can be achieved by using a waffled or coffered floor as shown in Fig. 20.4.

Fig. 20.4. Glass fibre reinforced polyester moulds laid on formwork to form a coffered effect when removed. Reinforcement is placed between the moulds.

Precast concrete beams and slabs (Fig. 20.5) for floors achieve their greatest economy when dimensions are standardised as far as practicable, and they can then be produced in a highly mechanised factory. A further advantage accrues when conditions are such as to render in-situ concreting impracticable or at least very expensive. In Russia this condition occurs for several months each year, and in consequence factory produced precast concrete is used extensively.

Fig. 20.5. Floor of precast concrete slabs, 6 ft wide, each containing four cavities, 12 in. wide

Among the advantages of the precast beam type of floor as against the in-situ type are (i) the saving in timber supports and shuttering, (ii) increased speed of construction, and (iii) the fact that the floor is ready for use immediately the beams are placed in position and grouted up. On the other hand, the in-situ floor is (i) usually cheaper to construct, (ii) better suited for distributing concentrated loads, (iii) not subject to cracking at intervals (at joints) as so often happens with a series of beams, and (iv) it provides better sound insulation than the lighter hollow floor.

Prestressed floor units

Prestressed floor units produced on the long line system described in Chapter 18, have the advantage of a saving in weight, an especially useful feature when long spans are envisaged. Their deflection may, however, be greater than that of ordinary reinforced concrete units, a point which requires consideration in design. An example of a possible source of trouble is that of a lightweight block partition, carried on a prestressed floor which is part beam and part cantilever, in which cracking may occur over the support adjoining the cantilever unless deflections are limited.

REFERENCES

20.1 Civil Engineering Code of Practice No. 4 (1950), *Foundations*. British Standards Institution (originally issued by Institution of Civil Engineers).

20.2 TOMLINSON, M. J., *Foundation Design and Construction*, 1963. Pitman, London.

20.3 *Treatise on Dams*. Various design standards published by U.S. Department of the Interior, Bureau of Reclamation.

20.4 *Symposium on Mass Concrete*, Special Publication No. 6, American Concrete Institute, Detroit, 1963.

20.5 STODOLA, P. R., O'ROURKE, J. E. and SCHOON, H. G., Concrete core block for Oroville Dam, *Proc. Amer. Concr. Inst.*, **62**, No. 6, June 1965.

20.6 Proposed Standard Recommended Practice for Concrete Floor and Slab Construction, Reported by A.C.I. Committee 302, *Proc. Amer. Concr. Inst.*, **63**, No. 1, Jan. 1966.

CHAPTER 21

Precast Concrete Products

THE term "precast concrete product" is used to describe the many types and varieties of concrete unit which are cast in moulds either in a factory or on the site, and are not built into the structure until they have fully hardened. Precast concrete products therefore include various structural members and such items as cast stone, concrete lintels, staircase units, fence posts, kerbs, lamp standards, railway sleepers, concrete blocks and bricks.

The factory production of concrete should offer certain advantages over site manufacture, chiefly in regard to ease of supervision and control. Compaction can be carried out more efficiently, as also can moist curing. However, the benefit of these advantages is often lost, and the user should check and test precast concrete products with just as much care as he gives to in-situ concrete. Against the advantages there is also the possibility of damage to the finished products during transport and the need for lifting equipment on site of sufficient capacity and range to lift units into position.

Precast concrete products may be divided roughly into five groups:

(1) Architectural cast stone and other types of ornamental concrete.

(2) Concrete steps, paving flags, etc., which have to be wear-resisting. Kerbs may be placed in the same class.

(3) Load-bearing structural units, and other concrete units such as lintels, beams, posts, railway sleepers, pipes, water tanks, and troughs. Many of these units are prestressed and in some cases the appearance is of secondary importance.

(4) Roofing tiles, which must be impermeable and weather resistant.

(5) Concrete blocks, slabs, and bricks. Air bricks.

Architectural cast stone

Architectural cast stone is used mainly in place of natural stone dressings on buildings, and is usually made to match a sample of natural stone or other material submitted by the architect. An example of cast stone masonry is given in Fig. 21.1. Well-made cast stone is dense and impermeable and is often indistinguishable from natural stone except by close examination. The various colours and textures are obtained by blending white and normal Portland cements with suitable inert pigments, and by the use of aggregates of the correct colour and texture. The material employed for the facing is expensive and it is the normal practice to use only $\frac{1}{2}$ in. to 1 in. with a backing of normal Portland cement concrete.

Aggregates for cast stone should comply with B.S. 882. They will not usually be larger than $\frac{3}{4}$ in. and for the facing may be limited to a $\frac{3}{16}$ in. maximum size. Materials commonly used are:

Crushed stone, such as Portland, York, Blue Penant, quartzite, whinstone and granites, Derbyshire spar. Crushed marbles.

Fig. 21.1. Entrance porch finished with cast stone in Portland stone colour

Leighton Buzzard and other light-coloured natural sands.

Two methods of making cast stone are in use, namely, the process by which an earth-moist mix is heavily rammed into the moulds, and the alternative of using a plastic mix with subsequent surface treatment to obtain the required appearance. The proportions of the mix depend on the maximum size of the aggregate, since it is necessary to use a richer mix with small aggregates in order to obtain a comparable strength. A typical facing mix is: 1 part of cement, 3 parts fine aggregate $\frac{3}{16}$ in. down. A richer mix is undesirable owing to the increased shrinkage and tendency to craze. The backing may be 1:1$\frac{1}{2}$:3 or 1:2:4, using $\frac{3}{4}$ in. maximum size aggregate if the section is large enough.

The advantages of using an earth-moist mix are that (1) a stonelike texture is obtained straight from the mould, (2) the casting can be removed immediately from the mould and any blemishes or loose corners atended to before the concrete sets, (3) the surface is free from blowholes, (4) the shrinkage is reduced, and (5) there is less tendency to craze.

Early removal of the moulds is especially useful if they can be used over and over again on repetition work. Care is necessary in the use of earth-moist mixes, however, to ensure that they are the wettest which will give a satisfactory surface texture and can be removed from the mould without sagging. Overdry mixes give weak corners and permeable concrete which soon blackens in town atmospheres.

Mixers of the open pan type with a rotating star of paddles are the most suitable for earth-moist mixes. Rotating-drum and tilting-drum mixers tend to cause "balling up," a condition which renders the mix unsuitable for the production of dense, close-textured cast stone.

Plastic mixes can be placed in accordance with the normally accepted principles of concrete making. Some method of removing the cement skin, such as scrubbing before the concrete has fully hardened, rubbing down with a carborundum stone or a flexible carborundum disc, or tooling, is necessary to obtain a satisfactory architectural finish. Blowholes must also be filled with a cement mortar grout. Thin skins of stone dust and cement rubbed on to the surface with a wooden float are unsatisfactory and prone to crazing.

Fig. 21.2. Cast stone as turned out of the wooden mould to the right

Moulds for cast stone may be of timber, steel or partly steel, plaster of Paris, gelatine, concrete, glass fibre reinforced plastic, or sand according to the number of repetitions to be made of any particular unit and the amount of ornamentation. Timber is the most commonly used material owing to the ease with which it can be worked. The moulds must be strong and rigid, especially for earth-moist mixes, and must retain their shape after repeated usage. Steel may be used for plain ashlar facing, and for reinforcing timber moulds, especially at corners likely to be damaged during ramming. Concrete moulds are very useful for some classes of repetition work, but are heavy to handle. Plaster of Paris and gelatine moulds are used for ornamental work. Sand moulds are sometimes used, and although they require a very wet mix, satisfactory cast stone can be produced from them.

Earth-moist mixes are usually rammed into the moulds by hand tampers or by pneumatic and electric tampers. Plastic mixes require less vigorous treatment, while the fluid mixes used in plaster of Paris, gelatine, and sand moulds need only judicious stirring with a rod or spade, taking care that the concrete is pushed into every detail. Stiff mixes are often compacted by holding the moulds down on electric or pneumatic vibrating tables, or by attaching vibrating units to the moulds themselves.

Many different surface textures may be obtained on cast stone by suitable treatment. Earth-moist mixes can be left as cast against the face of the mould, apart from final cleaning, and give a smooth and uniform texture. Otherwise, they can be wire brushed before they have fully hardened or etched to give a rough surface, bush hammered, combed, or rough tooled in a similar manner to natural stone. Plastic mixes can be wire brushed as soon as the mould is removed, giving an exposed aggregate texture. A retarder is sometimes used to facilitate the removal of the cement skin. Alternatively they can be tooled or rubbed down with carborundum stones or discs, and if necessary given a dull polish by rubbing with successively smoother stones, ending up with a soapstone.

Exposed aggregate finishes are also obtained on slabs by casting face upwards, and trowelling in extra coarse aggregate of whatever size is required to give the desired texture. As soon as the concrete has stiffened a light spray of water, aided as necessary by a soft brush, is used to remove the cement skin. The slab is tilted during this washing off. Except where this treatment is to be adopted it is advisable to form the facing of cast stone against the mould. Sometimes, however, casting is only practicable with the facing uppermost. In such cases it is important to keep the amount of trowelling to a minimum.

Cast stone should be well matured before being built into the structure in order to avoid cracking at the joints due to shrinkage. Most manufacturers coat light-coloured stones with a lime slurry before delivery in order to protect the surface and keep it clean. This should not be cleaned off until the stone is fixed in place. There should be no difficulty in removing it with a stiff brush or by a light rubbing with a carborundum stone. The joints should never be made in rich cement mortar; a mix of 1 : 1 : 6, cement : lime : sand gives satisfactory results.

Many cast stone units, particularly slender units such as cills and coping stones, contain reinforcement to withstand stresses due to handling. Such units are usually exposed to the weather and it is essential for the concrete mix used to be one which permits full compaction, otherwise a porous concrete will result and the unit will have a short life before spalling is caused by corrosion of the steel. Many instances are known of units which have deteriorated within a few months. These units are preferably made using a plastic mix rather than one having an earth-dry consistency.

Concrete units to resist wear

The principal requirement of concrete units required to resist wear by foot and other traffic is that they shall be made with a hard tough aggregate and shall have a dense strong wearing surface.

Concrete steps, floor tiles and other similar units are usually made with such aggregates as marble, granite, basalt, quartzite, and Hopton Wood stone, having a maximum size of $\frac{3}{8}$ in. In order to obtain a satisfactory grading, 2 parts $\frac{3}{8}$ in. to $\frac{3}{16}$ in. stone may be used either with 1 part of $\frac{3}{16}$ in. to dust stone or $\frac{3}{16}$ in. down washed sand. If $\frac{3}{16}$ in. to dust stone is used, care should be taken to ensure that the supplies do not contain an excess of crusher dust.

A rich mix of the order of 1:3 cement:aggregate is used for the facing, but a leaner mix may be used as a backing. The mix must be as dry as possible consistent with the production of dense concrete. Casting may be done with the tread either upwards or downwards; the former gives the densest and strongest surface if experienced trowel hands are employed. Too much emphasis cannot be placed on the necessity for good workmanship in producing trowelled wear-resisting concrete surfaces. As contrasted

Fig. 21.3. Manufacturing paving flags by hydraulic pressure

with the trowelling of plaster, considerable pressure must be employed to compact the surface without bringing up more fat than is necessary to fill the interstices between the larger pieces of aggregate. Any fat which is brought to the surface should be removed. Two or three trowellings may be necessary at intervals, allowing the concrete to stiffen and dry out to some extent between each. The period from casting to the final trowelling should not exceed about four hours. In order to increase the proportion of the larger pieces of aggregate in the surface, extra stones are sometimes trowelled into the surface. These are later exposed by polishing.

Steps and floor tiles cast face downwards are certain to have a number of pinholes in the surface which must be filled with neat cement. There is a tendency to use a wetter mix, and this added to the fact that the surface is not compacted to the same degree as a properly trowelled surface, accounts for the poorer wear resistance.

Floor tiles can be made very successfully by means of hydraulic presses similar in principle to those used for paving flags (Fig. 21.3).

Steps and tiles are often given a dull polish by means of carborundum stones, beginning with a coarse stone and ending with a fine one. Various types of machine are available for this work, all having rotating carborundum stones to perform the grinding. Terrazzo paving is made in a similar manner.

Non-slip finishes to concrete steps and tiles are obtained by the use of carborundum grit or by inserts of alundum. The carborundum grit may be trowelled into the surface or sprinkled in the bottom of the mould in the case of units cast face downwards. The former method is more effective. Small alundum inserts about 1 in. square and $\frac{3}{16}$ in. thick can be obtained in various colours for use as non-slip inserts in concrete steps and tiles. They are usually inserted to form a mosaic pattern 2 or 3 in. wide, and at a distance of about 2 in. from the nosing.

Paving flags are nowadays produced in large quantities in hydraulic presses. A mixture of 1:3 cement:granite chippings, $\frac{3}{16}$ in. to dust, is normally used, and is mixed in a pan-type mixer situated above the hydraulic press. Steel moulds are used, perforated gauze and a new sheet of paper being put in the bottom before filling. Similarly, a sheet of paper and a zinc plate are placed on the top of the filled mould. This arrangement allows drainage from both top and bottom, and the squeezing out of water increases the density of the pressed flag, so giving a high wear resistance.

The moulds are mounted on a rotating table so that each one is filled from a measuring apparatus situated below the mixer. The slab then moves round into position for pressing at a pressure exceeding 1000 lb/in^2. After pressing it moves round to a third position, where it is raised in the mould by means of a hydraulically operated discharge ram. A vacuum plate is then lowered into position over the slab, which is lifted by suction, and swung round on to a stillage, usually mounted on a trolley for convenience in handling. The top perforated zinc plate and the top paper are stripped immediately, but the bottom paper is left in place, and is stripped after the paving flag has hardened.

Paving flags are made in a number of standard sizes given in B.S. 368, 1956.[21·1] Details are given of tests which the flags are required to pass, covering transverse strength, wear resistance, and absorption. Paving flags should not be delivered until they have been properly matured.

Special sizes of paving flags, or those required in special colours, and in small quantities only, are made by hand in wooden or metal moulds. If wooden moulds are employed it is preferable to use a steel sheet as a facing in order to get a good finish. The mix used should be as dry as possible consistent with the production of dense concrete. The use of a vibrating table is a distinct advantage in this respect.

Kerbs are usually made to comply with B.S. 340, 1963,[21·2] for kerbs, quadrants and channels.

Concrete for kerbs must be of good quality, since kerbs are particularly liable to attack by frost if they are at all porous, and a low water/cement ratio should, therefore, be used, even if this results in a rather less smooth finish than that obtainable with a highly workable mix. The water/cement ratio should not exceed 0·6 and preferably be less than 0·55. B.S. 340 states in a Foreword "the water/cement ratio should not exceed 0·55 by weight immediately after moulding or pressing."

Air entrainment is an advantage in resisting frost attack in the presence of salt solutions, but reduces the transverse strength.

B.S. 340 requires that the aggregate, when tested in accordance with B.S. 812 shall not have an aggregate crushing value exceeding 30 per cent, nor a flakiness exceeding 35 per cent.

Standard sizes are given in the specification. Tests are given for transverse strength and absorption. Kerbs may be made by hand,

Fig. 21.4. Vibrating concrete kerbs

by vibration (Fig. 21.4), or by hydraulic pressure, a pressure of not less than 1,000 lb/in^2 being used. The moulds must have an internal surface of steel or other material which will give a smooth uniform surface.

Load-bearing structural units

Precast load-bearing structural units have been used increasingly in recent years, either to form the concrete structure, or in combination with in-situ concrete. The main advantage to be gained from the use of precast units is a saving in formwork and supports, so that it is those parts of the structure for which the formwork is most expensive, that the precast unit is most likely to be used. The advantage to be gained depends, within certain limits, on the amount of repetition; as a rough guide it may be assumed that the use of 50 times or more of moulds for precast concrete units gives reasonable economy. The saving on formwork must be sufficient to offset the fact that it costs much less to produce and place concrete on the site than it does to take the raw materials to a factory and then to have the precast unit transported to the site, together with attendant handling costs.

Fig. 21.5. Construction of multi-storey flats entirely in precast concrete

Other advantages of precast units accrue from the possibilities of reducing the formwork by casting them on their side, or the possibility of saving in concrete quantity and, consequently, the weight of the unit, by the use of more complicated shapes as, for instance, beams of "I" shape.

Accuracy of dimension is important in precast load-bearing units and, if a large number of repetitions is expected, soft wood is unsatisfactory for the moulds, and it is necessary to revert to hard wood or metal. Synthetic resin finishes are helping to protect mould surfaces and so prolong their life.

For load-bearing units, a high concrete strength is often required, necessitating the adoption of weigh batching of materials, low workability, and good control and workmanship to ensure uniformity, as described in detail in other chapters.

It is usually most economical to use the largest precast units which can be handled and transported without difficulty, and with which the erection tackle provided on the site can cope. As erection proceeds joints are made between units by bolting or by interlocking the reinforcement and infilling with concrete or cement mortar. In Russia where infilling with concrete would be impracticable in the long cold periods, steel plates are cast into the concrete units, and these are welded together. Jointing arrangements should be as simple as possible, and should not result in projecting concrete ribs or other details sensitive to damage during transport and handling. The infilling concrete or mortar must be well compacted and properly cured; otherwise ingress of water might lead to corrosion of the steelwork.

When precast units are used in conjunction with in-situ concrete, as for instance, to form the soffit of a floor, it is usual to form a mechanical key in the top of the unit in order in to provide a bond between the two. Reinforcement is also left projecting from the tops of units for the same purpose and to resist shear stresses.

Various concrete units

Such units as lintels, beams, columns, posts, water-tanks and troughs are made in accordance with normal concrete practice. The maximum size of aggregate may be limited by the size of the mould or by close spacing of reinforcement. Care should be taken to ensure that any reinforcement is placed correctly and is not displaced during casting, since in the small sections often made in precast concrete any movement may result in a serious reduction in cover. In such units as fence-posts the cover is sometimes reduced to $\frac{1}{2}$ in. or less. Cases of spalling of fence-posts due to insufficient cover are common everywhere, and it would appear advisable to limit the minimum size of post by the requirements of both strength and the minimum cover specified in the Reinforced Concrete Code of Practice.[7.5]

It is an advantage to reinforce all precast lintels and beams so that they can be used either way up. Some top reinforcement would probably be necessary in any case for handling, so that the extra cost is not likely to be serious, and provides the only practical insurance against failure resulting from fixing upside down.

Railway sleepers must be of high-grade concrete, adequately reinforced, if satisfactory service is to be obtained. B.S. 986, 1945,[21.3] for railway sleepers gives a minimum crushing strength for the concrete of 5,000–6,000 lb/in^2 at 28 days. In order to obtain this strength a mix containing about 625 lb of cement per cubic yard of concrete is normally used, with a low water/cement ratio. The sleepers are usually made in steel moulds, compacted on a vibrating table.

Concrete pipes may be made in a variety of ways. They may be cast in steel or wooden moulds either flat or on end, being compacted by hand or mechanical tampers, or they can be made in special machines. One such machine utilizes a revolving core which is slowly withdrawn as the concrete, which is of stiff consistency, is tamped into place. The revolving core brings sufficient fat to the surface to form an internal skin on the pipe and makes it more impermeable. Another method is to spin the mould so that the concrete is spread and consolidated against the mould by centrifugal force. The speed of spinning is critical, and must not be too fast, otherwise the centrifugal force may be sufficient to cause separation of the heavier particles. The coarse aggregate collects at the outer surface leaving cement mortar on the inside, and the pipe is weakened. The interior of the pipe is not formed against a mould by this method, and a round iron bar is used to smooth the surface while the pipe is still spinning.

Concrete pipes, the larger sizes in particular, should be carefully handled to avoid fracture.

The method of haunching concrete pipes is important since, except in the smaller sizes, they are not capable in themselves of sustaining a load from the overburden when given point support only. An illustration

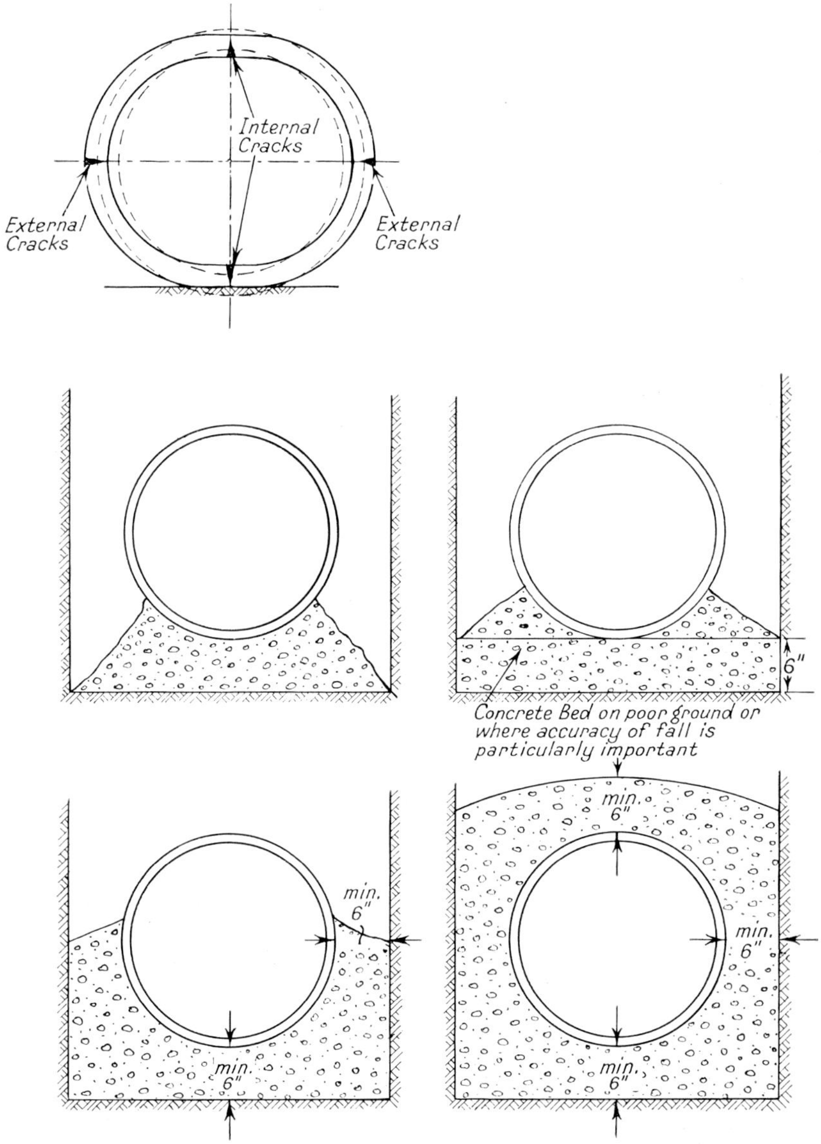

Fig. 21.6. Various methods of haunching concrete pipes

of the type of failure likely to occur is given in Fig. 21.6. Concrete pipes should therefore be adequately haunched if not completely surrounded with concrete. When the fall is small or the foundation is doubtful a concrete base should be used. Methods of haunching are shown in Fig. 21.6.

The most important point to remember is that the support given to the pipe should be continuous and uniform throughout its length.

Care is necessary to see that the concrete is worked well under the pipe, often a somewhat difficult procedure.

Uneven support results in unequal settlement along the length of the pipe with consequent failures.

Concrete pipes may suffer serious attack in sulphate-bearing or other aggressive soils, unless adequate precautions are taken, either by using a sulphate-resisting cement (possibly high-alumina cement) in the manufacture of the pipes themselves, or by providing a protective haunching in dense resistant concrete. External coatings of bitumen on the pipes are also useful.

Roofing tiles

Roofing tiles are made in a range of standard sizes given in B.S. 473, 1956[21.4] and B.S. 550, 1956.[21.5] Limits of variation in size and shape are given. Other requirements of a good roofing tile are that it shall be impermeable, weather resisting, and shall not fade; it should also be strong, in order to minimize breakages both in transit and after being placed in position. Tests designed to cover these points are given in the respective British Standards.

Roofing tiles are made in a variety of machines, designed to press the tile in a mould formed in the machine. Each tile is made on a separate pallet (shaped to the underside of the tile) placed in the bottom of the mould. The top of the tile is struck off by a combined screed and tamper, and a lifting device is used to push the tile out of the mould. The tile, still resting on its pallet, is then removed to curing racks.

The materials used for making roofing tiles are either normal or rapid-hardening cement and sand in the proportions of 1:3. The mix used is of earth-moist consistency, but must be the wettest which will allow

Fig. 21.7. Automatic machine for the manufacture of concrete roofing tiles

immediate removal of the tiles from the machine. Suitable pigments are added to the dry cement and throughly mixed, preferably by grinding them together.

Some tiles are given a surface finish after moulding, by brushing on a coloured cement:sand grout, by sprinkling with sand, by dashing with water to give a rough finish, or by using a roughened screed.

Precautions should be taken to see that the tiles do not dry out quickly, as this would result in loss of strength and high permeability.

Concrete blocks, slabs and bricks

Concrete units for masonry wall construction are usually turned out from machines in which an earth-moist mix is heavily rammed or pressed into a collapsible steel mould. A pallet is used in the bottom of the mould, and the blocks or bricks are removed immediately.

In the machine shown in Fig. 21.8, the pallet is first placed in the mould box at the front of the machine and rests on the extrusion carriage. The mould box is then filled with concrete by sliding a mixture box backwards and forwards over the aperture of the mould box. During this filling process the concrete is being distributed and partly consolidated by the jolting action of the bottom ram. After sufficient material has entered the mould box, the compression head of the machine is pulled down until it locates under the tension hooks, when the machine automatically applies pressure simultaneously to the top and bottom of the

Fig. 21.8. An automatic concrete block-making machine

concrete block. This is then extruded and carried off on the pallet to the stacking rails, placed adjacent to the machine.

Care must be taken to avoid the use of too dry a mix, as this would cause a loss of strength, friable corners, and a high permeability. Premature drying out should be avoided for similar reasons. The mix used for concrete blocks and slabs is usually between 1:6 and 1:8 of cement: aggregate, the maximum size of aggregate being $\frac{3}{4}$ in. A facing of 1:3, cement:sand is used if a good surface is required, as on "rock-faced" blocks. Concrete bricks are made from a mix of between 1:6 and 1:10 of cement:sand.

Blocks of odd shapes, and special bricks such as air-bricks, are normally made by hand either in wooden or metal-lined moulds, using the same materials as outlined above.

The following British Standards cover the manufacture of concrete bricks and blocks:

B.S. 1180:1944. Concrete Bricks and Fixing Bricks. (Amendment 1962)

B.S. 2028:1953. Precast Concrete Blocks (Amendment 1964).

Extracts from B.S. 2028 are included in Table 22.1. It will be noted that three types of block are included; the B.S. indicates the uses for which each type is intended. The requirements for crushing strength, drying shrinkage and moisture movements are important, and check tests should be made by the user from time to time. In the purchasing order a clause providing that the cost of testing shall be borne by the purchaser if successful, or by the manufacturer if unsuccessful, is not unusual.

In order to obtain the requisite strength and drying shrinkage values, it is usually necessary to mature concrete blocks and bricks for at least 4 weeks.

The first shrinkage should be completed before the bricks are built into the wall, and this means that they should be dried out before delivery to the building site. In winter it is necessary for the manufacturer to carry large stocks or to provide suitable dry accommodation under cover. Good air circulation is required to dry the units properly. See also Steam Curing, p. 325.

Troubles due to shrinkage cracking are not likely to be experienced if the blocks comply with the relevant British Standard, particularly the clause limiting the drying shrinkage, unless the mortar used for the blockwork is unnecessarily strong. A mix of cement:lime:sand of the order of 1:1:6 or 1:2:9 by dry volume is generally regarded as satisfactory.

Reference is made to the manufacture and use of lightweight blocks, including breeze and clinker blocks, in Chapter 22.

Colours

Only certain types of inert pigment give satisfactory results in concrete, the principal requirements being that the colour shall not be affected by lime, shall not fade when exposed to the weather, and shall not contain any substances such as gypsum, which are deleterious to concrete. Organic pigments do not fulfil these conditions and should not be used. Mineral colours are generally satisfactory, but are sometimes diluted with other cheaper constituents. However, to obtain the same colour shade, a greater quantity of the diluted pigment must be used, so that no advantage is gained, while in fact the increased proportion of fine material causes a loss in strength and increased drying shrinkage.

For general purposes the paler tints of reds, yellows, buffs and browns are the most successful. These pigments are usually oxides of iron. Pale green shades obtained by additions of chrome oxides are also quite good. Brighter colours are obtained by using white rather than grey cement.

Deep reds, blues, greens, and black are apt to come out patchy, and in any case soon lose their brightness due to the slight efflorescence which forms on most concrete exposed to the weather.

Dark grey may be obtained by using small quantities of manganese black or carbon black.

Pigments are not very different from silt in their effect on the physical properties of concrete and not more than 10 per cent of colour by weight of cement should be used. Larger percentages cause serious loss of strength and increased shrinkage. Pigments are not easily mixed with cement on the site, and if large quantities of coloured cement are required it is better to mix the pigment and cement in a ball-mill at a central depot, or to buy one of the coloured cements prepared by the cement manufacturers.

The effectiveness of concrete colouring depends to a considerable extent on the colour of the aggregate used. Unless a contrast is required between the aggregate and the cement, as in terrazzo, it is advisable to select an aggregate of similar colour to that required for the finished concrete. In some cases, aggregate may be selected, which, when mixed with cement, gives a suitable colour without additions of pigment.

Crazing

Crazing of concrete products results from differences in shrinkage between the surface and the interior. As has been shown in Chapter 2, this is a combination of differences in drying shrinkage and a shrinkage due to carbonation which affects only a layer up to $\frac{1}{2}$ in. or so thick near the surface. Cracks due to crazing rarely exceed $\frac{1}{2}$ in. or so in depth, and are therefore not serious, apart from their unsightliness.

Several methods of overcoming crazing have been advocated, but the best insurance appears to be either to use an earth-moist mix, or if a plastic mix is necessary use as low a water/cement ratio as is practicable and remove the cement skin to expose the aggregate. Aggregates which have low shrinkage in themselves are likely to produce less crazing than highly shrinkable aggregates. Proper attention to curing, especially during the early stages, is an added safeguard.

Trowelled surfaces are particularly prone to crazing, and trowelling should therefore be avoided whenever possible. If a surface must be trowelled, the amount of trowelling should be the minimum necessary to finish the surface.

Steam curing

Steam curing is used to obtain a high strength at early ages and so release the moulds more quickly than would be the case with normal methods of curing, especially during cold weather.

Steam curing may be effected either by the use of low pressure or high pressure steam, but, while the former is the cheaper, the use of high pressure steam results in an improvement of certain of the properties of the concrete which may be considered to offset the increased cost.

When high pressure steam curing is adopted the precast concrete units are stacked in steam chambers, or autoclaves, and the steam introduced for the required period of curing, usually a matter of a few hours. It is advisable to allow a short curing period at ordinary atmospheric temperatures before introducing the steam, and an excessive rate of heating should be avoided. There is considerable controversy on the question of the maximum rate of heating which may be allowed. Figures for the safe rate of increase in temperature quoted by various investigators range from $5\frac{1}{2}$°C to 66°C per hour. Rapid cooling at the end of the curing period should also be avoided.

High pressure steam curing encourages the growth of a crystalline bond and in the presence of finely divided siliceous aggregate, there is a reaction between the aggregate and lime from the cement which accounts very largely for the increased sulphate resistance. The shrinkage and moisture movement is also reduced.

Low pressure steam curing is effected either by stacking the precast concrete units in specially built chambers, or by using light portable boxings which are placed over the units wherever they are cast, the steam being supplied through flexible connections. This latter method has the advantage that small individual chambers take less time to warm up than large chambers containing a large number of units, and the amount of handling is reduced. Low pressure steam curing speeds up the rate of hardening of concrete units, but does not improve sulphate resistance or reduce shrinkage and moisture movements.

As with high pressure steam curing, excessive rates of heating and cooling must be avoided.

If hot concrete units dry out after having hardened in a steam curing chamber for only a few hours sufficient to facilitate demoulding, the loss in ultimate strength can be serious. Both with high pressure and low pressure steam curing, it appears that the ultimate strength is not fundamentally different from that of concrete cured at atmospheric temperatures, unless the rate of heating is too high. Under these conditions, further gain in strength on subsequent moist curing appears to be arrested.

Efflorescence

The chief cause of efflorescence on cast stone is the formation of crystals of calcium carbonate and calcium sulphate on the surface due to leaching and chemical reactions of the free lime (calcium hydroxide) with rainwater. This action is unavoidable in porous concrete, but is unlikely to be serious if the concrete is dense and impermeable.

Efflorescence is more apparent on dark or deeply coloured surfaces than on those which are white or tinted.

Treatment with waterproofers containing soluble salts, or which react with the formation of soluble salts, should be avoided. Aggregates containing alkalies or sulphates are liable to cause efflorescence, as also are aggregates containing sea salts.

REFERENCES

21.1 B.S. 368:1956, *Precast Concrete Flags*. British Standards Institution.

21.2 B.S. 340:1963, *Precast Concrete Kerbs, Channels, Edgings, and Quadrants*. British Standards Institution.

21.3 B.S. 986:1945, *Railway Sleepers*. British Standards Institution.

21.4 B.S. 473:1956, *Concrete Plain Roofing Tiles*. British Standards Institution.

21.5 B.S. 550:1956. *Concrete Interlocking Roofing Tiles*. British Standards Institution.

CHAPTER 22

Lightweight Concrete

LIGHTWEIGHT concrete is manufactured not only on account of its light weight, which is useful for floor thicknessing, screedings, and partition slabs, but also because of the high thermal insulation compared with normal concrete. The thermal insulation of dry lightweight concrete is mainly dependent on its density, and may be assessed roughly from the following equation. This may be used for densities between 20 and 100 lb/ft³.

$$K = 0.031(d - 20) + 0.5$$

where K = thermal conductivity in Btu/in. ft² hr deg F

d = density in lb/ft³

Some materials produce lightweight concretes which have generally better insulation than given by the equation. An indication of this is given in the sections in which the materials are discussed. It is important to realise that the high thermal insulation is only obtained while a lightweight concrete remains dry. It is therefore not always satisfactory for roof insulation because of practical difficulties in drying out before applying the waterproof finish, and because condensation may cause saturation later. This condensed water sometimes shows as damp patches on ceilings and is mistaken for rain penetration. Similarly lightweight concrete is of no use for thermal insulation in any position in which it may become saturated, for example underground duct work. The strength of lightweight concrete is generally much lower than that of dense concrete, but this is not usually of any great importance for the purposes to which lightweight concrete is put.

Structural "lightweight" concrete can be produced with a strength in excess of 2000 lb/in², but its weight is then usually 100 to 120 lb/ft³. The aggregates used include pumice or scoria, expanded clays, slates, shales, sintered fly ash, and foamed slag. As most of these aggregates absorb considerable quantities of water (up to 80 per cent by volume), the effect on workability within a few minutes of mixing is such that a wet mix can become too dry to work. It is therefore necessary to wet (but not saturate) the aggregate before mixing. A good portion of the mixing water is also best added before introducing the cement. Rich mixes containing 600 lb cement per cubic yard or more are usually required to give a satisfactory strength. Sand is often added to increase the strength, but also increases the density. The cover to reinforcement when using lightweight concrete should be adequate, say, 1 inch more than for dense concrete.

The increased cover is necessary because, besides being more permeable, lightweight concretes carbonate more quickly than dense concrete and the protection afforded to the steel by the alkaline lime is lost. (See also Expanded Minerals, p. 333.)

There are various methods of producing lightweight concrete but, they all depend on (1) the presence of air voids in the aggregate, (2) the formation of air voids in the concrete by omitting fine aggregate, or (3) the formation of air voids in a cement paste by the addition of some substance which causes a foam. In some types of lightweight concrete, methods (1) and (2) may be combined.

No-fines concrete made with gravel aggregate is not strictly a lightweight concrete, although its weight is only two-thirds that of dense concrete, but it is convenient to consider it with concrete made with other lighter aggregates.

Breeze and clinker

Breeze and clinker aggregates have been in use for many years in the production of blocks and slabs for internal partitions and other interior walls. These aggregates are cheap and plentiful, and provide a very useful product. The thermal conductivity of clinker concrete falls roughly in line with the equation given above, perhaps up to 10 per cent in excess depending on the clinker.

There is no definite demarcation between materials described respectively as breeze and clinker. In general, clinker is regarded as a well-burnt fused or sintered mass containing little combustible material, whereas breeze is a more lightly sintered and less well-burnt residue and therefore contains more combustible matter.

Increasing quantities of combustible matter cause corresponding increases in moisture movement. The following values may be quoted,[22.1] for clinker containing no dangerous coal.

Combustible content per cent	Moisture movement per cent
50	0·12–0·17
25	0·06–0·10
8–15	0·05–0·07
0–8	0·02–0·06

The combustible content is determined approximately by igniting a small sample of the clinker at specified temperatures and finding the percentage loss in weight. The specified limits for combustible content are varied according to the position in which the concrete is to be used. They are given in B.S. 1165:1966[22.2] as follows:

Class A1.	Not more than 10 per cent	General purposes, plain concrete.
Class A2.	Not more than 20 per cent	Interior work not exposed to damp conditions, cast in situ.
Class B.	Not more than 25 per cent	Precast clinker blocks. (Some British Standards call for a clinker aggregate which, by virtue of the test requirements in those standards, requires a lower limit on the percentage of combustible content.)

The presence of sulphur rarely causes failure in practice. However, sulphur may be dangerous if the compounds in which it occurs are present as a fine powder. B.S. 1165:1966 requires that the soluble sulphate content, as determined by the method laid down in B.S. 3681,[22.3] shall not exceed 1·0 per cent when expressed as sulphur trioxide.

In the light of experience the soundness test, which appeared previously in the British Standard, has been dropped.

A large proportion of fine material necessitates a high water content for the production of the concrete, and retards the setting.

Clinker and breeze aggregate is quite unsuitable for reinforced concrete due to its porosity and absorptive properties, which maintain a more moist condition than the surroundings. The sulphur content is also a factor in accelerating the corrosion of embedded steel.

Breeze and clinker blocks

Precast breeze and clinker blocks are used extensively for the internal partitions of houses and other buildings, chiefly on account of their cheapness. They are sometimes used for load-bearing walls, but unless arrangements can be made for regular testing for compressive strength and drying shrinkage, their use for this class of work may entail some risk of subsequent trouble. Precast blocks should comply with B.S. 2028:1953, Precast Concrete Blocks and later amendments. Amendment No. 2, 1964 deals with blocks of greater strength and with moisture movements in Type B blocks. The specification includes strength requirements and limits for drying shrinkage and moisture movement, together with limits on the tolerances for dimensions and definitions of types and uses of the blocks.

The widespread small-scale manufacture of breeze and clinker blocks without appreciation of the ill-effects of lack of control of mix proportions, and delivery within two or three days of manufacture, is reflected in the low compressive strength, high shrinkage and excessive moisture movement, often revealed by testing, with the risk of subsequent cracking and shelling of plaster coatings. However, cracking in clinker block partitions should not be ascribed to their shrinkage without investigation of the structural details. Partitions are commonly built on timber joists which may shrink $\frac{1}{8}$ in. and more in their depth, quite sufficient to cause cracking at junctions with more rigid supports.

Methods of use of precast concrete blocks are described in the British Standard Code of Practice C.P. 122, Walls and Partitions of Blocks and Slabs, and for design purposes reference should be made to C.P. 111, Structural Recommendations for Load-bearing Walls, which includes data on working stresses for load-bearing walls constructed with blocks.

The Building Research Digest No. 52, Lightweight Concrete Blocks for Housing, gives data on manufacture, site storage, and laying of lightweight precast blocks. Information on design of structures using these blocks is given and the following points are emphasised:

(*a*) Long uninterrupted lengths of lightweight concrete block walling should be avoided by providing breaks or flexible mastic-filled joints in the construction at intervals of about 20–30 ft in the length, and at points of marked change of section.

(*b*) Relatively weak mortars should be used so that any shrinkage cracking will tend to occur in the mortar joints rather than in the blocks. Mixes of 1:2:9 cement, lime and sand are recommended, or for winter conditions or severe exposure, a 1:1:6 mix is satisfactory.

(*c*) When blocks are used externally and rendered, a relatively weak porous rendering should be used. This recommendation should be tempered by consideration of the exposure conditions. In Britain, where renderings may remain wet for several days and may be frozen, it is unwise to use a mix leaner than about 1:1:6 of cement:lime:sand, except in sheltered situations.

Fly ash concrete

A lightweight concrete can be made by the use of fly ash, cement or lime, and a foaming agent. Sand is also sometimes added to the mix or may be used on its own. The mixed material is poured into moulds and autoclaved. The strength of fly ash concrete is dependent on the mix proportions, the density, and the extent to which the fly ash and/or sand are ground together in a ball mill before mixing with the foaming agent, which is usually aluminium powder. The autoclaving is essential, not only to accelerate the hardening of fly ash concrete, but also to reduce the drying shrinkage which would otherwise be too large for most building purposes.

It is doubtful whether fly ash concrete possesses advantages over ordinary clinker concrete, to justify the increased cost of production. However, the diminishing supplies of clinker have resulted in an increasing use of fly ash concrete.

Pumice concrete

Pumice is the only natural lightweight aggregate in common use. Provided it is free from fine volcanic dust and materials not of volcanic origin such as clay, pumice produces a satisfactory lightweight concrete with a density of between 45 and 70 lb/ft^3. Other properties are given in Table 22.1.

Pumice provides better thermal insulation than other types of lightweight concrete, its conductivity when dry being only about $\frac{3}{4}$ of that given by the equation at the beginning of the Chapter.

Foamed slag

Foamed slag is made by rapidly quenching blastfurnace slag produced in the manufacture of pig iron. Its texture and strength are dependent on the chemical composition and the method of treatment, but in general the structure is similar to natural pumice.

Foamed slag aggregates are supplied in two sizes, coarse $\frac{1}{2}$ in.–$\frac{1}{8}$ in., and fine $\frac{1}{8}$ in. down. They should comply with B.S. 877, 1939.[22.4] Foamed slag may be regarded as satisfactory if it fulfills the following requirements:

(1) Freedom from contamination by heavy impurities, including air cooled slag.
(2) Freedom from volatile impurities such as coke or coal.
(3) Freedom from an excess of available sulphate.

The properties of foamed slag concrete are given in Table 22.1. It will be noted that light weight is achieved by using relatively low ratios of cement to aggregate. The thermal conductivity when dry is only about

TABLE 22.1 *Properties of Lightweight Concretes*

Type of lightweight concrete	Mix cement: aggregate by volume	Water-cement ratio	Crushing strength: lb/in²	Transverse strength: lb/in²	Drying shrinkage: per cent	Density air dry: lb/ft³	Working properties	Nail and screw holding
Clinker	1:6 1:10	— —	150–600 —	75–250 —	0·03–0·20 —	50–105 95	Easily worked Easily worked	Satisfactory Satisfactory
Pumice	1:6 1:10	— —	200–550 —	100–150 —	0·04–0·08 —	45–70 41	Easily worked Easily worked	Satisfactory Satisfactory
Foamed slag	1:6 1:10	— —	800–2,000 350–750	200–300 100–250	0·04–0·08 0·03–0·07	80–95 60–95	Easily worked Easily worked	Satisfactory Satisfactory
Expanded slate, clay and shale. Sintered fly ash.	1:6 1:10	— —	600–1,100 250–400	200 100–150	0·04–0·08 0·03–0·07	40–80 35–60	Easily worked Easily worked	Satisfactory Satisfactory
Aerated cement	—	—	200–500	100–240	0·05–0·18	35–60	Easily worked	Satisfactory
No-fines concrete (*a*) gravel aggregate (*b*) lightweight aggregate	 1:8 1:6	 0·40 0·60	 500–1,000 350–450	 Low Low	 0·02–0·03 Depends on aggregate used	 100–110	 Difficult Easily worked	 Fixing blocks necessary
Solid concrete	1:2:4	0·60	4,000 (av.)		0·03–0·05	145–150	Hard	Good when plugged
B.S. 2028: 1953 requirements, type A, dense aggregate			500 min.† 375 min.††	No requirement	0·04 max.	Not exceeding 100		
B.S. 2028: 1953 requirements, type B lightweight aggregate, load bearing			400 min.† 300 min.††	No requirement	0·08 max. (moisture movement 0·05)**	Not exceeding 100		
B.S. 2028: 1953 requirements, type C, lightweight aggregate, non-loadbearing			No requirement	140–244† depends on size of block	0·08 max.	Not exceeding 100		

† Average of 12 blocks for compressive test or 6 blocks for transverse test.
†† Lowest individual block.
** For blocks made with furnace clinker aggregate.

$\frac{3}{4}$ of that given by the equation at the beginning of the chapter, being almost as low as pumice concrete and considerably better than clinker concrete.

In common with other lightweight concretes, foamed slag has a higher moisture movement than ordinary concrete. The value is, however, as good as the best of other lightweight concretes.

Expanded minerals

Porous material of cellular texture may be produced by heating clays, shales, and slates, to a suitable temperature. After crushing and screening to the desired size, these materials form good lightweight aggregates.

The properties of sintered fly ash, expanded slate, clay and shale concretes are given in Table 22.1 and their thermal conductivities are roughly as assessed by the equation given, being a little above if anything.

Expanded clay aggregate is produced on the Continent and at one plant in Britain. It is generally porous, and in structural concrete extra cover is required to reduce the risk of corrosion of the reinforcement. Extra reinforcement is also necessary, the modulus of elasticity being only 2 to $2{\cdot}5 \times 10^6$ lb/in^2, while the saving in weight is not great, with the result that the use of such aggregates (costing more than natural aggregates) does not normally result in a saving in cost in the structure.

Expanded shale is also produced on a relatively small scale, and has similar properties to expanded clay, except that it is more porous, having an ultimate water absorption of 50 per cent or more by volume.

A somewhat similar lightweight aggregate is produced by pelletizing fly ash and burning it on a sinter strand. A bed of pellets, on a moving steel grid, passes under a furnace and the pellets become sintered into a clinker-like mass. Light crushing and grinding to size follows. Its properties are very similar to those of expanded shale.

Exfoliated vermiculite is very light in weight, but is so soft that the action of mixing with cement and water is sufficient to cause collapse of particles. Vermiculite concrete is therefore likely to be of uncertain density and to have little compressive strength.

No-fines concrete

No-fines concrete is composed of cement and coarse aggregate only, the fine aggregate being omitted in order to leave uniformly distributed voids throughout the mass. The aggregate may be gravel or crushed stone, blastfurnace slag, crushed brick, or one of the light aggregates already mentioned. The quality of the aggregate should accord with the requirements already given, and the relevant British Standard.

Recommended practice for the use of no-fines concrete cast in-situ for one-to-three-storey housing was originally given in Post-War Building Studies, No. 1.[22.5] The following comments are based on experience in the production of no-fines concrete on a large scale for houses and multi-storey flats, both in Britain and abroad.

(1) *Grading of aggregate.* The aggregate should be as nearly single sized as practicable. In Great Britain it is usual to use an aggregate, of which not more than 5 per cent is retained on a $\frac{3}{4}$ in. mesh sieve, and not more than 10 per cent passes a $\frac{3}{8}$ in. mesh sieve.

(2) *Recommended proportions.* Heavy-weight aggregate, 1 cwt cement to 10 ft^3 aggregate; light aggregate, 1 cwt cement to $7\frac{1}{2}$ ft^3 aggregate.

(3) *The water/cement ratio* should be the minimum necessary to ensure that each particle of aggregate is coated with cement grout. Too little water gives a friable appearance with a large proportion of uncoated particles. Too much water causes the cement grout to run and separate from the aggregate.

(4) *Mixing.* It is most important that the aggregate is used in a damp condition or wetted before adding the cement. Mixing should continue until the aggregate is evenly coated with cement grout. With rotary-drum mixers it is generally advisable to add some of the water before raising the loading skip and then to add the remainder of the water after the materials have entered the drum. In this way a more uniform mix is obtained. This difficulty does not occur with open-pan mixers fitted with a rotating star of blades.

(5) *Placing.* No-fines concrete will not segregate and can therefore be dropped from a greater height than ordinary concrete, and greater lifts can be placed in one operation. Pressures on the formwork are not high and a comparatively light timber framework serves to support plywood, hardboard, sheet metal, or expanded metal coverings. No-fines concrete should not be rammed but careful rodding should be done wherever obstacles occur, since bridging is more likely to occur than with ordinary concrete.

(6) *Joints.* Since the bond between new and existing work is weaker than in ordinary concrete, construction joints should be horizontal and as few as possible. Vertical joints weaken the structure, especially if they occur near an external angle, and except where necessary for the provision of expansion joints they are better omitted altogether. Expansion joints should be used at suitable spacings to take care of shrinkage and thermal movements.

(7) *Cube strength.* Experience has shown that with good site control a minimum crushing strength of 600 lb/in^2 or more at 28 days is easily practicable with the mix given under (2) above, enabling the building of unreinforced structures 5 or 6 storeys in height. Greater strengths can be obtained by using richer mixes and by compacting to a higher density. In Germany, no-fines concrete has been used without reinforcement for structures up to about eight storeys. The total height of these structures is, in some cases, greater than eight storeys, but this is about the maximum number of storeys in no-fines, as defined above. In Great Britain one system of construction, utilizes no-fines panels to form the moulds for reinforced beams and columns which eventually provide the supporting framework. By this method multi-storey structures have been produced.

The B.S. Code of Practice, C.P. 111:1964, (Amendment 1966) gives minimum cube strengths and maximum permissible stresses, as in Table 22.2. These are for walls with a slenderness ratio of less than 15. For greater slenderness ratios it is necessary to apply a reduction factor.

Other characteristics. No-fines concrete presents some difficulty in the fixing of various fittings and it is necessary to embed nailing blocks of timber, sawdust-cement, or foamed slag. Cutting away for services should be avoided by forming suitable openings and chases at the time of placing.

TABLE 22.2 *Recommendations for Cube Strengths and Permissible Stresses in No-Fines Concrete*

Nominal mix	Volume of coarse aggregate per 112 lb of cement ft³	Cube strength* within 28 days after mixing lb/in²	Maximum permissible stresses lb/in²
Special mixes	10 or less	1000†‡	190
		500	90
1:8	10	400	70

The recommendations cover the use of natural aggregates to B.S. 882 and air-cooled blastfurnace slag to B.S. 1047, mixed with Portland cement, Portland blastfurnace, or other cement included in C.P. 114.

* These requirements may be deemed to be satisfied if two-thirds of the value is obtained at 7 days, except in the case of high-alumina cement.

† The average cube strength for mix design purposes should be 300 lb/in² in excess of the cube strength specified unless more definite information is available.

‡ The attainment of this strength increases the density of the concrete to an extent where the thermal-insulation properties may be impaired.

No-fines concrete has little resistance to the penetration of water, but there is also very little capillary action. Thus there is no tendency for the water to be drawn into the wall; the provision of a rendering, with care in arranging and fixing flashings at various openings, is sufficient to waterproof the structure satisfactorily.

The rendering should preferably have a rough surface texture, dry dashings being particularly suitable. A 1:1:6 or 2:1:9, cement:lime: sand rendering is normally considered suitable, except perhaps in cold climates where a somewhat richer mix may be used to provide increased resistance to disintegration by frost and continually wet conditions.

The thickness of no-fines concrete is more often determined by the requirements of thermal and acoustic insulation than stability. For instance, in order to obtain insulation similar to that of an 11 in. closed-cavity brick wall, plastered internally, it is necessary to use 10 to 12 in. of no-fines concrete rendered externally and plastered internally.

A general indication of other important properties of no-fines concrete is given in Table 22.1.

Aerated cement

Aerated cement is made by adding powdered zinc or aluminium to cement, and gauging with water. About 1 part of powdered metal to 1,000 parts of cement are thoroughly mixed dry, and then water added to form a paste. Within a few minutes hydrogen gas begins to evolve causing the paste to rise, the action continuing for an hour or so. The cement then hardens to form a material consisting of a multitude of closed bubbleholes surrounded by hardened cement. The density of aerated cement depends on the quantity of powdered metal and the temperature at the time of manufacture. For most purposes densities within the range 35–60 lb/ft³ are satisfactory.

In making aerated cement blocks, the paste is poured into the mould to fill it about one-third full. As soon as the paste has risen fully, any surplus is struck off, and the block set aside to harden.

While aerated cement can be relatively impermeable to moisture, owing to the closed type of pores, it has the disadvantage of a high drying shrinkage and moisture movement. Precast units, which can be matured well before use, thus materially reducing effective shrinkage movements when the units are built into a structure, do not suffer from cracking to the same extent as aerated cement cast in situ. Curing in an autoclave, using steam under pressure, considerably reduces drying shrinkage and moisture movements.

Aerated mortars and aerated mortar concrete

Aerated mortar of very similar texture to aerated cement can be made by using a foaming agent, such as a resin soap. The foaming agent is either mixed with the cement and water and the air entrainment achieved by whisking in a high speed mixer, or is turned to foam by the use of compressed air using a special foam producing apparatus, and the foam then added to a cement:sand mortar in an ordinary concrete mixer. The latter method provides a more uniform density provided the foaming time is carefully controlled.

Foamed cement can be made with a density as low as 20 lb/ft^3 but then has no appreciable strength, and is only of use for insulation purposes in dry situations.

Some foaming agents are sufficiently stable to permit the use of cement: sand mixes for the production of foamed concrete at densities above about 50 lb/ft^3. The average crushing strength varies with the density as indicated in Fig. 22.1. As with other lightweight concretes the relatively high drying shrinkage presents problems in the use of this material in situ, and ample provision of "expansion joints" is essential. Nevertheless aerated mortars cast in situ have been used successfully in the construction of some thousands of dwellings in countries such as Kuwait and Borneo,

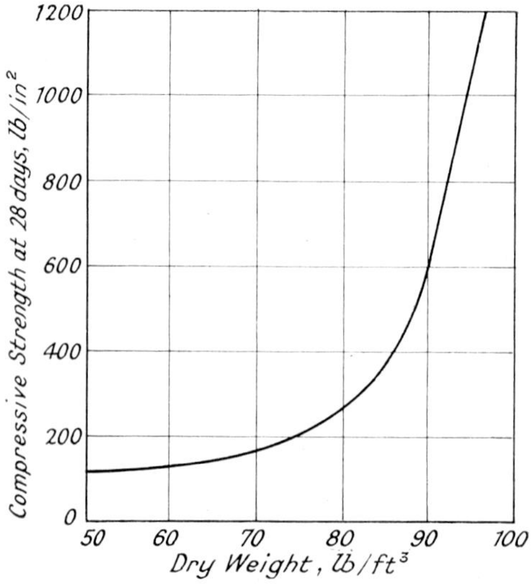

Fig. 22.1. Compressive strength-density relationship for aerated cement and concrete

where sand has been in ample supply in the vicinity of the construction site. Precast blocks must be cured for several weeks before use unless steam curing is adopted. Even then it is advisable to build in a cement: lime:sand mortar not richer in cement than 1:1:6 if visible cracking is to be avoided.

Coarse natural or lightweight aggregate can also be used in conjunction with aerated cements and mortars. Natural aggregates increase the density, and to some extent reduce the moisture movements. Lightweight aggregates do not cause the same increase in density, but the moisture movements are more nearly those of aerated mortars.

A process for making blocks similar in texture to aerated cements and mortars is based on the use of silica sand and lime, often with a little cement added. All or part of the sand and lime are ground together in a ball mill, and are then mixed with water and a foaming agent, usually aluminium powder. The fluid mix is then poured into large moulds and allowed to stand for about two hours during which aeration occurs so that the mixture rises to fill the moulds. By this time the mixture stiffens sufficiently for it to be cut by wire slicers to blocks or slabs of suitable size. These are then autoclaved to bring about the chemical reaction between the lime and the silica which causes hardening. The process and the product are very similar to that of fly-ash:lime mixes (see Fly Ash Concrete, p. 331).

Wood-cement mixes

Wood in the form of fibers, chips and sawdust is sometimes used in various types of block, partition slab and floor surfacings. The high drying-shrinkage and moisture movement is apt to cause cracking, and the use of these materials should not be considered unless freedom of movement of each individual unit can be arranged.

REFERENCES

22.1 LEA, F. M., *The Properties of Breeze and Clinker-Aggregates and Methods of Testing their Soundness*, Building Research Bulletin No. 5 (revised edition 1936). H.M.S.O.

22.2 B.S. 1165:1966, *Clinker Aggregate for Concrete*. British Standards Institution.

22.3 B.S. 3681:1966, *Methods for the Sampling and Testing of Lightweight Aggregates for Concrete*. British Standards Institution.

22.4 B.S. 877:1939, *Foamed Blastfurnace Slag for Concrete Aggregate* (Addendum 1947). British Standards Institution.

22.5 *House Construction*, Post-war Building Studies, No. 1, Ministry of Works, 1944. H.M.S.O.

CHAPTER 23

Concrete Floor Surfaces

CONCRETE floors may be finished either by trowelling up the concrete surface or by applying a special finishing course of the granolithic type. The former are suitable only for relatively light wear as, for instance, foot traffic and rubber-tyred vehicles, and in positions not requiring a smooth hard surface, and therefore require little comment.

Heavy-duty floors are usually of the latter type and a carefully controlled technique is essential if good results are to be achieved. The following considerations therefore apply to granolithic and similar finishes.

Aggregates

The aggregate should consist of a hard tough natural rock, and of suitable grading to comply with B.S. 1201:1965, Aggregates for Granolithic Concrete Floor Finishes (included with B.S. 882:1965)

The specification limits the type of rock to one of seven trade groups. An essential property of aggregates for granolithic floors is wear resistance, for which no direct test is available, although the ten per cent fines value gives an indication of hardness; the test is made in accordance with B.S. 812, the limit being ten per cent fines under a load of not less than 15 tons.

B.S. 1201 also gives maximum desirable quantities of clay, silt and fine dust.

Cement

Normal Portland cement is used almost exclusively for granolithic floors, coarsely ground cements being preferred as against the more finely ground rapid-hardening cements. Experience suggests that the latter result in an increased tendency to crack.

Additions

Coloured floors are not entirely satisfactory, as variations in colour intensity are difficult to avoid, and efflorescence is often the cause of white patches. Colours have the same effect in a concrete mix, as the presence of clay and silt, reducing strength and wear resistance while increasing drying shrinkage. Large percentages of colour have been the cause of failures; in one case investigated it was found that the use of dusty granite aggregate and 10 per cent (by weight of the cement) of red pigment had increased the drying shrinkage to twice that of the undercourse.

Metallic aggregates are widely used in the belief that they improve the wear resistance of floors subjected to heavy traffic. They are mainly composed of particles of cast iron of varying degrees of fineness. The finer materials have a large portion passing a 100-mesh sieve and their value, at least, is open to doubt. They may be either incorporated in the mix or thrown on to the surface and trowelled in.

Large proportions of metallic aggregates have caused spalling due to rusting and expansion of the iron. The safe limit depends on the density of the concrete, but about 15 lb/yd^2 is suggested as a maximum.

Metallic aggregates may improve well-laid floors but are valueless in badly-laid floors. They are more evenly distributed when incorporated in the mix than when applied by throwing on the surface and trowelling in.

Discoloration by rust during damp curing of the floor is difficult to avoid.

Carborundum grit trowelled into the surface is useful for preventing slipperiness, but being brittle does not improve the wear resistance under conditions of heavy impact and abrasion by iron-tyred trucks and similar vehicles.

Calcium chloride is sometimes added to accelerate the setting and hardening. The precautions necessary when using calcium chloride have already been described in connection with cold-weather concreting.

Mix

The mix used should be 1:3 cement:aggregate by weight. The aggregate should be measured accurately by weight or by volume, and the cement by weight. The water content should be the minimum necessary to give sufficient plasticity for laying and compacting and does not generally need to exceed a water/cement ratio of 0·42. The influence of cement content and the cement/water ratio (converse of water/cement ratio) on wear resistance are indicated by Fig. 23.1 (*a*) and (*c*), which is plotted from the results of tests made by Schumann and Tucker.[(23.1)]

In two-course work the bottom course may be 1:2½ to 1:6 by weight and of somewhat drier consistency than the top course, which is mixed in the proportions given above. Care should be taken to ensure that the bottom course is not so dry that a porous, friable layer having little or no bond to the base concrete, results. This point cannot be emphasized too much, as many floors have failed because of the use of a dry bottom course which has failed to bond to the base concrete. The consistency should be such that a dense layer is obtainable with the compacting method employed.

Preparation of base concrete

The topping may be laid before the base concrete has completely hardened, thus forming a monolithic slab, or it may be necessary to lay the topping on set concrete. At whatever stage the topping is to be laid the surface of the base should be freed from laitance, and the best way to achieve this appears to be to brush the concrete with a stiff broom just before it sets. This method also leaves a roughened surface which forms a good key.

If measures have not been taken when laying the concrete, hardened bases should be thoroughly roughened; a few chips hacked out here and there is not sufficient and mechanical means are strongly recommended if a good bond between the hardened concrete and the topping is to be achieved. The base should then be thoroughly cleaned and wetted before

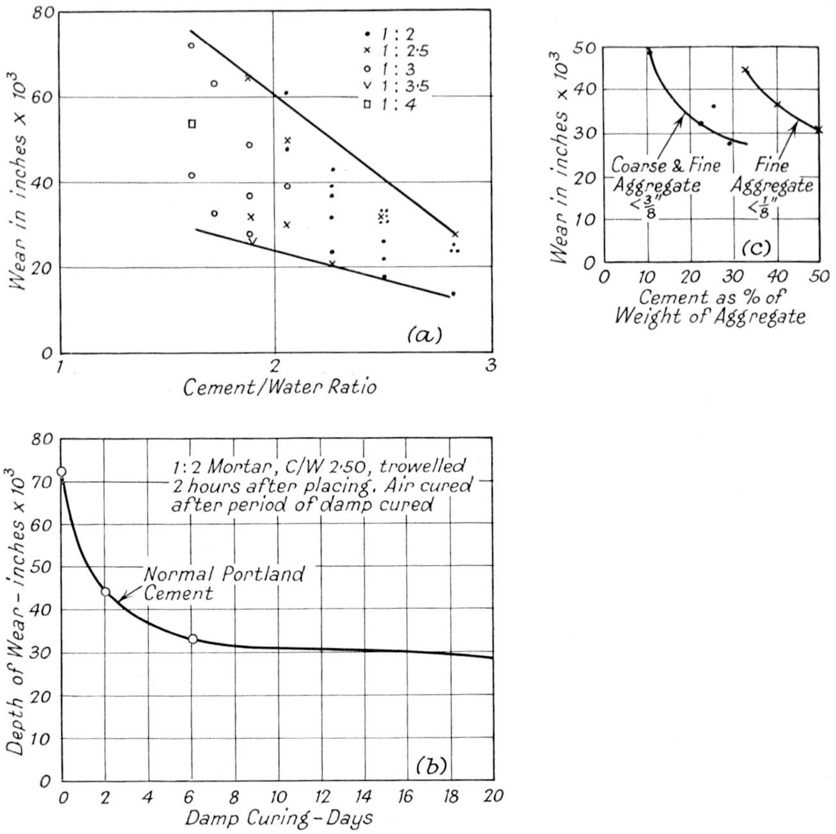

Fig. 23.1. Relation between wear resistance and cement/water ratio, length of curing, and cement content

Note. Points are average of all specimens at each mix, irrespective of water content, types of aggregate or method of finishing.

commencing to lay; surplus water should, however, be removed from the surface before placing the topping.

Lifting and curling of granolithic floors due to failure to clean away dust, chippings and other loose debris from the base, has commonly occurred. Plaster droppings are specially detrimental to the bond, and even small quantities may cause sulphate expansion under damp conditions. Careful cleaning is essential, particularly in the corners.

When laying on an oily or otherwise contaminated base, it becomes necessary either to chop out all the oil-soaked concrete, or, if conditions of height and level make it possible, to overlay a thickness of, say, 3 in. of topping, possibly reinforced.

American practice favours the use of a detergent on bases which have been laid a week or more previously, followed by commercial hydrochloric acid broomed over the surface. When the acid stops foaming the surface is flushed with water and all loosened material removed with a stiff broom. Rubber boots and suitable protective clothing are necessary.

Grouting base

After cleaning and wetting, the hardened base concrete should be grouted with a cement:sand (in about equal parts) grout scrubbed into the surface, keeping just ahead of the topping.

In lieu of grouting, certain epoxy and epoxy-sulphide resins may be used to bond fresh to hardened concrete. Their use is likely to be restricted to special circumstances by the added cost and the more involved technique necessary. Their effective life is unknown, but should be several years at least, if used properly.

Thickness of floor surfacing

The minimum thicknesses of floor surfacing are usually specified to be as follows:

(1) If laid before the base has hardened, minimum $\frac{3}{4}$ in.

(2) If laid on a hardened base to which good bond can be expected, minimum $1\frac{1}{2}$ in.

(3) If laid on a hardened base to which little or no bond can be relied upon, due to contamination or other causes, minimum 3 in.

Unfortunately as the thickness increases, the tendency to curl upwards at the edges, and to break the bond with the base concrete is increased. Attention to those details which improve the bond with the base concrete, is probably more important than thickness. Entirely satisfactory granolithic surfacings, $\frac{3}{4}$ in. thick, have been laid on a hardened base carefully prepared. Wire fabric may be used in the thicker surfacings to resist shrinkage and to assist in reducing crack width.

Laying

Granolithic surfacing is laid in bays which may be up to 20 ft square although areas about 10 ft square are generally advisable, particularly for the thinner surfacings, as less likely to crack within the bay.

The bays are usually laid alternately in order to reduce the opening at joints due to shrinkage. Wooden battens are used at the edges, and act as guides to the finished level. Joints should coincide with those in the base concrete, otherwise cracks are certain to form over all joints in the base. The new concrete is butted against the old, no special jointing medium being used. It is an advantage to chamfer the edges slightly at the joints.

(*a*) *One-course work.* The surfacing mix should be placed on the grouted base, spread level, and then compacted by means of a punner, a heavy straightedge or a roller.

(*b*) *Two-course work.* The underlying course is usually of a somewhat drier mix than the top course, and is compacted by punning or rolling. A top course of at least $\frac{1}{2}$ in. thickness of a wetter finishing mix is then spread roughly and either struck off or tamped down with a straightedge. Excess water from the top course is absorbed by the bottom course.

There is a danger that the use of too dry a bottom course may result in poor bond with the base concrete. Many failures have resulted from the use of over-dry, and consequently weak and friable, bottom courses.

Trowelling

Trowelling is the most important factor in the laying of a good granolithic surfacing, and too much emphasis cannot be placed on the necessity for experienced trowel hands for this work. Trowel hands, although classed as "plasterers," should be specially trained in floor laying and need not be experienced plasterers in the normal sense. Conversely, experienced plasterers are not necessarily good trowel hands.

Finishing is carried out with the trowel and the float, the important points being the period at which the trowelling is done, and the pressure used by the trowel hand. Immediately after screeding off, the surface should be floated just sufficiently to level it while not bringing too much laitance to the surface. The surface should then be allowed to dry and

Fig. 23.2. Use of power float in finishing granolithic floor surfacing

stiffen until all the excess water has disappeared before any further trowelling is attempted. The next trowelling should be accompanied by considerable pressure, just enough work being done to tighten and smooth the surface. Further trowelling should be given at intervals as the mix stiffens so that a hard compact surface is produced without bringing laitance to the surface. Experienced trowel hands obtain such a surface with no more than a thin skin of cement covering the aggregate.

Machine tools are available for floating and finishing the surfaces of concrete or granolithic concrete floors. One such tool, known as a power float, which employs a heavy rotating steel disc for compacting and floating out the concrete, is illustrated in Fig. 23.2.

The power float is used to best advantage in finishing large surface areas such as aircraft hangars, or factory floors. It may not be economical for small isolated areas, such as floors for dwelling houses or flats. The manufacturers of this machine claim that lean mixes of low water-content can be laid at a speed of 5 to 10 times faster than that of hand floating. In easing the physical effort involved in trowelling, and facilitating the use of relatively lean stony mixes, e.g. 1:4 to 5, cement: aggregate, the use of a power float results in several advantages from the technical point of view. The higher stone content of the surface gives increased resistance to surface wear, and a lesser likelihood of dusting. The reduced water content provides a higher strength and lower shrinkage, thus reducing the risks of cracking and curling. It is important, however, that a good bond is obtained with the underlying concrete, and the grading of the aggregate and the cement content of the mix must be carefully adjusted to ensure that the compacted surfacing is dense and in full contact with the roughened surface beneath.

Dustings of cement intended to dry up the surface during trowelling are dangerous and liable to produce a neat cement skin which later peels off. They should therefore not be tolerated. Artificial drying can be effected very satisfactorily by laying cotton sheets on the floor and covering them with an inch or two of dry cement; this absorbs water and the sheets can be lifted after, say, ten minutes, when trowelling can proceed. The damp cement can usually be used for other concrete work which may be proceeding at the time.

This risk of peeling is also present when dry mixtures of cement and iron filings, with or without pigment, are trowelled in to produce a hard wearing skin or to give a coloured surface.

All trowelling should be finished on the same day as the laying operation.

Builder's traffic should be prevented from damaging the floor soon after laying. Attempts are often made to hide such damage by trowelling in neat cement slurry, but this is useless and invariably flakes off or becomes dusty.

Curing

Granolithic floor surfacings should be moist cured for 7 days after laying by hosing, flooding, or by covering with a p.v.c. or similar plastic film, a suitable thickness being about $\frac{20}{1000}$ in. The film should be laid over the surfacing while still wet and as soon as it is hard enough not to be marked. No further watering is then necessary. Other methods such as coverings of wet hessian or wet sand require frequent watering in dry weather. Wet sawdust often causes staining. The value of moist curing is not generally realized and reference should be made to Fig. 23.1, which gives the results of tests on the wear of a 1:2 mortar after different lengths of curing, and illustrates the large reduction in wear resulting from moist curing during the early stages of hardening. Furthermore, prolonged curing delays the start of drying shrinkage and enables the topping to develop sufficient tensile strength to withstand the shrinkage stresses.

Surface hardeners

The most commonly used surface hardener in this country is sodium silicate, but magnesium silicofluoride, other silicofluorides, and certain other salts are also in use.

A properly laid and trowelled granolithic floor surfacing will be hard wearing and durable without the use of surface hardeners. The value of surface hardeners is therefore a matter of controversy. Sodium silicate is generally believed to produce a hard skin which may reduce the wear of new floors at early ages. On old floors it is regarded as effective in reducing dusting. In either case the effect only appears to last for a limited period, perhaps six or twelve months.

Floor hardeners are not effective on weak and friable floors.

Common causes of failure

Common causes of failure in the past have been:

(1) The use of aggregates containing a large proportion of dust.

(2) Overwet mixes coupled with unsatisfactory trowelling.

(3) Lifting due to poor bond with the base concrete resulting from omission to clean and roughen it properly. When laying between previously constructed bays, failure to clean out loose dust near the edges has been a common cause of lifting.

(4) Possibly, insufficient thickness of surfacing.

(5) Use of a dry, friable mix for the bottom course in two-course work, failing in bond with the base concrete or even in itself.

(6) Unsatisfactory trowelling.

(7) Dustings of cement. A neat cement skin is formed even up to $\frac{1}{8}$ in. thick, and this later flakes off.

(8) Trowelling-on of a skin of neat cement slurry to hide poor finishing or damage. This invariably flakes or dusts off.

(9) Oversized bays.

General

Granolithic floor surfaces can be made hard, dustless, and wear-resisting provided the technique and workmanship are good. Some comment is, however, necessary concerning floors subject to exceptionally heavy abrasion and chemical attack.

(1) *Dairy floors.* Rapid deterioration, particularly where heavy impact or abrasion is combined with the action of milk wastes, is liable to occur unless particular attention is paid to all the details that ensure a hard and dense surface.

Where heavy trucking or churn rolling occurs, heavy cast iron or steel grids embedded in dense granolithic concrete give good service, see Fig. 23.3. They are, however, expensive and the floor is noisy. The best size of opening seems to be about 2 in. square; grids with small openings become slippery, while grids with larger openings permit potholes to form.

(2) *Food factories.* In food factories where the floor is likely to be affected by sugar, vegetable and fruit juices, acid solutions and fats, very similar remarks apply as in the case of dairy floors. Sugar solutions are more detrimental than dry sugar.

In repairing floors in sugar factories care is necessary to prevent any contamination of the mix by traces of sugar, otherwise the concrete may be spoilt or prevented from setting.

High-alumina and super-sulphated cements are more resistant than normal Portland cement, providing attention is paid to curing at early ages.

(3) *Engineering workshops.* Lubricating and cutting oils which have a deleterious effect on concrete are usually present and soften the concrete wherever they penetrate the dense surface skin. Such penetration may

Fig. 23.3. $\frac{3}{8}$ in. thick cast iron perforated plates laid on granolithic base course. After levelling further granolithic is placed and compacted to fill the perforations. Note the dry texture of the granolithic in the background

occur where the floor is accidentally chipped and where there is heavy trucking by iron-tyred vehicles.

Again, the effective life of the floor is dependent on the hardness and denseness. Cast iron or steel grids are effective in trucking corridors.

REFERENCE

23.1 SCHUMANN, L. and TUCKER, J., Tests of wear resistance of concrete, *Nat. Bur. Stand. J. Res.*, **25**, No. 5, 1939, pp. 549–570.

Further useful Reference

SHACKLOCK, B. W., The design and construction of granolithic floor toppings, *Structural Engineer*, **41**, No. 9, Sept. 1963.

CHAPTER 24

Durability of Concrete

ATMOSPHERIC pollution in the neighbourhood of large towns is a major cause of the decay and discolouration of concrete. Most of the substances which pollute town atmospheres are derived from coal, with lesser amounts of gases and soot from vehicular exhausts and acid gases from factory chimneys.

Analyses of the grime on buildings in towns have shown that it consists mainly of chimney smoke, soot from motor vehicles, and street dust. Structures exposed to railway smoke are coated mainly with soot, ash, partly oxidized iron from brake-shoes, wheels, and rails, and small quantities of other mineral and organic matter. Moisture is also present. Stains in rural surroundings contain only a small quantity of soot (from motor vehicles), while the quantity of firmly adhering oily or waxy pollen and fragments of decaying vegetable matter is much larger.

Stains due to weathering cannot be entirely removed without rubbing off the stained particles from the surface, and since penetrations of 0·3 in. to 0·5 in. have frequently been noted in columns and walls, this is extremely difficult.

An instance of the darkening and discoloration of a concrete surface by a polluted atmosphere is illustrated in Fig. 24.1. Except for the reeded portion in the centre, the surfaces were wire-brushed while the concrete was green. Differences in porosity at each lift have been accentuated by the deposits of grime, and the construction joints show as dark lines.

Of the gases present in the atmosphere, only carbon dioxide and sulphur dioxide cause material damage to concrete surfaces. Carbon dioxide dissolves in rainwater forming carbonic acid, which has a solvent action on any lime compounds present in the concrete. Calcium carbonate is removed from the surface in this manner and is replaced by diffusion of calcium hydroxide from the interior. When the calcium hydroxide is precipitated, water held in the pores becomes unsaturated and more calcium hydroxide is taken into solution. The depth to which the calcium hydroxide is changed to calcium carbonate by carbon dioxide depends on the porosity of the concrete, but even for porous concrete it rarely exceeds half an inch.

The white deposit of efflorescence so commonly seen on concrete surfaces is the result of leaching and subsequent carbonation and evaporation.

The sulphuric acid present in rainwater reacts with lime compounds, forming calcium sulphate. This is only slightly soluble in water and is deposited on or near the surface, forming a skin. Crystallization behind this skin, and differential movements between it and the concrete underneath, loosen it and cause disintegration.

The concentration of sulphur dioxide in the atmosphere is highest in industrial areas and varies somewhat with the time of year, December, January and February being the worst months. Concrete exposed at,

Fig. 24.1. Darkening and discoloration in a polluted atmosphere. Differences in porosity are accentuated by deposits of grime

say, Manchester or Sheffield, would be expected to weather very differently from that in a country district well away from towns. The former would get very dirty with a sulphate-bearing skin which would eventually disintegrate, while in the country district the concrete might be expected to remain clean and any wear would be due to erosion.

The effect of erosion in a polluted atmosphere is shown in Fig. 24.2. The surface was originally smooth as it left the mould, but much of the cement and fine material has weathered away, especially on the coping,

Fig. 24.2. Erosion of concrete in a parapet wall. Reinforcement which had been given insufficient cover has also corroded and become exposed

leaving the coarse aggregate exposed. Spalling due to corrosion of the reinforcement has also occurred. Other examples of the weathering of concrete are given in Figs. 24.3 and 24.4.

Erosion by water action

Serious erosion may result from the action of flowing water. It is well known that under certain conditions of hydraulic flow, a persistent free space or cavitation pocket will occur at some given location, which may under some conditions contain nothing but water-vapor and flying particles of water. If this occurs in contact with a solid object such as a concrete surface, damage results from the hammering and penetrating effect of the extraordinary high water pressure generated. Serious erosion results from cavitation.

The American Bureau of Reclamation[3.4] have found that, under atmospheric conditions, cavitation and destructive erosion begin downstream of relatively small offsets when water velocities reach about 40 ft per second.

Fig. 24.3. Weathered concrete to a crane gantry in which a sloping pour plane is in evidence, and also spalling of concrete over reinforcement, due to insufficient cover

Frost action

Concrete surfaces may be damaged by frost action either before or after the concrete has set. If the temperature falls below zero before the final set, the expansion of water while freezing exerts a force sufficient to destroy the cohesion between particles of the green concrete. The surface of concrete frozen before the final set has taken place scales badly and becomes pitted. The imprint of frost crystals may often be seen on the broken surface. Reference should be made to Chapter 14 for further information on concreting in cold weather.

Disruption of set concrete by frost is caused by the expansion of water in the pores during freezing. Pressures are developed within the material progressively as ice formation continues.

Slow rates of freezing are the least harmful since they are accompanied by appreciable extrusion from the surface with only small extrusional strains. Extrusion into unfilled pores helps to relieve strains in materials which are not saturated, and it is generally found that if the percentage saturation of the pores is below 80 per cent little damage is done, but the effect varies with different materials and with the rate of freezing.

Concrete in exposed damp positions such as road surfaces, reservoirs and dams, which is liable to suffer disintegration from frost, should be made with the lowest practicable water content which will give sufficient workability for placing. Concrete with a water/concrete ratio of less than 0·5 may be regarded as safe from frost damage; between 0·5 and 0·6 there is a risk in isolated cases; over 0·6 the risk becomes progressively greater. Examples of frost failure are shown in Figs. 24.5 and 24.6.

Where salt and calcium chloride have been used during the winter months to remove ice from road surfaces, spalling and flaking of the surface have proved troublesome.

Fig. 24.4. Weathering of a bridge abutment in Scotland. Efflorescence has occurred at two construction joints due to seepage of water through the joints

Additions of small quantities of such admixtures as resins, fats, detergents and other organic compounds entrain air in the concrete and give somewhat increased frost resistance. Air contents of 2 to 6 per cent are normally used. Air entrainment is more effective at the lower water/cement ratios, which in any case give improved frost resistance. Air entrainment appears to be of greatest value when frost attack and the effects of salt used for de-icing are combined. There is some doubt about the long term resistance if the air voids become saturated. Further reference is made to air-entraining under Admixtures, Chapter 5.

Sea action

Sea attack is due not only to chemical action and the mechanical action of tides and frost, but to the growth of plant life and the activities of marine borers. The relative severities of these actions cannot be determined, but it seems probable that during the first few years while the

Fig. 24.5. Disintegration of concrete kerbs due to frost action on porous aggregate[10.2]

concrete is new, the only actions are chemical and mechanical. As the concrete ages its surface becomes rougher and cracking occurs, so that salt-laden water reaches the reinforcement, resulting in corrosion with consequent expansion and disruption of the concrete. After 30 years of observations the Sea Action Committee of the Institution of Civil Engineers[24.1] concluded that the deterioration of concrete piles in sea water was primarily due to the corrosion of the steel reinforcement. Salts in the water contributed to the incidence of corrosion but were not the dominant factor.

The chemical constituent of seawater that is mainly responsible for the attack on concrete is magnesium sulphate. This substance reacts with cement to form calcium sulphate and calcium sulphoaluminate. A molecular increase in volume takes place during the reaction adding a physical action to the process, and on crystallizing the further increase in volume results in a disruptive force not only at the surface but at progressively greater depths.

The attack of marine borers is not important unless soft aggregates such as (some) limestones or sandstones are used in making the concrete.

Reliance should not be placed solely on the use of cements or cements and admixtures thought to be immune from attack by seawater. Callet[5.4] quotes the case of large blocks of concrete made with pozzolanic cement, forming part of the original quays at Le Havre, which were burst open by explosions during the war. The broken concrete, which had been in contact with seawater for about four years, showed advanced chemical decomposition.

Fig. 24.6. Failure of over-wet concrete in a bridge. Failure was caused by frost action. With a water/cement ratio below 0·6 such failure would be unlikely

In order to minimize chemical action and rusting of the reinforcement it is essential to prevent infiltration of water. A well-made and properly designed concrete should be sufficiently impermeable, but trouble is likely to occur at construction joints and these should be avoided as far as possible. The concrete should contain at least 650 lb of cement per cubic yard and the cover to the reinforcement should be at least 2 in. and preferably 3 in.

Crack formation in plastic concrete

Cracks occurring before the concrete hardens are usually due to the settlement of the plastic concrete around a rigid obstacle, such as a piece

Fig. 24.7. Complete disintegration of a concrete breakwater by sea action. Failure occurred primarily at the construction joints, with the result that the concrete broke up into a series of independent masses(19.2)

of aggregate or reinforcing bar, or excessive evaporation of water from the concrete surface.

Settlement cracks, which tend to occur more in thick than thin sections, particularly when top reinforcement is present, can best be prevented by using a well proportioned concrete mix with as low a slump as possible, and by adequate compaction of the concrete.

Plastic shrinkage cracks, caused by water evaporating from the surface of the concrete faster than it bleeds from beneath, are apt to occur when hot drying winds are prevalent. Ground and road slabs are susceptible to this type of cracking, as described in Chapters 13 and 19.

The formation of plastic cracks results in a weakened section of concrete possibly unable to resist subsequent drying shrinkage stresses, with the result that cracks may develop through the full depth of the section.

Cracks in hardened concrete

The formation of cracks in hardened concrete is not often attributable to any single cause, but besides being unsightly they may allow the leakage of water and lead to corrosion of any reinforcement. Factors which lead to a tendency to crack on the part of concrete are:

(1) drying shrinkage of restrained members.

(2) changes in length due to changes in temperature. This applies, particularly, to the larger masses of concrete in which there is a fairly considerable temperature rise during setting and hardening, and a subsequent shrinkage as the concrete cools. Cracking in dams results from these temperature changes.

(3) Movements in the structure due to settlement of the foundations, resulting in local overstressing.

(4) Expansion of metal, partly embedded in the concrete, or firmly fixed to the concrete. This is especially liable to happen in industrial structures subject to heat from boilers or furnaces.

(5) Overloading, whether due to inaccuracies of construction, or from the imposition of a greater load than that for which the structure was designed.

The risk of cracking can be reduced by the inclusion of expansion joints in the structure, as outlined in Chapter 10. It is important, however, that such joints should provide a complete severance, otherwise there is some danger that the joints may cause sympathetic cracking in other members,

Fig. 24.8. Badly stained and cracked reinforced concrete warehouse in London. The surface was originally cast against a hardboard form lining to give a smooth finish. Crazing may be seen about the middle of the column on the right

or that the joints may not function because they are not the weakest part of the structure. The effect of increasing the amount of reinforcement in a concrete member is to reduce the width of any particular crack along its length, but to increase the number of cracks. With sufficient reinforcement, it is possible to reduce the width so that the cracks are no longer visible, and this method is often applied when the risk of crack formation is foreseen, and would be difficult to avoid by other means, see also p. 260. In roadwork, reinforcement is used principally to hold together any cracks which form due to movement in the subgrade and so preserve the road from further deterioration.

Crazing

Crazing, or the formation of fine hair cracks on concrete surfaces, is a common cause of disfigurement. Crazing is caused mainly by the accentuated drying shrinkage of the surface layers of concrete, to which must be added the effects of carbonation, and differences in temperature and moisture content between the surface and the interior. This leads to the formation of very fine cracks rarely more than about a quarter of an inch in depth.

Hair cracks form a convenient route by means of which disintegrating agents may carry their attack into the interior of the concrete. It is unfortunate that dense and impermeable concrete surfaces which normally have the best weather resistance, are more liable to craze than porous surfaces, and are thus weakened.

Sulphate attack on concrete

Sulphates in solution attack concrete and cause expansion, deterioration, and eventual disintegration. In the early stages of attack there is little visual evidence of any change in the concrete, although there may be some reduction in strength as the chemical action proceeds. This is accompanied by slight expansion, which may not be apparent in the concrete itself, but may cause trouble at points of restraint. As the attack proceeds there is usually some change in colour from the normal cement grey, and cracking or spalling occurs, starting from the surface in dense concrete, but penetrating much deeper in more porous concrete. The rate of attack depends on a number of factors, the principal of which are:

(i) *Form in which the sulphate occurs.* Easily soluble sulphates such as those of sodium, potassium, magnesium, and ammonium, react more vigorously than calcium sulphate or gypsum. Insoluble sulphates do not cause any deterioration.

(ii) *Concentration.* The higher the concentration of sulphates in solution the more serious the attack. The severity of attack is increased in circumstances in which a flow of sulphate-bearing water brings a continuous supply of the salt into contact with the concrete.

(iii) *pH value of the soil or groundwater.* Its temperature may also be a factor at ground level, the activity of sulphate solutions increasing with temperature.

If the pH value is below 6, i.e. in acid soils, the rate of attack is likely to be increased.

(iv) *Permeability of the concrete.* Impermeable concrete is attacked only at the surface and the deterioration is therefore slow, whereas a permeable concrete, particularly if one side is in contact with moist soil containing sulphates while the other side is open to the air so that evaporation occurs, or if subjected to a one-sided water pressure, is attacked throughout its thickness resulting in an overall expansion and complete disintegration.

(v) *Formation of cracks.* Attack proceeds along the lines of cracks, particularly when the movement of moisture along any cracks is encouraged by one-sided water pressure or evaporation from a free surface.

Examples of sulphate attack

The severity of sulphate attack depends mainly on the quality of the concrete, the type of cement, and the varying conditions which surround it. Before considering precautionary measures to minimize the effects of sulphate attack, two examples of severe attack may serve to illustrate the inter-relationships of the various factors involved. The first example is of concrete sewers in sulphate-bearing ground which have failed in a number of instances within a year of construction, one-sided water pressure being the cause of attack throughout the thickness of the pipes, although the concrete was of reasonably good quality.

These failures have occurred even in clay soils which might have been thought to be impermeable. However, the disturbance in digging the sewer trench, and probably some porosity in the backfilled clay is known to have resulted in the flow of ground water necessary to carry sulphates in solution into the concrete. It is probable that the failures would not have occurred if the pipes had been of dense impermeable concrete made with sulphate-resisting Portland cement, or if they had been surrounded by a haunching of such concrete. Other precautions might have been to wrap the pipes in a p.v.c. membrane, or to coat them with bitumen. In some cases, the failures might not have occurred if nothing more had been done than to set the pipe on dense concrete in order to avoid problems in compacting clay underneath the pipes, and then to pun the clay well in backfilling the sides and over the top of the pipe.

The second example is of failures of oversite concrete forming the ground floor of a number of houses, resulting in lifting by as much as 3 inches, and pushing outwards of the external walls due to expansion. This was attributed to movement of sulphates from the underlying shale fill into the concrete. This movement was aided by cracking and evaporation from the upper surface before completion of the floor surfacing with pitch mastic. Due to the high concentration of sulphates, it is probable that, in the particular instances examined, it would be necessary to use supersulphated cement in a dense concrete. However, in cases such as this it is usually simpler to provide a p.v.c. membrane at least 20/1000 inch thick immediately below the concrete.

It will be noted that concrete is attacked by sulphate salts only when they are in solution, and that the rate of replenishment is important. It follows that attack is most severe when the soil conditions allow a free movement of ground water, as in the case of sands and gravels, and porous or fissured rocks. Clay is relatively impermeable when undisturbed. Examples where moisture movements result from evaporation at an exposed surface are: oversite concrete laid directly on the soil surface, porous concrete in foundations through which moisture is sucked up into brickwork above, and especially permeable concrete below fireplaces, boiler houses, etc.

Precautionary measures

An essential condition in the production of a concrete resistant to chemical attack is that the concrete itself should be of high quality. A lean mix or badly graded aggregate, will not give a resistant concrete whatever type of cement is used. If the conditions are such that in-situ concrete is immediately exposed to the aggressive agent, it is more susceptible to attack than precast units provided they are moist cured for a month or more before being placed in a position subject to attack. The compressive strength of concrete is a reliable index of its resistance to attack, since strength, density, and low permeability are all interrelated; the stronger the concrete the more durable it is likely to be.

The importance of dense well-compacted concrete of high quality cannot be over-emphasised, but the resistance to attack can be supplemented by other precautions. As, however, the probable severity of attack is largely a matter of judgment based on general guidance, such as that given in Table 24.1, any decision concerning precautions to be taken is likely to be a compromise between what may be accepted as a reasonable risk and economic savings. In cases of doubt, particularly where any failure in the concrete could have disastrous consequences or would be difficult to repair, it is wise to err on the side of excessive precautions. Special consideration should also be given to thin concrete sections, site conditions which may make the production of dense concrete impracticable, and the chances of one-sided water pressure. Changes in groundwater level occur with the seasons, and may also result from changes in conditions as a result of the building of a structure or adjacent structures. Footings and mass concrete buried completely in clay are unlikely to suffer seriously unless the concrete is of poor quality and the sulphate concentration of the groundwater high. Little is known of the possibility of attack on concrete piles, mainly due to the difficulty in obtaining evidence. It seems probable, however, that attack does at times occur near the head and when such piles penetrate sulphate-bearing soil through which there is a continual flow of water.

An assessment of the probable severity of attack is primarily based either on a determination of the percentage of sulphates (as SO_3) present in the soil, or on a determination of the quantity of sulphates (as SO_3) in parts per 100,000, present in the ground water. The analysis should also indicate the metal radicals present in conjunction with the sulphate and the pH value of the soil or groundwater. A guide to the precautions necessary may be obtained by reference to Table 24.1.

The table is provided only as a general guide. From the paragraph headed Sulphate Attack on Concrete, it will be noted that the presence of gypsum is less important than that of the easily soluble sulphates. For mass concrete surrounded by undisturbed and relatively impermeable clay, it is suggested that the presence of gypsum crystals may be ignored provided dense well-compacted concrete is used. Whenever practicable, footings in clay should be cast without shuttering. This avoids backfilling, which may leave a porous zone likely to become saturated by surface water even if the general groundwater level is below foundation level. If shuttering cannot be avoided, but mass concrete footings are isolated there may not be much movement of water, so that the precautions given for concrete buried in clay should suffice. For small concrete sections, the guidance given in Table 24.1 should give a reasonable insurance against failure; if spaces between the member and the soil are backfilled the precautions adopted should be those for buried concrete in permeable ground.

If the soil or groundwater is acid, i.e. with a pH value of less than 6, it is suggested that the precautions taken should be equivalent to those outlined under class 4, i.e. supersulphated cement should be used. When the SO_3 content of the soil is high, particularly if the soil is permeable, other precautionary measures, such as those described on p. 359, should be employed. For an SO_3 content above 1·0%, unless supersulphated

TABLE 24.1 *Classification of Sulphate Soil Conditions Affecting Concrete and Recommended Precautionary Measures*

Classification of soil conditions			Precautionary measures	
Class	Sulphur trioxide in ground-water parts SO_3 per 100,000	Sulphur trioxide in clay % SO_3	Buried concrete surrounded by undisturbed relatively impermeable clayey soils.	Concrete exposed to one-sided water pressure or evaporation from one face. Buried concrete in permeable ground, including all backfill materials.
1	Less than 30	Less than 0·2	No special measures except that the use of lean concretes (e.g. concrete containing less than 460 lb cement per cubic yard) inadvisable if SO_3 in water exceeds 20 parts per 100,000. In latter case use good well-compacted concrete made with ordinary Portland cement and a maximum w/c ratio of 0·55.	Dense fully-compacted concrete made with ordinary Portland cement (see Table 24.2 for cement content), and a water/cement ratio not exceeding 0·50.
2	30 to 100	0·2 to 0·5	Dense fully-compacted concrete made with ordinary Portland cement (see Table 24.2 for cement content), and a water/cement ratio not exceeding 0·50. If sulphates are in highly soluble form use sulphate-resisting Portland cement.	Dense fully-compacted concrete made with sulphate-resisting Portland cement (see Table 24.2 for cement content), and a water/cement ratio not exceeding 0·45. Supersulphated cement may be used for the higher concentrations if in doubt.
3	100 to 250	0·5 to 1·0	Full compaction to give the densest possible concrete essential. Use sulphate-resisting Portland cement (see Table 24.4 for cement content), and a water/cement ratio not exceeding 0·45.	Full compaction to give the densest possible concrete essential. Use supersulphated cement (not less than 650 lb/yd^3) and a water/cement ratio not exceeding 0·45.
4	Above 250	Above 1·0	Full compaction to give the densest possible concrete essential. Use supersulphated (or high-alumina) cement (not less than 650 lb/yd^3) and a water/cement ratio not exceeding 0·45.	Preference should be given to protection by impervious membrane, see p. 359. Otherwise use of supersulphated cement as for buried concrete in undisturbed clay, with lowest practicable water/cement ratio for full compaction.

See text for influence of pH value for soil, type of sulphate, and size of section.

cement is used, such measures are a wise precaution in any soil conditions. The sulphate content of the ground water is a more reliable guide than that of the soil, since the latter is more subject to local variation.

Sulphate resistance and type of cement

Cements vary considerably in their resistance to attack even when nominally of the same type. This is illustrated in Fig. 24.9,[4.7] which shows the rate of deterioration of concrete made from two ordinary Portland cements, when exposed to sea water. The two samples of cement used in the experiments had very much the same strength at 1 year, and

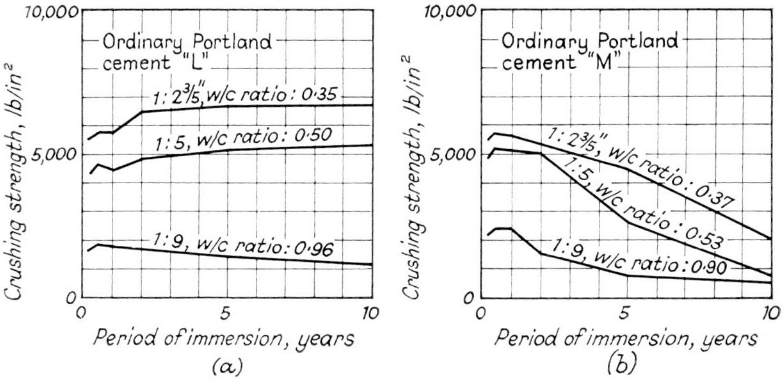

Fig. 24.9. Deterioration of concrete made from two ordinary Portland cements, when exposed to sea water

yet cement L appears to have been highly resistant, whereas cement M was much less resistant.

Sulphate-resisting Portland cement is more resistant than ordinary Portland cement, but must be used in dense concrete to be effective, otherwise deterioration may occur almost as rapidly as with ordinary Portland cement. Recommended minimum cement contents are given in Table 24.2 for the more usual irregular and angular aggregates and varying water/cement ratios and workabilities.

Supersulphated cement is particularly useful in hot climates where high-alumina cement cannot be used. It is resistant to sulphates other than ammonium sulphate which causes severe attack. This cement is also resistant to attack by hydrochloric and sulphuric acids in concentrations up to 15 g per litre.

Reference is made to the properties of high-alumina cement and pozzolanic cements on page 65.

Other precautionary measures

As an alternative to the use of one or another of the special cements, an impervious coating to the outer surface of the structure may be considered. In the case of a basement to a building constructed in water-bearing ground, an asphalt tanking is normally provided to ensure

TABLE 24.2 *Recommended Minimum Cement Contents to Give Resistance to Sulphate Attack for Different Types of Aggregates, Water/Cement Ratios and Workabilities*
(Use this Table in conjunction with Table 24.1)

w/c Ratio (effective)	Aggregate	Estimated average 28 day compressive strength lb/in²*	Cement contents per cubic yard of compacted concrete for different workabilities			
			C.F. 0·78 Slump 0–½ in.	C.F. 0·85 Slump ½–1 in.	C.F. 0·92 Slump 1–2 in.	C.F. 0·95 Slump 2–4 in.
0·55	¾ in. irregular	4600	460	520	570	590
	¾ in. angular	4800	500	560	610	630
0·50	¾ in. irregular	5300	520	580	630	650
	¾ in. angular	5600	560	620	670	690
0·45	¾ in. irregular	6000	590	660	710	730
	¾ in. angular	6500	640	700	750	770

* Based on average strength: w/c ratio relationship.
Table 24.1 gives maximum values for w/c ratios. These should be used to indicate (from Table 24.2) the minimum cement content.

watertightness of the basement. This tanking provides an adequate protection from attack by sulphates on the concrete in the basement structure, and no additional precautions are normally needed. Concrete foundations not subjected to one-sided water pressure can be protected against attack by thick coats of bituminous paint on the outer face, or by a built-up coating using two or three layers of bituminous felt with a bituminous paint tack coat on the concrete surface and between each layer. Bituminous emulsion does not provide a satisfactory coating.

Resistance to other chemical attacks

Gases such as sulphur dioxide and carbon dioxide may, in damp situations, attack concrete. These gases are prevalent in gasworks, railway tunnels, and at various chemical factories. Sulphuric acid from coke, coal or furnace ash is liable to attack concrete in bins, bunkers, or retaining walls. The chemical attack is in these cases likely to be accompanied by abrasion, and to produce results similar to that shown in Fig. 24.10.

Sulphates of sodium, potassium, magnesium and ammonium may cause serious damage to Portland cement concrete if moisture is also present. This begins by expansion within the concrete, which may be enough to cause a general expansion in the member. Cracking and disruption follow.

Inorganic acids generally are destructive to concrete. These may also be released from some salts, such as ammonium chloride and ammonium nitrate[24.2] by interaction with lime. Leaching then follows.

Organic materials vary in their rate of attack on concrete and their effect may roughly be summarized as follows:
Petroleum oils—none.
Coal tar distillates—none or very slight.

Fig. 24.10. Abrasion due to coke at Beckton Gasworks(17.5)

Organic acids such as stearic, oleic, lactic and tannic—all attack concrete.
Lactic substances, which are derived from dairy produce, have a most destructive effect.
Vegetable oils—slight or very slight attack.
Molasses, sugar, syrup, glucose—fair degree of attack.
Acetic acid which occurs in vinegar, and tartaric acid which occurs in some fruit juices all attack concrete.
Oxalic acid—little effect.

Sewage normally has an alkaline reaction and is harmless, but it may become acid by contamination with factory wastes and will then attack concrete.

Hydrogen sulphide gas, which is evolved from stale sewage, may cause severe attack along the tops of sewers.

Resistance to abrasion

Concrete is sometimes called on to resist abrasion as in the case of road surfaces, silos and bunkers, retaining walls to storage areas, etc. The abrasion may prove serious, as in the case of columns in Fig. 24.10, which were subject to the abrasive action of coke stored in the area. The best insurance is good dense concrete made with a hard tough aggregate in accordance with the principles already described.

REFERENCES

24.1 *The Durability of Reinforced Concrete in Sea Water*, National Building Studies Research Paper, No. 30. H.M.S.O.

24.2 LEA, F. M., The Action of Ammonium Salts on Concrete. *Mag. Concr. Res.*, **17**, No. 52, Sept. 1965. Cement and Concrete Association.

CHAPTER 25

Quality Control Inspection and Testing

THE quality control of concrete should aim at producing a uniform material which provides those properties particularly desirable for the work envisaged. At the same time, a careful balance must be maintained to ensure that the quality of concrete is satisfactory for the purpose, without being of unnecessarily high strength, the achievement of which inevitably results in a marked increase in cost. Thus, lean mixes of low strength are usually adequate for mass concrete in foundations and in the cores of dams where, in fact, they are advantageous because the heat generated during hydration is less than that for richer mixes. On the other hand, concrete pipes and haunchings used in sulphate-bearing ground must be of rich dense mixes if rapid and costly failures are to be avoided.

The choice of plant and procedures may have a considerable influence on the results achieved, and if effective quality control is to be obtained, discussions between the design engineer and the contractor should be arranged at the planning stage. From this point the details of control can be developed.

It is fortunate that most of the desirable properties of concrete are enhanced by increasing the compressive strength, and this, therefore, affords a simple measure of the general standard of quality. It should be noted, however, that it is not only necessary to control the quality of the concrete at the mixer, but to provide experienced supervision of all the processes which follow. This cannot be a matter for individual effort only, and any measures envisaged for the control of concrete quality must provide for co-operation and general instruction in good or improved techniques, and in the use of tools and machinery. It is essential that the educational methods should be such that their influence is apparent from the site agent down to each operative involved.

In order to achieve good results, it is necessary to adopt several methods of approach,[7.2] such as:

(1) Direct control by means of a concrete engineer, and assistants if necessary, specially trained in the work and resident on the site, where the quantity of concrete is large, or for some special reason an unusually high standard of concrete quality is desired. This concrete engineer may be backed by a site laboratory. He should be responsible not only for the production of good quality concrete at the mixer, but for supervision of the many operations which ensue.

The principal duties of the concrete engineer or other person responsible for concrete quality may be defined as follows:

(i) He should be familiar with the plans and specifications, especially those sections which will deal with the concrete work under his control, and should see that the work is executed accordingly.

(ii) If any sections of the specification are impracticable, or require modification, he should report to his immediate superior, or to the employing authority, in order that suitable alternative measures can be discussed.
(iii) He should see that the materials are of satisfactory quality, the proportioning of the materials is accurate, the water content is within the limits allowed, and the concrete is of uniform and satisfactory consistency.
(iv) He should acquaint himself with the details of the formwork, reinforcement, handling, placing, consolidation, finishing and curing, and he should point out any errors which occur in the techniques. He should be able to advise on the best method to adopt in any particular circumstance.
(v) He should take samples of the materials, either for site tests or tests at a central laboratory, make slump tests, and prepare concrete cubes for testing.
(vi) He should keep a record of work he has done and should chart results of tests, in order that any deterioration over a period can be observed and measures taken to effect improvements.

The concrete engineer must be certain before making a complaint that it is of sufficient importance to be justified. If it is, he should insist that the defective work or method is remedied within a reasonable period. At the same time he should make every effort to maintain control and co-operation with the agent, site engineer, foreman, ganger, and workmen.

Delegation of the duty of the control of concrete quality on the site must depend on the quantity of work involved. On small contracts it is obviously uneconomic to employ an engineer solely for the purpose of controlling quality of concrete, and it is usual for the agent to delegate the duties to the general foreman or one of the site engineers. In such cases it is advisable to make every effort to impress on the responsible person the importance of the correct technique in making concrete and the correct method of making control tests. Control on large contracts may require one or more concrete engineers, with assistance to help with the preparation of test specimens, working in conjunction with a site laboratory.

(2) Instructions and advice on the site, on both specific and general questions, by a visiting engineer with wide experience of concrete technology. This form of approach is particularly valuable on the smaller sites, provided the concrete engineer is prepared to talk to the site staff, foreman, and gangers in non-technical language which they can understand.

(3) Education of site staffs by means of pamphlets dealing with concreting techniques, copies of technical papers and encouragement to attend any meetings at which they can join in discussions on concrete technology.

(4) Testing of materials and concrete by experienced laboratory engineers. The emphasis on experience is important, since the observations made during testing and the interpretation of the results are just as important as the figures obtained from the testing machine.

The production of a good concrete at the mixer is not sufficient to ensure a good concrete structure since, for instance, a small percentage of voids in the concrete due to air-bubble holes or honeycombing, causes a reduction in crushing strength of 4–6 times that percentage. Further, it is necessary in most concrete structures built in situ to provide construction joints or lift planes at intervals and, unless carefully treated, these provide a plane of weakness at which disintegration may begin, or through which water may seep. It is therefore necessary in considering quality control measures not to place too much reliance on the results of slump tests and cube tests, and to arrange for supervision of all the other aspects of concrete production besides proportioning and mixing. Much depends on the skill and experience of the foreman carpenter, the reinforcement fixer and the placing ganger, or foreman, and unless they fully understand the basic principles of good concreting techniques, they may run into difficulties when faced with some new shape of member or the use of concrete of different mix proportions from that to which they have been accustomed.

TABLE 25.1

Concrete strength category	Minimum specified cube strength at 28 days: lb/in²	Standards of control for quantities (yd³)		
		Large: over 10,000	Medium: 1,000 to 10,000	Small: up to 1,000
1	Over 4,000	A	A	A
2	2,500 to 4,000	A	A or B[1]	B
3	1,000 to 2,500	B[2]	B or C[3]	C
4	Unspecified	D	D	D

NOTES:
[1] The control would be dependent on the nature of the work.
[2] For large quantities it may be economical to use standard A if the saving in cement justifies it.
[3] The control would depend on the strength specified.
See also Table 25.2 for description of standards of control.

The Committee appointed by the Institution of Civil Engineers to prepare a Report on Quality of Concrete in the field,[(25.1)] divided concrete into four arbitrarily chosen strength categories. It then considered small (up to 1,000 yd³), medium and large (over 10,000 yd³) works, and indicated the standards of control desirable in each case, in relation to the strength category. The standards needed are summarized in Table 25.1[(25.1)] and a general indication of the quality control requirements and number of tests may be obtained from Table 25.2.[(25.1)]

Diagnosis of the causes of blemishes and defects of concrete is not always easy, since there may be several possible causes of any one blemish or defect, and the necessary or possible remedies may be even more numerous. This is illustrated in Table 25.3, in which some major causes of blemishes on concrete structures and their possible remedies are tabulated. The table is by no means exhaustive, but an examination of the remedies illustrates the dependence of the finished product upon

TABLE 25.2 *Requirements for the Control of Concrete Quality*

Standard of Control	Aggregates				Concrete				Personnel required for testing
	Separation		Tests		Batching		Tests		
	Fine	Coarse	Grading	Moisture determination	Cement	Aggregate	Workability	Strength (see note 2)	
A	One size normally, but two sizes may be required to obtain correct and uniform grading	Single sizes as B.S. 882: e.g. $\frac{3}{8}$ in., $\frac{3}{4}$ in., $1\frac{1}{2}$ in.	After standard has been established, one per week or when variation is suspected	Frequently at first and then as for Standard of Control B	By weight or by bag	By weight	Frequently each day if used to control added water	After standard has been established 3 cubes for each 500 cu. yd. or part thereof in each day's work	Skilled laboratory technician (full time or majority of his time)
B	One size	$\frac{3}{4}$ in. to $\frac{3}{16}$ in. graded aggregate plus $1\frac{1}{2}$ in single-sized aggregate when required	After standard has been established, when variation is suspected	At beginning of each day's work and after moisture is likely to have changed, e.g., after rain	By weight or by bag	By weight	Frequently each day if used to control added water	After standard has been established 3 cubes each week	Resident Engineer or Assistant (part-time)
C	One size Well-graded all-in aggregate may also be used	One size	None	None	By weight or by bag	By volume	None	If required	—
D	All-in or unscreened aggregate		None	None	By volume approximately		None	None	

NOTES:
1. The aggregates used in conjuction with Standards of Control A, B, and C should comply with B.S. 882, Coarse and Fine Aggregates from Natural Sources for Concrete.
2. Where appropriate, beam or cylinder splitting tests may be used in place of cube tests.

TABLE 25.3 *Some Major Causes of Blemishes on Concrete Structures and Possible Remedies*(25.2)

Blemish	Possible causes of trouble	Possible remedy
A. Concrete alignment unsatisfactory.	1 Unsatisfactory formwork design.	1 Remedy design.
	2 Faulty erection of formwork.	2 Tighten bolts, see that wedges are in correctly, bracings tightly fixed, etc.
	3 Placing of concrete at too fast a rate for design of formwork.	3 Alter formwork design or adjust rate of production of concrete.
	4 Dropping of concrete from a height into formwork.	4 Improve plant arrangement and placing technique.
	5 Use of vibration in formwork not designed and erected with vibration in mind.	5 Improve design of formwork or erection technique.
B. Construction joints in random positions.	1 No attention to formwork design.	1 Show joint positions on formwork detail.
	2 Concreting stopped for a period during lift.	2 (*a*) Improve concreting gang technique to avoid breaks. (*b*) Adjust quantity of concrete between joints to plant output. (*c*) Eliminate plant breakdowns by regular maintenance.
C. Construction joints badly made, including honeycombing just above joint.	1 Failure to remove laitance from previously placed concrete.	1 Remove laitance by scabbling, wet grit blasting, etc.
	2 Failure to clean up	2 Clean up thoroughly by air or air-water jets.
	3 Formwork not tight or gives under pressure, allowing loss of cement-sand grout.	3 Remedy design or erection technique of formwork.
	4 Mix too harsh.	4 (*a*) Revise concrete mix design. (*b*) Check proportioning of materials at the mixer.
	5 Concrete allowed to flow along formwork.	5 (*a*) Instruct concreting gang to place evenly round formwork. (*b*) Improve transport method to simplify even placing.
D. Honeycombed patches.	1 Loss of cement grout through formwork.	1 Check formwork.
	2 Shortage of sand in some batches.	2 (*a*) Use more care in proportioning. (*b*) Mix more thoroughly.
	3 Change in grading of aggregate.	3 Check grading of coarse and fine aggregates.
	4 Segregation of mixed concrete.	4 (*a*) Change plant used for transport. (*b*) Alter mix design to decrease tendency to segregate.
	5 Segregation during placing.	5 (*a*) Improve placing technique. (*b*) Improve plant used in placing (if any).
	6 Lack of consolidation.	6 Improve compacting technique.

everyone concerned, including the designer, the agent, engineers, foremen, gangers and others.

Equipment

Specifications usually require that certain testing equipment shall be provided by the contractor. The following items are among those commonly required:

Straightedge. Foot rule. Tape.
Thermometer (maximum and minimum).
Slump cone and rod.
Cube moulds, base plates and rammer.
1 cubic foot measure.
Set of B.S. sieves, $1\frac{1}{2}$ in. down to 200 mesh.
Cylinder, 100 cm for field silt test.
Primus, hot-plate or gas-ring for quick drying of aggregates. An oven may be used on larger contracts.
Frying-pan, metal trays, etc.
Pycnometer, or other apparatus for measuring the moisture content of aggregates. Tins for sampling aggregates, cement, etc.
Scales, 5 kg $\times$ 1 g, and weights.
Storage tank for storing concrete cubes. This should be thermostatically controlled on larger contracts.
Trowels, shovels, spanners and other tools.

Testing of materials

Methods of testing aggregates and cement for quality have already been described in Chapters 3 and 4. Samples of the aggregate, and in some instances of the cement, should be sent to a central laboratory for comprehensive tests before concreting commences. It is often advisable, and sometimes imperative, that one or more trial mixes be prepared at the laboratory; the aggregate samples should therefore be sufficient for this purpose. It is necessary to give the central laboratory as much information as possible concerning the source of supply of the aggregate, the purpose for which the concrete is to be used, and the method by which the concrete is to be compacted, in order that a suitable and practicable mix can be designed. The nominal mix, when specified, should be quoted. When submitting cement for testing, the brand, situation of works, date and time of sampling, method of storage, length of storage, and other relevant details should be quoted.

Whether or not trial mixes have been made in the laboratory, it is essential for trial mixes to be made on site using the materials and the plant which will be used for the construction. Adjustments to the mix proportions may be necessary to obtain both the workability and the average compressive strength for which the mix was designed.

Determination of surface moisture on the site

Aggregates are usually delivered to the site in a saturated condition and the water drains away gradually after depositing. This results in a saturated layer about a foot or so thick at the bottom of the stockpile. It is advisable to avoid using material from this layer and to be careful that samples for moisture content are not taken near the bottom of the heap.

Small differences in moisture content coincide with changes in the grading of the aggregate, especially the sand. Samples for testing should, therefore, be collected from different parts of the stock-pile and mixed. At least three samples should be taken and tested and the average value found.

In large hoppers drainage occurs from the bottom. It has been found that this results in a low moisture content at the bottom, becoming wetter towards the middle. Allowance must be made for this by adding rather more water than that indicated by the tests for the first few mixes after any prolonged break.

Practical difficulties occur on the site for the various reasons discussed above, and it is rarely economical to make frequent tests for surface moisture when the quantities of concrete produced are small. On those larger works where strict control can be maintained, adjustments to the water added at the mixer on the basis of frequent surface moisture tests, should make it possible to maintain the water-cement ratio within the limits $\pm 0{\cdot}02$. It is suggested that when the quantity of water to be added to the mix has been estimated, a tolerance of ± 1 gallon in 20 should be allowed, in order to take care of short-period changes.

An alternative method of strictly controlling the uniformity of the concrete is by frequent checks on the workability, preferably by means of the compacting factor test (see below). While this is only an indirect way of controlling the water/cement ratio, it has been found to be satisfactory, and is particularly valuable in road construction, for which the plant demands very uniform workability. However, checks by surface moisture tests on the aggregates must still be made periodically.

Mention has already been made of the necessity for determining the moisture content of the aggregate in order that the water/cement ratio may be controlled. Several methods have been devised for this purpose, but of these only two will be described in detail.

(*a*) *Pycnometer Method*

A full description of this method is given in Appendix B. It is necessary to determine the bulk specific gravity of the material from a saturated surface dry sample. This needs careful manipulation, and it is suggested that at least three determinations should be made and the average found. The surface moisture can then be estimated without repeat determinations of the specific gravity, unless material is delivered from another source with the possibility of a change in the specific gravity.

It should be noted that all calculations when using a pycnometer are based on the saturated surface dry condition, and to obtain results based on a dry condition, a correction for absorbed water must be applied as indicated in Fig. B.1, p. 387.

Precautions to be taken in the use of the pycnometer are listed below:

(1) Weighing must be very accurate.

(2) The outside of the bottle must be carefully wiped free of any drops of water.

(3) It is not easy to get rid of air bubbles, and any carelessness in this respect upsets the results.

(4) Use care when testing coarse aggregate since the bottle is easily broken by it. (It is necessary to use a glass bottle in order to see when the air bubbles are getting away.)

The pycnometer has the advantage that it can be used when no sources of heat for drying are available.

(b) Drying Methods

The surface moisture can be found by drying in two different ways. Either the aggregate can be dried until it is in a saturated surface dry condition, or it can be dried right out and an allowance made for the loss of absorbed water. Full descriptions are given in Appendix C.

It is difficult to decide just when the aggregate is saturated surface dry, especially for fine aggregates and careful manipulation is essential if reliable results are to be obtained.

The second method is recommended for use on the site since it is easy to dry the aggregate completely. The absorption can be determined by preliminary tests in the laboratory, and for all practical purposes it may be assumed at a constant figure.

The method of drying the aggregate depends on the facilities available. The most accurate results are obtained by drying a sample weighing 1 kg to 5 kg, according to the grading, in an oven maintained at 105–110°C. As this takes several hours, the method is too slow for the site control of concrete. Drying can, however, be speeded up by using 1 kg to 2 kg of aggregate, according to grading, in a frying-pan over a primus or other stove. This takes about 10 to 15 minutes, but care must be taken not to overheat the aggregate, and to prevent any small pieces from jumping out of the pan.

It is necessary to make at least two tests, and to find the average, if reliable results are to be obtained for the moisture content.

(c) Other Methods

Various sets of equipment are available which employ other methods of determining the surface moisture. Some of them are not very accurate or are less convenient than the methods already described. Others use far too small a sample.

The methods include measurement of the change in specific gravity of acetone or a saline solution when mixed with the wet aggregate, addition of methylated spirit and ignition to dry off the water, use of small cups on finely balanced scales reading off the moisture content directly, a saturation method using a graduated cylinder, the syphon-can method, and measurement of the electrical resistivity.

Sampling of concrete

The sampling of concrete for testing should be done with care in order to ensure that the test results are representative. Concrete may not be homogeneous for a number of reasons, such as, errors in proportioning, non-uniform mixing, segregation during handling and placing, change of consistency during handling, or loss of moisture from the surface or from absorption by contact with an absorbent material.

Samples should be taken either at the mixer or as close as possible to the point of final deposit, in which case they should be taken immediately after the concrete is deposited and before it is compacted. Occasionally tests for consistency should be made on the same concrete both at the mixer and at the point of placing in order to establish the extent of any loss of workability during transport.

The procedure to be used for obtaining samples of fresh concrete is described in B.S. 1881 : 1968 to which reference should be made. When taking a sample from the mixer it is specified that three approximately equal increments totalling 1 cu ft shall be taken at about the time when one-quarter, one-half and three-quarters of the batch have been discharged. When obtaining a sample at the point of deposition it is required that concrete shall be collected from not less than five well distributed positions, avoiding any segregated material at the edge of the mass. In both cases the composite sample should be mixed to ensure uniformity and used immediately for the required tests.

Several other points are worthy of note:

(1) Samples should not be taken at the beginning or end of a period of concreting without making special note of the fact.

(2) It is advisable to take samples at irregular intervals and without prolonged preparation which is evident to the mixer operator.

(3) Note should be made of the appearance of the mix, including any tendency toward segregation.

(4) Note should be made of the time and date, mix proportions, the batch number, the mixer from which the concrete was delivered, the position in the structure, the method of sampling and the temperature.

Testing for consistency and compressive strength

The tests normally adopted on the site for quality control are the slump test, or on larger jobs the compacting factor test, for workability, and the cube test for compressive strength. The cube test is, however, no more than a measure of the quality of concrete produced at the mixer, and as the amount of concrete used in cube making is only a very small proportion of the quantity of concrete placed in the structure, and is only taken from a few of the batches mixed, individual tests should only be regarded as giving a general indication of the quality of the concrete. The variability of cube test results is partly due to factors outside site control, but the efficiency of the control measures at the mixer, particularly the accuracy of proportioning, the maintenance of a uniform consistency, and the limits within which the sand grading can be maintained, have a marked effect in reducing the extent of the variations.

Cube test results should be examined in relation to the average cube strength which might reasonably be expected from a consideration of the mix proportions, water/cement ratio and workability required. Limits suggested for cube strengths, given different site conditions, are tabulated in Table 7.4.[7.2] When cube strengths are well within the limits, it may be accepted that the concrete being produced at the mixer is of reasonable standard, but when the cube strengths approach or exceed the upper or lower limits, then an investigation should be made of the conditions at the mixer. Inaccurate proportioning, or an excess of water, are the most likely sources of trouble, and a check should be made on these first.

It may, however, be found that the water/cement ratio has been increased beyond that intended in order to increase the workability of the concrete, either to facilitate placing in a difficult position, or possibly because the plant being used will not deal satisfactorily with the original consistency. An instance of this is the attempt to use a transporting hopper with too narrow an opening for the discharge of stiff concrete, with the result that the concrete gang quite rightly calls for wetter concrete in order to get on with the job. A study of such plant difficulties can often lead to a substantial improvement in concrete quality, and sometimes to a saving in cost at the same time.

Compression tests

Compression tests are usually made on 6 in. cubes in Great Britain, whereas in the United States of America cylindrical specimens, 12 in. high and 6 in. in diameter, have been adopted as standard. Unfortunately, the height of the test specimen in relation to its width greatly

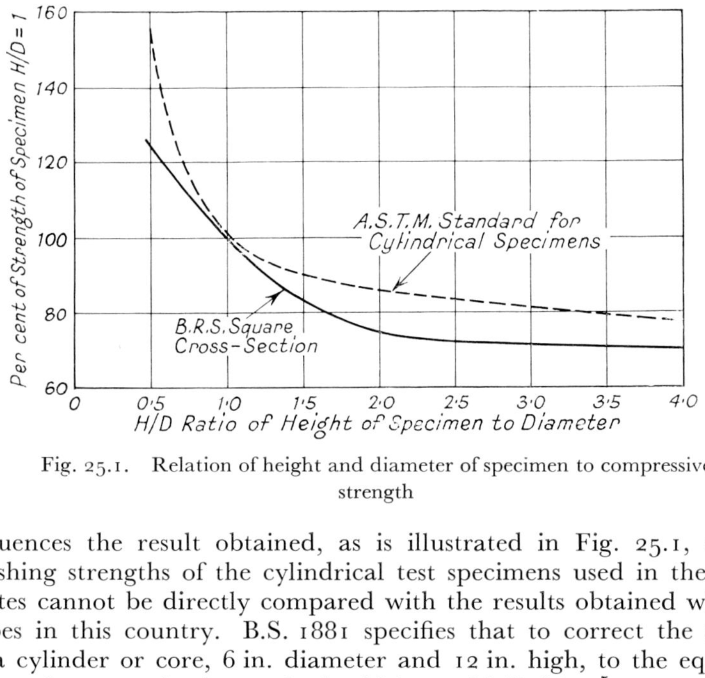

Fig. 25.1. Relation of height and diameter of specimen to compressive strength

influences the result obtained, as is illustrated in Fig. 25.1, and the crushing strengths of the cylindrical test specimens used in the United States cannot be directly compared with the results obtained with 6 in. cubes in this country. B.S. 1881 specifies that to correct the strength of a cylinder or core, 6 in. diameter and 12 in. high, to the equivalent 6 in. cube strength, the result should be multiplied by $\frac{5}{4}$.

The size of specimen also affects the crushing strength; for instance, compared with a 6 in. cube, a 4 in. cube of the same concrete is approximately 4 per cent stronger. Again, an 8 in. cube is approximately 8 per cent weaker, and a 10 in. cube 10 to 12 per cent weaker than a 6 in. cube. The relation between diameter and strength of concrete cylinders is given in Fig. 25.2 by a curve based on the results of a large number of tests made in the United States.

The advantage of the cube form of specimen used in British practice is that the sides can be made accurately plane and parallel by using carefully machined steel or cast-iron moulds. The cubes can then be tested on their sides without any special preparation of the test specimen or any necessity for bedding material, which results in lower strengths, and unless carefully done, leads to unreliable results. For this reason the use of wooden moulds is unsatisfactory, since it is impossible to maintain their accuracy, thus necessitating some form of bedding material. As an illustration of the influence of end packings on the crushing strength, a reduction of 30 to 50 per cent may be obtained by the insertion of lead or soft millboard.

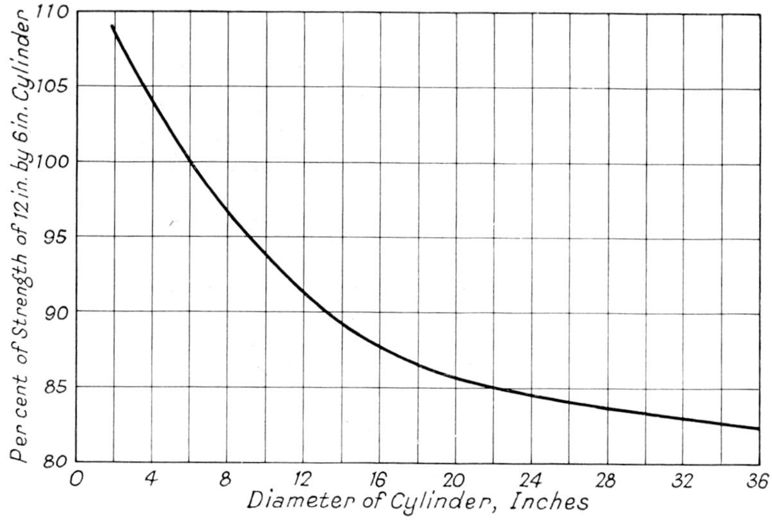

Fig. 25.2. Effect of cylinder diameter on crushing strength

In the United States the difficulty of trowelling accurately the top surface of the cylindrical specimen is overcome by capping with neat cement an hour or so after filling with concrete. However, it does appear that this method introduces an additional source of error.

The rate of applying the load to the specimen affects its crushing strength, especially for high rates of loading. Below a rate of loading of 2,500 lb/in^2 per minute, however, there is little change. It will be noted that a rate of 2,000 lb/in^2 per minute is specified in B.S. 1881 : 1968.

A point which may be missed is that cubes should be tested in a saturated condition in order to obtain uniformity. Dry cubes give higher results, but since there are different degrees of apparent dryness, it is not possible to ensure agreement.

Crushing tests are made in various machines either of the lever-arm type or adaptations of hydraulic presses. A machine of the hydraulic-press type with a maximum capacity of 200 tons, is illustrated in Fig. 25.3. The machine consists of two parts, namely, the press itself and the dynamometer which measures the load. The press consists of a ram, a crosshead, and upper and lower compression plates between which the specimen is crushed. The lower compression plate is scribed to facilitate centering of the specimen, and rests on the ram of the press. The beam seen in the photograph is incidental and is intended for tests on beams and slabs.

The upper compression plate is itself provided with a spherical seating, and is thus enabled to rotate sufficiently to take up any lack of parallelism in the sides of the specimen. In some machines the upper compression plate is supported at the end of a screw, enabling the operator to bring it down into contact with the specimen. In the machine shown this is dispensed with, and the ram provided with a large up-and-down movement. When the ram, the diameter of which is 12 in., rises in the cylinder the lower compression plate presses the specimen against the stationary upper plate.

Fig. 25.3. Hydraulic testing machine with a capacity of 200 tons

The ram is raised by pumping oil into the press cylinder from a reservoir at the base of the dynamometer. The pump is operated electrically, and is of the three-throw type. Provision is made to raise the ram quickly until contact is made with the specimen, and then by movement of the control valve the rate of movement of the ram can be controlled at any desired speed. On opening a release valve the press cylinder is unloaded and the weight of the ram drives the oil back into the reservoir.

The principle of the dynamometer is that the oil pressure acting on a small piston raises a weight attached to the end of a rotating arm (not seen in the photograph). The movement of the arm is proportional to the pressure, and is transmitted eventually to the pointer on the dial. After the test specimen is broken the pointer returns gradually to zero. A second pointer having a light friction fitting to the glass front, is engaged by the indicating pointer as it ascends, and remains in the highest position, thus recording the maximum load obtained.

Some hydraulic machines are hand operated, but this results in a jerky rate of loading, and is tiring if a large number of specimens have to be tested. Ordinary pressure gauges are also commonly used to indicate the load instead of a dynamometer. They are less expensive, but are not particularly reliable or accurate.

Inaccuracies may occur in making crushing tests for various reasons, such as errors in centering the cubes, wear of the platens, friction in the spherical seating of the upper compression plate, and inaccurate calibration of the machine itself. It has been found that considerable discrepancies occur between the results of crushing tests made with different designs of machine, although when calibrated with standard gauges these machines agree with one another to within 0·5 per cent. The cause of these discrepancies is not yet fully understood, although they have been the subject of much research in recent years.[25.3]

It appears advisable that all cubes from a particular site should be tested in the same machine in order to get comparable results.

Preparation of specimens for compression test

The preparation of test cubes from a representative sample of the concrete, obtained in the manner already described, should be done in accordance with the method laid down in B.S. 1881 : 1968, Methods of Testing Concrete, and it should be noted that the prime requirement is

Fig. 25.4. Compaction of 6 in. cube using a standard rammer

Fig. 25.5. Use of electric hammer to compact a stiff mix which could not be compacted by hand

for the concrete to be fully compacted. Whereas the minimum number of blows per layer on a 6 in. cube is specified as 35, mixes of low workability with less than 1 in. slump are likely to require more than the minimum. Very dry mixes with zero slump may require over 100 blows per layer to achieve full compaction, and are best compacted using vibratory methods. Poker vibrators are not suitable.

Particular attention should be given (*a*) to the ramming of the concrete into the moulds to obtain full compaction, Figs. 25.4 and 25.5, and (*b*) to protection to prevent premature drying out.

Cubes may either be "standard cured," i.e. in a tank of water maintained at 50–70°F in accordance with B.S. 1881, or "field cured." "Standard cured" specimens are a measure of the quality of the concrete itself and its inherent strength. "Field cured" specimens are an indication of the degree to which the strength is affected by the actual conditions of temperature and moisture, and they may be used as a guide to the safe time for removal of the forms, or opening of a road to traffic; they should not be used to determine the concrete strength in relation to the minimum required by the specification.

Standard-cured specimens should be cured at the proper temperature, commencing immediately after casting. A substantial variation in compressive strength may occur even within the limits of curing temperature of 50–70°F, given in B.S. 1881. Reference to Fig. 4.10 shows that for normal Portland cement, concrete cured at 70°F may be expected to have a strength of roughly one and a quarter times that cured at 50°F when tested at 7 days. In the Laboratory a temperature of 64°F has been generally accepted as standard in the United Kingdom and it is suggested that wherever possible a thermostatically controlled tank should be provided on the site. The variation in temperature should not then be more than ±2°F.

On the site it is not easy to ensure that cubes will not become uncovered after casting, and it is preferable to take the sample of concrete to the site laboratory or a small shed, and make the cubes there. They should be covered with damp matting, or hessian over which a polythene sheet should be placed to prevent loss of moisture. As soon as they are demoulded they should be immersed in water in a tank as specified in B.S. 1881. Concrete cubes should not be allowed to dry out from the time they are cast until the time they are tested. They should never be placed near a fire.

A difficulty arises with the curing of cubes for testing at 7 days when these have to be transported to a distant laboratory for testing; it is that the cubes must be packed two or three days after casting and may be in transit for the next two or three days. During transit the temperature cannot normally be controlled, and the effect on the crushing strength may be considerable in very cold weather.

Specimens intended to be cured under field conditions should be placed in contact with the forms and given the same protection as the concrete structure with a little additional insulation to compensate for the fact that they are cast in steel moulds and have greater relative surface area from which heat losses can occur. Some indication of the quality of the concrete can be obtained by making careful note of the temperatures during the curing period, and then making an allowance for any reduction in early strength using the curves given in Fig. 4.10. A fair approximation can be obtained in the following manner:

Assume average daily temperatures are as tabulated below:

Day	1	2	3	4	5	6	7
Average daily temperature, °F	40	40	30	50	50	60	40
Ordinary Portland cement in use							

Then, start off on the 40°F line and continue for 2 days, then move horizontally to the 30°F line (2—2) and move up this line for a one day period, reaching point 3. Move horizontally back to 3 on the 50°F line, and move along this for a two day period (days 4 and 5), to point 5. Continue to point 7 on the 40°F line. A strength of 53 per cent of that of concrete cured at 64°F for 28 days is indicated, compared with 70 per cent at 7 days on the 64°F line. Thus, concrete cubes cured for 7 days at the temperatures indicated, might be expected to reach $\frac{53}{70} = 76$ per cent of that of similar concrete cured at 64°F for 7 days.

Cores, usually about 6 in. diameter, may be cut from the finished concrete by diamond drills or by the grinding action of steel shot under a rotating steel tube, a supply of water being used to wash away the slurry formed.

Information on the drilling, examination and testing of cores is given in B.S. 1881:1968.

Beam tests

Tests on beams to determine the flexural strength of the concrete are described in Chapter 2. Two point loading is called for in B.S. 1881, which also provides for compression testing of the broken halves of the beam to obtain the equivalent cube strength.

Wet analysis

A method of analysing samples of fresh concrete is given in B.S. 1881. It is too elaborate, and is not intended, for ordinary site use; but it can be valuable under experimental conditions, particularly for measuring the efficiency of concrete mixers.

Number of tests

The number of tests must depend on circumstances, and no definite figure can be given. Slump tests can of course be made as often as any change in consistency becomes discernible. Concrete cubes are best prepared in pairs or even three or four at a time, so that reliance is not placed on the result of a single cube. It is a good plan to make three cubes together, one for test at 7 days as a guide, and the other two for test at 28 days as the final check.

An indication of the number of tests required has been given by a Committee of the Institution of Civil Engineers[25.1] and is summarised in Table 25.2. For strict compliance with C.P. 114 (1957), three cubes should be tested at any one age, which, because designs are based on 28 day strengths, is usually at that age. It is, however, desirable that an early indication of strength should be obtained for control purposes. It is therefore suggested that at least one extra cube is made on each occasion for testing at 7 days.

When sampling cement for testing it is advisable in most cases to obtain samples from a number of different loads and to submit them separately in airtight tins.

Despatch of materials and specimens for testing

Samples of material or specimens of concrete should be accompanied by complete information regarding the conditions under which they were obtained or made. The name of the site and other means of identification should be clearly marked on strong labels attached to each article.

Cement should be packed in airtight tins holding not less than 12 lb.

Samples of aggregate should be sent in strong polythene or closely woven jute bags, since the loss of fine particles in transit can materially change the grading, more particularly with sands.

Strong labels should be used with an identification number which can be quoted on an information form sent separately.

Concrete cubes should be packed in boxes and surrounded with wet sand or preferably placed in a strong polythene bag with a little water and the top folded over to prevent loss of moisture. The information to be sent with them should include the date of casting, the source of supply and sizes of the aggregate, the brand of cement, the proportions of the mix, the consistency, the location from which the sample was taken, the curing temperatures, the method of curing, and the age at which the test is to be made. A suitable form to use for such a purpose is shown in Fig. 25.6. It is preferable for the completed form to be sent by post, and for similarly numbered counterfoils, with the name of the site and the date on which the cubes were cast, to accompany the cubes.

DETAILS FOR COMPRESSION TESTS ON CONCRETE CUBES

CONTRACT
Name ....Ledger No. ....
Full address ....
MATERIALS
Coarse agg. Supplier ....Size ....
Pit ....
Fine agg. Supplier ....
Pit ....Zone ....
Cement type and trade name ....Works ....
MIX DETAILS
Ready mix supplier (if applicable) ....
Ready mix plant location ....
Class of concrete ....Mix proportions (wt/vol)* ....
Specified strength.............................. @ 7 days............................... @ 28 days
Mixer size....................................Slump (in).......................
Where concrete placed ....
Batch quantities: Cement ......lb. Fine agg. ......lb. Coarse agg. ......lb. Water ...gal.
CUBE DETAILS (dry/damp)*
Date cast.............................Cubes made by ....Status ....
Cube numbers ....Despatch date ....
Testing Laboratory ....
Age at which to be tested...
Site remarks...
....

* Delete whichever is inapplicable.

Fig. 25.6

Records

Tabular forms greatly simplify the work of recording and reporting, and also form a check of the items covered. Forms may be prepared for calculations of the batch quantities of aggregates, the recording of information when submitting aggregates or cubes for testing, records of slump tests, and records of cube test results. Charts such as that shown in Fig. 25.7 give a useful summary of test data in a form which can be readily appreciated, and from which the trend of results can be quickly assessed.

Variation in cube strengths

It is well known that variations occur in the strengths of dense concrete test cubes due to a variety of factors. Except for testing errors, which should be small, and variations in the rate of hardening of the cement,

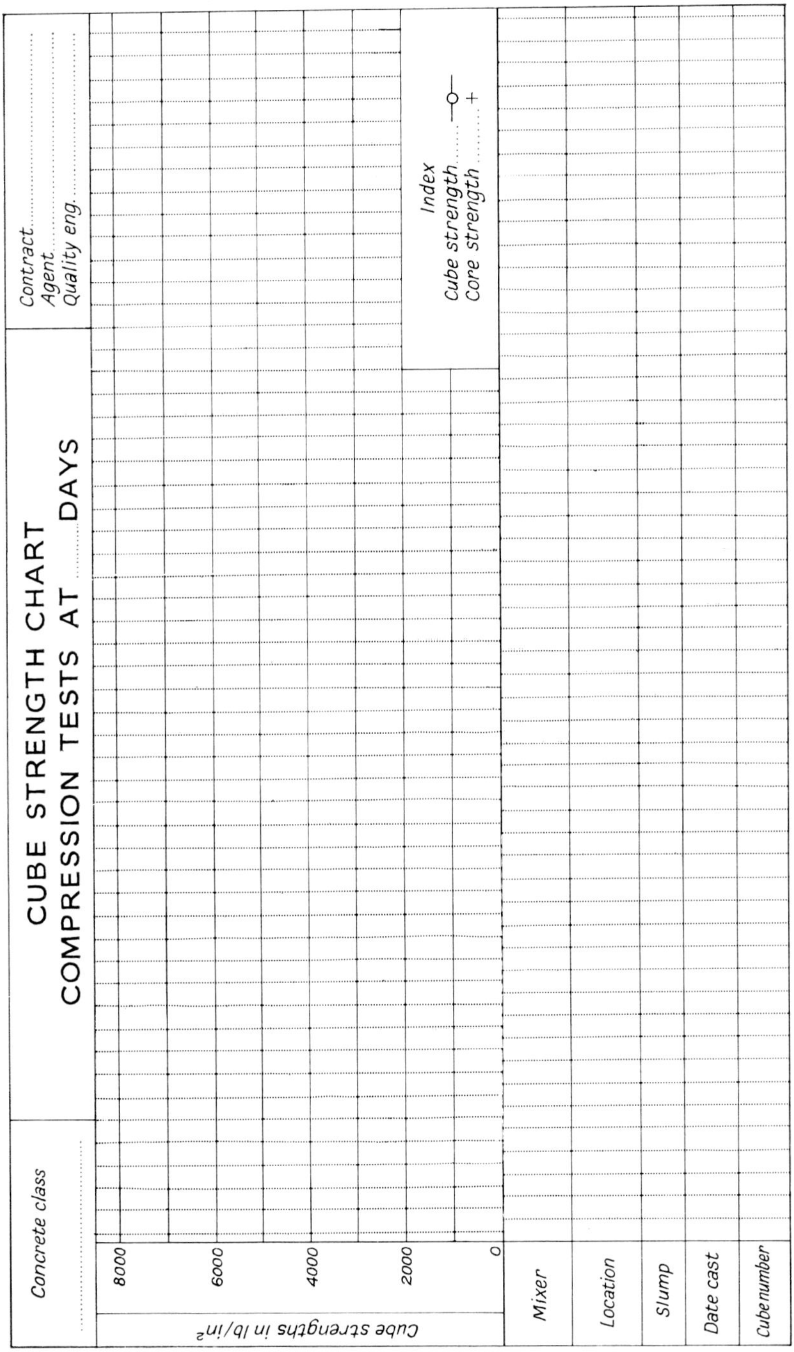
CUBE STRENGTH CHART
COMPRESSION TESTS AT .... DAYS
Contract..........
Agent..........
Quality eng..........
Index
Cube strength ....
Core strength ....
Concrete class ....
Cube strengths in lb/in²
8000
6000
4000
2000
0
Mixer
Location
Slump
Date cast
Cube number

Fig. 25·7

the variations are dependent on the care and attention given to the manufacture of the concrete and treatment of the specimens on the site. The main causes of variation in concrete cubes may be summarized as follows:

(1) Inaccuracy in the proportioning of the gravel, sand and cement. This is probably the largest single cause of site variations.

(2) Variations in the water/cement ratio. Such variations are aggravated by the need to obtain a reasonable workability for placing when using a mix in which the proportions of gravel, sand and cement are varying widely.

(3) Variations in the grading of aggregates which necessitate changes in the water/cement ratio if a uniform workability is to be maintained.

(4) Insufficient compaction. Very small percentages of air voids cause a large reduction in strength.

(5) Unsatisfactory curing. If the cubes are allowed to dry out during the first 24 hours, the loss in strength may be up to 50 per cent, which is not completely recovered by subsequent wetting.

(6) Variations in temperature. These have a most marked effect at early ages.

(7) Variations in cement quality.

Fortunately these variations are not additive, since some factors are tending to improve the concrete, while others are tending to cause a lowering of quality. The result is that concrete cube strengths vary about the average in such a way that while most of the values show little deviation above or below, a gradually decreasing number vary by greater amounts. If the number of cubes differing from the average by certain amounts, say, 0–500, 500–1,000, 1,000–1,500, and so on, lb/in^2 are plotted a histogram is obtained such as that given in Fig. A.1, p. 382. An envelope of such a histogram has the form shown for a Gaussian or Normal distribution.

It is sometimes desirable to summarize the test data obtained over a period. This may be done by taking an average and also indicating the maximum and minimum values, but such figures give little indication of the way in which most of the results departed from the average. An index which provides a measure of the extent of the variations from the average is therefore useful. The usual index is the "standard deviation" or, expressed as a percentage of the average value, the coefficient of variation. These terms are defined in Appendix A.

On most sites, changes in the grading of the sand, differences in workability required for differences in placing conditions, and other factors make some changes in the water/cement ratio necessary from time to time, and these are not normally recorded with sufficient accuracy for them to be taken into account in any statistical analysis of the cube strengths. The number of cubes made is also too small on most sites to be of much use statistically; the standard deviation calculated on a few results is likely to depart considerably from the true value, which can only be obtained with reasonable accuracy from a large number of cube results, say forty or more. With such variants unaccounted for, the standard deviation would be expected to be least on works requiring large quantities of concrete of uniform workability, such as road or airfield pavings, and this is borne out in practice. Even then, an upward or downward trend in the concrete strengths over a period, whilst increasing the standard deviation or coefficient of variation, may be caused by a similar trend in cement strengths, and not by a weakening of quality control.

Although these limitations on the value of statistical analyses must not be overlooked, such analyses, if carefully interpreted, are useful as an indication of the variations in cube strengths which might be deemed reasonable under a variety of site conditions. This knowledge is of importance both in mix design and in day-to-day control on the site.

The information given in Table 7.4 is based mainly on an analysis of cube results from nearly two hundred sites, ranging from small works on which only a few cubes were made, to London Airport where the number tested by the contractor was nearly 30,000. In the Table it will be noted that the variation in strength is taken to be proportional to the average strength for strengths below about 3,000 lb/in^2. Above this value the variation is indicated to be independent of the average strength, and is related more to the standard deviation than to the coefficient of variation. There is considerable evidence[7.2,25.4] to support this view as being approximately correct, both theoretically and from analysis of site cube test results.

REFERENCES

25.1 Report by Institution Research Committee, Quality of Concrete in the field, *Proc. Inst. Civ. Engrs*, **4**, No. 3, May 1955.

25.2 MURDOCK, L. J., Methods of achieving control of quality, *Symposium on Mix Design and Quality Control of Concrete*, London, 11–13 May 1954. Cement and Concrete Association.

25.3 NEWMAN, K., Concrete control tests as measures of the properties of concrete, *Proc. Symposium on Concrete Quality*, Nov. 1964. Cement and Concrete Association.

25.4 ERNTROY, H. C., *The Variation of Works Test Cubes*, Report No. 10. Cement and Concrete Association.

APPENDIX A

Variations in Cube Strengths

VARIATIONS in cube strengths result from many different factors such as the strength characteristics of the cement, the water/cement ratio, changes in grading of the materials, air-voids content of the compacted concrete, curing, temperature, and testing errors. For any particular sample of concrete, some of these factors will be tending to reduce the compressive strength, while others will be tending to increase it. The compressive strength actually obtained will then depend on the balance between the positive and negative influences, and the extent of deviations from an average value will depend on the amount by which the total of the positive influences exceeds the total of the negative influences, or *vice versa*. In general, it is found that the bulk of the results lie near the average value with a decreasing number showing greater deviations, following the distribution shaded in Fig. A.1.

Analyses of cube strengths obtained on large numbers of sites have shown that the distribution of values is in accordance with the Theory of Probability, i.e. they have a Normal or Gaussian distribution which is represented by the curve in Fig. A.1. A Gaussian distribution has the property of being wholly defined by the average value of the set of results and their standard deviation.

The standard deviation of a set of numbers is defined by:

$$\delta = \sqrt{\frac{\sum (x - \bar{x})^2}{n}}$$

where δ is the standard deviation of the set of numbers,
n is the number of values in the set,
x is any value in the set of numbers,
$\bar{x}$ is the average of the set of numbers.

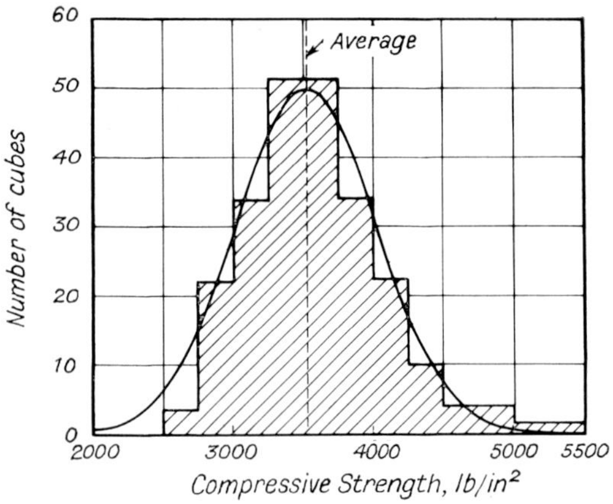

Fig. A.1. Example of histogram and a normal probability curve

However, when examining concrete cube results the main interest is in an estimation of the variation of the compressive strength of the whole consignment of concrete, rather than the variation in the small number of samples which have been taken and tested from the entire consignment. A better estimate of the standard deviation for the whole consignment is given by the following formula:

$$\sigma = \sqrt{\frac{\sum (x - \bar{x})^2}{n - 1}}$$

The coefficient of variation is obtained by dividing the standard deviation (σ) by the average ($\bar{x}$) and is usually expressed as a non-dimensional percentage:

$$v = \frac{\sigma}{\bar{x}} \times 100$$

The normal distribution curve may be used to calculate the proportion of cubes differing from the average by any multiple of the standard deviation. Figure A.2 has been plotted to show the relationship in convenient form.

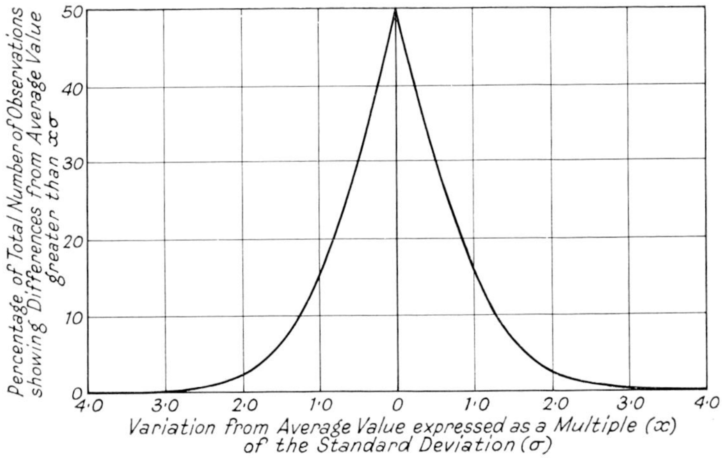

Fig. A.2. Distribution of observations deviating from the average value due to operation of numerous chance causes

The standard deviation provides a useful basis for quality control when large numbers of cubes are made from one quality of concrete and are tested in a reasonably short period, particularly during the early stages of construction. As an example, assume that thirty or more cubes are made and tested during the first few days of construction (or if economically justifiable from trial mixes before commencing concreting operations), and that the standard deviation is calculated from the results obtained. It is then possible to form an estimate of the future distribution of results. Should these show greater variability as time goes on, it is obvious that either the site control is showing some relaxation, or other factors such as increased variation in the quality of the cement, decreased mixing efficiency, poor cube-making, etc. are having an adverse effect. Whatever the cause, the increased variability should be investigated.

An example will make this method of control clear. Assume the specified minimum strength required to be 3,000 lb/in² and that to ensure that not more than 2 per cent of the cubes shall fail, the aim has been to achieve an average strength of 4,400 lb/in². (See Table 7.4 for suggested allowances.)

Suppose now that compressive tests have been made on the first 40 cubes cast on the job, and two have failed; the average strength is 4,200 lb/in² and the standard deviation (σ) 580 lb/in².

Referring to Fig. A.2, it will be seen that the condition for not more than 2 per cent cubes failing, is that about twice the standard deviation shall not exceed the allowable variation from the average, in this case 4,200–3,000 = 1,200 lb/in². As twice the standard deviation is $2 \times 580 =$ 1,160 lb/in² this condition is just fulfilled and the cube strengths may be accepted as fulfilling the specification requirement. The fact that two cubes of the forty tested have already failed may be ignored, since the analysis suggests that they were chance failures, and that if an infinite number of cubes were tested less than 2 per 100 would fail.

Suppose now that testing continues and the results of the next 20 cubes are as follows in the order printed:
3,460, 4,460, 4,890, 5,250, 3,210, 3,300, 5,700, 4,120, 4,770, 4,610, 6,000, 4,040, 4,600, 3,090, 5,310, 4,700, 4,540, 4,260, 5,910, 3,890.

The average strength for these 20 cubes is 4,500 lb/in² but the standard deviation has increased to 860 lb/in². Although no failures occurred in the batch of 20 cubes the increase in standard deviation alone is sufficient to suggest that the quality control measures have not been so effective and an investigation should be made. Twice the standard deviation is 1,720 lb/in² which is larger than the difference between the average strength and the minimum required, $4,500 - 3,000 = 1,500$ lb/in² and if the trend continued it would be expected that more than 2 per cent of cubes would fail. This fact would not be apparent from a consideration only of the average cube strength which had, in this example, increased.

This example of the use of the statistical approach to quality control indicates the usefulness of the method in assessing whether cube strengths are being maintained at a sufficient level to comply with specification requirements. It should be noted, however, that an analysis based on only 20 or 30 results is not usually sufficiently accurate for investigational purposes, and it is desirable to base any conclusions on a very much larger number, say, a hundred or more cube tests.

The curves in Fig. A.2, may also be used to estimate the proportion of cubes differing from the average by multiples of the standard deviation. For instance, 16 per cent of the cubes are likely to deviate from the average by a value equal to the standard deviation, x being equal to 1. It will be noted that as many cubes are above the average as are below it, and that the deviation above the average is likely to be as great as that below it.

An example of a histogram plotted to show the numbers of cubes falling into 500 lb/in² intervals in cube strength is given in Fig. A.1. The outline

TABLE A.1 *Example of Calculations for Standard Deviation and Coefficient of Variation*

No.	Crushing strength of cubes, lb/in²	Diff. from average $(x - \bar{x})$	$(x - \bar{x})^2$
1	4,100	0	0
2	3,990	−110	12.100
3	3,180	−920	846,000
4	5,010	910	828,000
5	4,500	400	160,000
6	4,110	10	100
7	3,680	−420	176,500
8	4,910	810	656,000
9	4,220	120	14,400
10	3,260	−840	705,000
Average $(\bar{x})$	4,100	$\Sigma(x - \bar{x})^2 =$	3,398,100

$$\frac{\Sigma(x - \bar{x})^2}{n - 1} = \frac{3{,}398{,}100}{9} = 377{,}566$$

$$\text{Standard deviation} = \sqrt{377{,}566}$$
$$= 614 \text{ lb/in}^2$$

$$\text{Coefficient of variation} = \frac{614}{4{,}100} \times 100 = 15{\cdot}0 \text{ per cent.}$$

of such a histogram should generally follow the form of the Normal distribution curve. Any substantial departure should be a matter of investigation. In the example given in Fig. A.1 it will be seen that there is a somewhat sudden cut-off at the lower end, whereas there are a few more high results than would be expected from a normal distribution. It would be expected that the tails of the histogram would stretch to roughly the same range above and below the average. The sudden cut-off should be investigated. Low cube strengths may not have been reported, cubes of poor appearance may have been "lost," or there may have been some error. Other anomalies should also be investigated; for instance, too many high results may be due to selection of concrete from batches of lower workability than the average for the purpose of cube making.

For further information on the subject, reference should be made to standard textbooks.(A.1) Much useful information on the utilization of statistical methods for the control of quality is given in British Standard 2564:1955.(A.2)

REFERENCES

A.1 McINTOSH, J. D., *Concrete and Statistics*, 1963. C. R. Books Ltd., London.

A.2 B.S. 2564:1955, *Control Chart Technique when Manufacturing to a Specification*, P. B. Dudding and W. J. Jennett. British Standards Institution.

APPENDIX B

Specific Gravity and Moisture Content Pycnometer Method

General

THIS method is intended to cover the determination of specific gravity and surface moisture content for sand and gravel up to 1½ in. in size.

Apparatus

(1) A pycnometer consisting of a 1 quart glass fruit-jar to which is screwed a cone attachment of the dimensions shown in Fig. B.1.

(2) A balance weighing up to 2 kg, and accurate to at least 1 g, preferably 0·1 g.

(3) Towels, trays, buckets, scoops, etc.

Determination of bulk specific gravity

(1) Weigh the pycnometer full of water (P g).

(2) Weigh 700 g (B) of the aggregate, saturated and surface dry. (See note below.)

(3) Place the sample in the jar and then fill with water. Screw on the top, so that the match marks are in line, and then roll the jar with the top hole covered by one finger to remove all air from the sample.

(4) Completely fill the pycnometer with water and weigh. Any froth which rises to the surface should be carefully blown off, and a few drops more water added so that the water is level with the orifice in the top of the cone. Call this weight (P_s).

Bulk specific gravity on a saturated surface dry basis $= G = \dfrac{B}{P + B - P_s}$

The single line in Fig. B.1 may also be used to find the specific gravity, reading off on the right-hand scale.

Determination of surface moisture

The same procedure is used as for determining specific gravity except that the 700 g sample consists of the moist material to be tested. The surface moisture content is calculated by the following formula or determined by use of the graph in Fig. B.1.

$$M = \frac{100S}{P_s - P}\left(\frac{G - 1}{G}\right) - 100$$

Where M = surface moisture in the sample, expressed as the percentage by weight of the saturated surface dry aggregate,

S = weight of the wet sample (700 g),

P_s = weight of pycnometer plus wet sample plus water.

The surface moisture can, if desired, be read off from the curves given in Fig. B.1, using the left-hand scale.

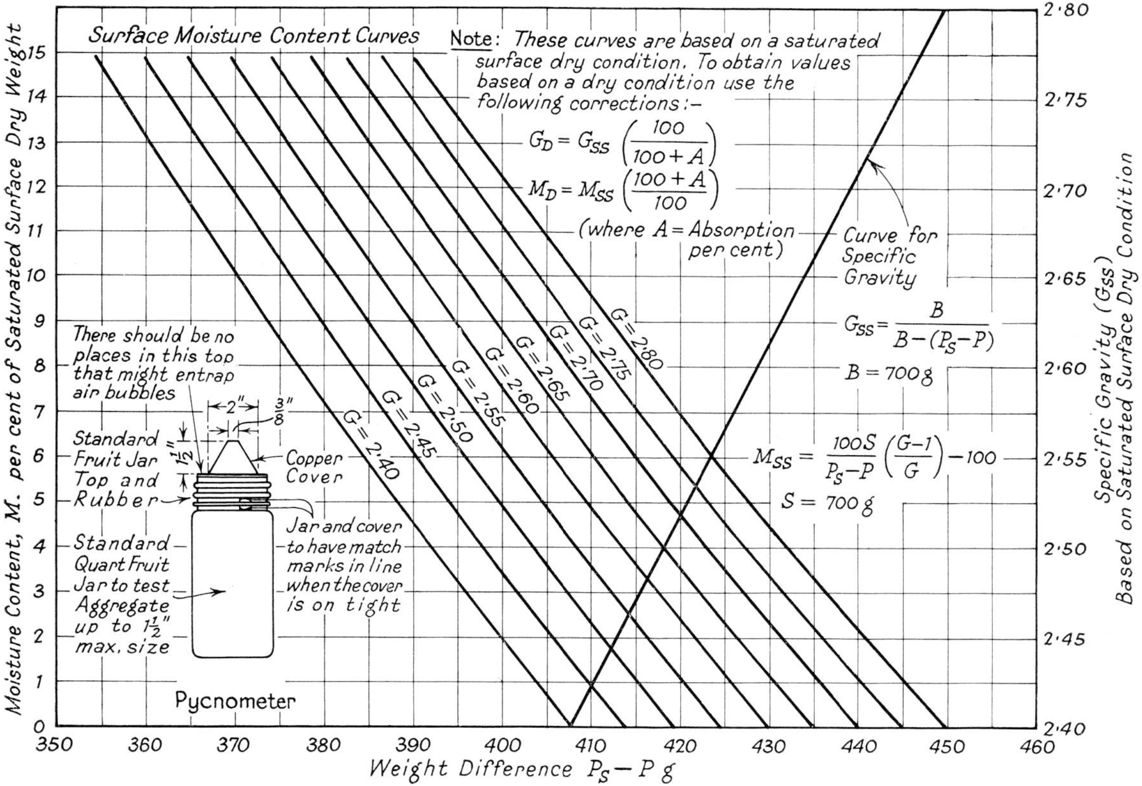

Fig. B.1. Curves for use with pycnometer when determining bulk specific gravity and surface moisture content

Precautions necessary in making the tests

(1) Match marks on the cone and on the jar must be in line before the pycnometer is weighed.

(2) The sample must be completely submerged in water before shaking and rolling to remove entrained air.

(3) The outside of the pycnometer must be thoroughly dried before weighing, and care taken to see that no water is trapped under the edge of the screw top.

(4) The weights P and P_s should be measured with the pycnometer filled with water at approximately the same temperature, say within ± 2°F.

Preparation of a saturated surface dry sample

The sample of aggregate should be placed in a tray or dish, covered with water, stirred to remove air bubbles, and allowed to stand for 24 hours.

Aggregate coarser than $\frac{3}{8}$ in. B.S. sieve should then be taken from the water and dried rapidly with a damp cloth until no water is visible on the surface. With most aggregates at this point the surface just loses its shininess and becomes dull.

Aggregate finer than $\frac{3}{8}$ in. is freed from water by decanting through a filter paper to avoid loss of the finer particles. The aggregate or sand is then exposed to a current of warm air, such as might be obtained by fanning in front of an electric fire, to evaporate the surface moisture. The sample must be stirred continuously to ensure even drying. On the site when a primus or a coke stove is the only means of heating, the surface moisture can be removed by placing the sample in a frying-pan and letting it warm gently while stirring continuously. This method requires more care than that previously described if reliable results are to be obtained. The surface moisture is considered to have disappeared when the aggregate loses its shiny wet appearance and becomes dull, or when it just attains a "free running" condition. Another test sometimes suggested is that the sand shall just cease to stick to a glass rod. Whichever end-point is adopted, care must be taken that this is not passed.

NOTE

If the surface moisture content is required as a percentage of the dry aggregate, an allowance must be made for the weight of absorbed water. Unless the aggregate is porous, the difference between the two figures will, for practical purposes, be negligible.

It is desirable to make at least three tests and find the average, if reliable results are to be obtained for the moisture content.

APPENDIX C

Absorption and Moisture Content Drying Methods

Method 1

DETERMINATION of surface moisture and absorption by drying.

(1) Weigh out the following quantity (A) of the wet material:

Sand	1 kg
or $\frac{3}{16}$ in.–$\frac{3}{4}$ in. aggregate	2 kg
or $\frac{3}{4}$ in.–$1\frac{1}{2}$ in. material	3 kg

(2) Place in a frying-pan and heat carefully on a stove as described in Appendix B, "Preparation of a saturated surface dry sample." If a better form of heating is available so that a current of warm air can be created, this is preferable.

(3) As soon as the material is saturated surface dry, weigh again (B).

(4) Continue drying until all the moisture is evaporated.

(5) Weigh the dry material (C).

$$\text{The percentage surface moisture} = \frac{A - B}{C} \times 100$$

$$\text{Absorption as percentage of dry weight} = \frac{B - C}{C} \times 100$$

Method 2

It is assumed that the absorption has previously been determined under laboratory conditions by a procedure similar to that given in Method 1.

(1) Weigh a similar quantity of wet material to that given in Method 1 (weight A).

(2) Place in a frying-pan on a stove until it is quite dry.

(3) Weigh the dry material (C).

The surface moisture as percentage of dry weight =

$$\frac{A - C}{C} \times 100 - W_A$$

where W_A = absorption as percentage of dry weight.

PRECAUTIONS NECESSARY

Care must be taken not to overheat the aggregate as this may cause some of the small pieces to jump out of the pan. Any material lost between the first weighing and the second will count as moisture, and every effort should be made not to lose any.

British, American and Metric equivalents

1 lb	= 454 g	1 ton (Br.)	= 2,240 lb
1 ton (Br.)	= 1,016 kg	1 ton (Am.)	= 2,000 lb
1 in.	= 2·54 cm		
1 ft	= 0·305 m		
1 in^2	= 6·451 cm^2		
1 ft^2	= 0·0929 m^2		
1 in^3	= 16·387 cm^3		
1 ft^3	= 0·0284 m^3		
1 yd^3	= 0·765 m^3		
1 gal (Br.)	= 4·543 litres	1 gal (Br.)	= 10 lb water
		1 gal (Am.)	= 8·33 lb water
1 lb/in^2	= 70·3 g/cm^2		
1 lb/ft^2	= 4·882 kg/m^2		
100 lb/ft^3	= 1,602 kg/m^3		
100 lb/yd^3	= 59·3 kg/m^3		

1 bag of cement (Br.) = 112 lb
1 sack of cement (Am.) = 94 lb

Subject Index